Springer Earth System Sciences

More information about this series at http://www.springer.com/series/10178

Alexandros Theodoridis • Athanasios Ragkos
Michail Salampasis
Editors

Innovative Approaches and Applications for Sustainable Rural Development

8th International Conference, HAICTA 2017, Chania, Crete, Greece, September 21–24, 2017, Selected Papers

Springer

Editors
Alexandros Theodoridis
Laboratory of Animal Production Economics
School of Veterinary Medicine
Aristotle University of Thessaloniki
Thessaloniki, Greece

Athanasios Ragkos
Agricultural Economics Research Institute
HAO Demeter
Athens, Greece

Michail Salampasis
Department of Informatics
TEI of Thessaloniki
Thessaloniki, Greece

ISSN 2197-9596 ISSN 2197-960X (electronic)
Springer Earth System Sciences
ISBN 978-3-030-02311-9 ISBN 978-3-030-02312-6 (eBook)
https://doi.org/10.1007/978-3-030-02312-6

Library of Congress Control Number: 2018966713

This Springer imprint is published by the registered company Springer Nature Switzerland AG
The registered company address is: Gewerbestrasse 11, 6330 Cham, Switzerland

Foreword

This collective volume is based on papers presented at the "8th International Conference on Information and Communication Technologies in Agriculture, Food & Environment." The Conference was held in Chania, Crete, Greece in September 2017 and was organized by the Hellenic Association for Information and Communication Technologies in Agriculture Food and Environment (HAICTA), the Mediterranean Agronomic Institute of Chania (CIHEAM), and the University of Macedonia, Greece. HAICTA is the Greek Branch of the European Federation for Information Technology in Agriculture (EFITA). The series of HAICTA conferences is an international venue for research and development in ICT in agriculture, food, and environment and welcomes contributions from various topics and disciplines relevant to these fields. The HAICTA conference in 2017 received 124 paper submissions, of which 55 (44.4%) were accepted as full papers. Submissions were received from authors from 45 countries. The top five countries in terms of accepted papers (according to contact author affiliation) were Greece, Poland, Italy, Czech Republic, and Spain. The top scientific areas covered by the submitted papers (as indicated by the primary keyword assigned by the authors) were Decision Support Systems, Information Systems, Environmental Impact Assessment, Precision Farming Systems, and Environmental Design and Policy.

The manuscripts originally submitted to the Conference Program Committee were reviewed by the members of the scientific committee of the HAICTA 2017 Conference in order to be published in the conference proceedings. The versions included in this volume are at least 50% different from the papers presented in the conference. The selection of the studies appearing here was based on their relevance to the scope of the book and on the evaluation score of the conference papers. All manuscripts have passed a new round of double-blind review process before being accepted for publication.

Thessaloniki, Greece — Alexandros Theodoridis
Athens, Greece — Athanasios Ragkos
Thessaloniki, Greece — Michail Salampasis

Preface

Agriculture is confronted continuously with new challenges, among which are: the globalization of food markets, which intensifies competition in the agricultural sector; the increase in input prices; the unstable economic environment, which entails financial stress; climate change; etc. In addition, increased consumer awareness regarding food security, environmental protection, animal welfare, social equity, and abolishment of poverty have serious impact on the operation of markets of agricultural products. This environment shapes a dualistic situation where small farms need to adjust their production activities to global demands as well as to local needs. Although the notions of "rurality" and "territory" nowadays seem to be increasingly excluded from discussions regarding the economic efficiency of agriculture, increasingly more theory and practice tend to converge toward the necessity for a new economic pattern in rural areas. Although agriculture constitutes their main source of income and employment, the current policy framework encourages the emergence of alternative economic activities to support development. A broad array of activities, policies, initiatives, and tools has been proposed to ensure income and employment for these areas and to avert depopulation as well as the adverse effects of climate change and unexpected turbulences in the external environment.

Within this framework, sustainable rural development is nowadays central to the discussions regarding policies and economic development. In fact, sustainability in rural areas is regarded as a concept with numerous facets including environmental (with the provision of ecosystem services without reducing the availability of natural resources), economic (allowing systems to function autonomously, without needing external inputs), and sociocultural dimensions (ensuring high quality of life for rural dwellers without altering their habits and traditions). Sustainable rural development is nowadays linked to close relationships among farms, agri-food industries, retailers, and other related actors across the supply chain, thus generating resilient networks. Indeed, by examining the reasons behind the resilience of agricultural systems and family farms, new rural development models can be built based on this accumulated knowledge. Innovation—either in the form of technological innovations or as social innovation—constitutes a central element in this development process. Effective governance of such issues is another key factor in the development process, as local populations require new ways to manage the commons in an efficient manner, mitigating potential conflicts and maximizing economic outcomes.

Innovative Approaches and Applications for Sustainable Rural Development will seek to provide insights regarding these issues. The main objective of this book is to raise the awareness of the importance of sustainable management in agriculture, food, and the environment. In total, the book includes 21 original contributions relevant to the various aspects of the key objective of this book. Most of these papers constitute updated and/or expanded versions of those presented at the "8th International Conference on Information and Communication Technologies in Agriculture, Food & Environment." Innovation is central to all contributions in the book, including real industry cases, sustainable management practices, new forms of rural cooperation and entrepreneurship, as well as models and techniques to face crucial decision-making problems in rural areas. The papers have been systemized into four parts, which are presented below.

Part I. Contemporary Views of Rural Development, Sustainability, and Cooperation

Rural development in the EU has been highly supported by policy measures and development projects and measures have been widely proposed. This part includes five contributions in total, each one of which provides an alternative view of the development of rural areas and/or proposed new pathways toward this direction, under alternative models. The chapter by Ragkos et al. refers to the relatively novel approach of "territorial development" and demonstrates how local breeds can be developed as a local asset toward the development of particular territories. On the other hand, Koulelis shows the role of forestry in the development process, especially under the economic crisis that Greece faces, which operates within an outdated framework, and proposes regulatory amendments to the policy framework in order to boost forestry-based development. Valsamidis describes the role of innovation, along with novel notions such as "ecopreneurship," at the business level and how this can be transformed into development opportunities within a wider scope. The contribution of Srdjevic et al. shows the applicability of a multicriteria analysis framework in the decision-making process for water management, by providing a new prospect including a collective approach, which pertains to the public-good nature of water. Another view of cooperation is provided by the work of Pisanelli et al., who present the framework of agroforestry and demonstrate how linking independent networks under a common umbrella can lead to novel solutions with environmental benefits.

Part II. Novel Technologies and Innovative Farming Systems to Support Rural Development

This part includes various examples of farming systems and innovations to support their overall sustainability. Two contributions deal with animal production. The first is by Papageorgiou et al., who demonstrate how an epidemiological investigation can contribute to disease eradication and thus lead to more sustainable production patterns. The second contribution is by Rose et al., who proclaim the role of innovation for the livestock sector and pinpoint key innovations necessary to boost the European sheep and goat sector as a whole. Zaralis' and Padel's work lies between livestock production and the environment describing how a particular type of sheep and goat grazing can bring about significant environmental benefits and, at the same time, support high levels of animal productivity. At the crop sector, the review by Symeonaki et al. of background technical information and challenges regarding the Internet of Things shows how this type of technology can support climate-smart agriculture in the context of rural development. Masner et al. present a new method to support the widespread adoption of Precision Agriculture, by showing how the systematic collection and analysis of users' perceptions of the method may serve as an important decision-making tool for farmers, suppliers, academics, and other stakeholders.

Part III. ICT Applications in Agribusiness and New Business Models in the Agri-food Sector

This part is the largest of the four included in this collective volume and demonstrates a wide range of ideas and applications of ICT in the agribusiness sector. It is true that ICT plays a key role in promotion of entrepreneurship and there are numerous relevant applications. The contribution of Kartsonakis et al. shows how the optimization of supplier selection can play an important role in entrepreneurship, especially when noncost attributes—such as quality, delivery, and collaboration—are included in the choice of suppliers. In the same spirit, Skordoulis et al. provide a description of how existing methodological approaches, such as a differential equations framework, can be used to redefine competitive advantage of alternative business models such as green entrepreneurship. Misso et al. demonstrate another important role of the Internet in the support of agribusiness, through the applicability of crowd funding for the support of entrepreneurship oriented toward socioeconomic and environmental well-being. In the context of alternative entrepreneurship, Sturiale and Scuderi show how e-marketplaces are tending to become mainstream in Italy and how they contribute to the transformation of the agri-food sector by providing fresh products at same-day delivery. A similar perspective is provided by the work of Očenášek et al., who compare farmers' e-shops in three countries to conclude that, while there is still room for development in this sector, these e-shops provide significant services to potential customers. Di Pascale et al. investigate the role of ICT at the farm level and specifically contribute to the adoption literature by showing that regardless of the clustering method, farmers' adoption patterns of ICT methods still follow the pattern of early and late adopters. Unlike the previous authors, Jagtap and Rahimifard do not focus on a specific type of actors within the supply chain and describe the applicability of ICT to broader areas by presenting the example of Internet of Things and its applicability across whole food supply chains and not to specific links separately.

Part IV. New Methods for High-Quality Agricultural Products and Environmental Protection

Alternative views of rural development are important, but their practical usability is closely linked to the existence or the emergence of production systems and patterns that will support novel development paradigms. Part IV includes four contributions, which describe such examples, combining innovation at the farm or system level and development implications. The work of Karelov et al. is an example of the sort, as it describes how genetics techniques can be applied to a commodity (wheat) in order to produce disease-resistant varieties, which can be cultivated in adverse environmental conditions, thus promoting economic activities in less favored areas. Papaefthimiou et al. focus on an equally important Mediterranean product, i.e., olive and olive oil. Their approach involves the development of an application (OLEA) that focuses on all links of the supply chain, starting from primary production and leading to branding and commercialization of products, with clear economic benefits for all actors involved. Muradova et al., on the other hand, present an innovative system, which can support the installation and operation of widely used flexible sprayer booms, thus leading to cost savings, environmental benefits, and also to prospects for more complex irrigation systems. For environmental protection, Czekała shows how solid biomass can be converted to biogas, providing an example of circular economy with multiple economic benefits from its more widespread application.

We believe that this book can be a useful tool for scientists and practitioners related to ICT, rural development and innovation in agriculture, food, and environment. Since it approaches and discusses various topics and concepts, the book should reach wider interdisciplinary audiences and stimulate further research or yield practical ideas for implementation. Finally, we trust that the contributions here will be of interest to young scientists who are just starting to be engaged in these research fields and need to be introduced to the new concepts in this book.

Thessaloniki, Greece — Alexandros Theodoridis
Athens, Greece — Athanasios Ragkos
Thessaloniki, Greece — Michail Salampasis

Contents

Contributors

Garyfallos Arabatzis is Professor at the Democritus University of Thrace. Most of his research analyses concern forest, rural, and regional development; forest economics; and forest valuation and accounting. However, part of his recent work also looks into socioeconomic evaluation of RES, operational research and natural resources, and green finance and development. He has published over 200 articles in international journals and proceedings. He is reviewer in 54 international journals while his work has been cited more than 1000 times.

Georgios Arsenos is Professor at the School of Veterinary Medicine, Aristotle University of Thessaloniki, Greece. His research fields include behavioral nutrition, animal growth, meat quality and sustainability of small ruminant production systems.

Konstantinos G. Arvanitis is Professor in the Agricultural University of Athens. He holds an Electrical and Computer Engineering Diploma and a PhD in the field of automatic control and automation systems. He has participated in several research and development projects and has authored numerous publications. His current interests include among others applications of artificial intelligence in agriculture, automation and electrification in agriculture, image processing and remote sensing in agriculture, ICT in agriculture, precision farming, etc.

Francesca Camilli is a researcher at CNR-IBIMET, Italy. She holds a PhD in Commodities Sciences and her research interests include in rural development, recovery of multifunctional agro-environmental resources, local knowledge assessment, and implementation of local supply chains.

Gian Paolo Cesaretti is President of Simone Cesaretti Foundation and Professor at the University of Naples "Parthenope." He has held various institutional roles in both the agri-food and the environmental sectors, and he is author of numerous articles and books concerning well-being sustainability.

Miltiadis Chalikias is Associate Professor at the University of West Attica. He holds a PhD in Statistics from the National and Kapodistrian University of Athens. He is the author of several scientific articles, books and chapters. His principal research interests are in the area of business mathematics, business statistics, qualitative methods, and multivariate analysis.

Angelos Chatzimparmpas started as an undergraduate student at the University of Western Macedonia, Greece, in 2012. Angelos is currently a PhD student in Visual Analytics for Engineering Smarter Systems at the Department of Computer Science and Media Technology of Linnaeus University, and member of Data Intensive Sciences and Applications (DISA), and Information and Software Visualization (ISOVIS) groups.

Marco Ciolfi is a research fellow at CNR-IBAF, Italy. He holds a PhD in Forestry. He has research interests in remote sensing, GIS—geostatistics, stable isotopes in ecophysiology, and ecological modeling.

Daniel Martin-Collado is an agricultural engineer currently employed as a researcher at the Agrifood Research and Technology Centre of Aragón. His research fields include extensive livestock farming systems, animal genetic resources conservation and development, and livestock breeding and genetics focusing on farmers' preferences and socioeconomic aspects of livestock sector development.

Claudia Consalvo is a research fellow at CNR-IBAF, Italy and holds a PhD in Forest Ecology. Her scientific expertise includes the environmental constraints of the main physiological processes, at tree and stand levels; the acclimation responses to climate change and water scarcity; the techniques and methodological approaches used in tree's ecophysiology; and the implementation of agroforestry systems.

Wojciech Czekała is an environmental protection and agricultural engineering specialist currently employed at the Institute of Biosystems Engineering of the Poznan University of Life Sciences. His research fields include waste management, agricultural engineering, biofuels production, production of organic fertilizers, food waste management, and sustainable development.

Gianluigi De Pascale is a PhD student in Health Food Innovation and Management at the Department of Economics of the University of Foggia, Italy. His PhD thesis focuses on the evaluation of the Short Food Supply Chain. His field of research is focused on applied economics and politics economics by considering lean approaches in organizing resources.

Nicola Faccilongo is a researcher in the Department of Economics, University of Foggia, Italy. His field of research is applied economics. His research interests are focused on integrated supply chain projects, supply chain management and the use of ICT, agro-food chain, and public policies assessment and environment improvement.

Georgios Filioussis is Assistant Professor in Laboratory of Microbiology and Infectious Diseases in the Faculty of Veterinary Medicine, Aristotle University of Thessaloniki, Greece. George is an experienced microbiologist with a principal interest in Veterinary Microbiology. He has been employed by the Greek Army from 1992 until 2010 and by the Aristotle University of Thessaloniki since 2011. He has published widely in the field of veterinary microbiology and infectious diseases.

Spyros Galatsidas is Associate Professor at the Democritus University of Thrace. His articles are published in several scientific journals and conference proceedings. His principal research interests concern sustainable management of forest ecosystems, planning forest resources production, inventory and analysis of forest ecosystems, assessment and valuation of ecosystem services, database and GIS systems in forest management, analysis of forest inventory data, exploitation of forest biomass in energy production, and certification of sustainable forest management.

Evangelos Grigoroudis is Professor in the School of Production Engineering and Management of the Technical University of Crete, Greece. He serves as Vice President of the Hellenic Operational Research Society. He has coauthored/coedited more than 17 books in operational research, service quality measurement, and corporate strategy and published more than 200 articles in scientific journals, books, and conference proceedings.

Ioannis Grivas is Assistant Professor in Anatomy, Histology, and Embryology in the School of Veterinary Medicine, Aristotle University of Thessaloniki. His research fields include neuroscience, comparative anatomy, designing animal research protocols, and bioengineering.

Sandeep Jagtap is a research assistant at the Centre for Sustainable Manufacturing and Recycling Technologies (SMART) at Loughborough University. He is involved in research to improve the environmental sustainability of the food supply chain through the application of the Internet of Things.

Jan Jarolímek is Assistant Professor, secretary, and researcher in the Department of Information Technologies FEM CULS, Prague. Jan has been involved in the development and use of ICT in rural areas for more than 20 years and has led many national and international projects.

Anatolii Karelov is a biotechnologist, currently employed as a researcher at the Institutes of Plant Protection and Food Biotechnology and Genomics, National Academy of Sciences of Ukraine. His research fields include molecular studies in wheat, nucleic acid purification, and bioinformatics.

Stratos Kartsonakis is a graduate student of the School of Production Engineering and Management at the Technical University of Crete. He obtained his master's degree in Technology and Operations Management at the University of Groningen. His research fields include multicriteria decision analysis, supply chain management, and quality control.

George Kokkonis received his PhD diploma from the Department of Applied Informatics, University of Macedonia, Greece. He received an Engineering Diploma from the Department of Electrical and Computer Engineering, Aristotle University of Thessaloniki and his MSc in Information Systems from the University of Macedonia. He has been working as a contract lecturer since 2006 in the Department of Computer Applications in Management and Economics, Western Macedonia University of Applied Studies, Greece.

Sotirios Kontogiannis graduated from Democritus University of Thrace, Department of Electrical and Computer Engineering. He received an MSc in Software Engineering and a PhD from the same department in the research area of algorithms and network protocols for distributed systems. He is currently a scientific staff member and Director of the Distributed Micro-computers Laboratory at the Applied Mathematics and Engineering research section of the Department of Mathematics, University of Ioannina.

Panagiotis Koulelis is a researcher at the Institute of Mediterranean Forest Ecosystems, Hellenic Agriculture Organization "Demeter." He conducts research and applied work on forest sector modeling, the economics of forest management, and the climate change/policy implications estimating the effects of the increased demand for wood and environmental services in natural resources sustainability.

Georgia Koutouzidou is an agricultural economist, now a PhD student. She has published papers in peer-reviewed journals and conference proceedings and has participated in two EU scientific projects. Her research fields include agricultural economics, agricultural policy, cost accounting for agricultural products, project appraisal, and statistical analysis.

Stavriani Koutsou is Professor in the Department of Agricultural Technology at the Alexander Technological Educational Institute of Thessaloniki. She is a rural sociologist with many years of experience in research regarding rural transformations, collective action, and social capital. She has many publications in peer-reviewed journals and has coordinated research projects in fields relevant to her expertise

Natalia Kozub is a molecular biologist, employed as senior researcher at the Institutes of Plant Protection and Food Biotechnology and Genomics, National Academy of Sciences of Ukraine. She is currently working on the biochemical analysis of quality in wheat (bread and durum), hulled wheat and triticale, and the molecular detection of biotic resistance in the aforementioned species.

Spyridon K. Kritas is Professor of Microbiology and Infectious Diseases, working in the School of Veterinary Medicine, Aristotle University of Thessaloniki, Macedonia, Greece. His research fields include microbial pathogenesis, epidemiology and prevention of infectious diseases (vaccines), as well as the efficacy of probiotic bacteria in animals both under controlled and field conditions.

Piermichele La Sala is a researcher in Agricultural Economics and Assistant Professor in Agri-Food Economics at the Department of Economics, University of Foggia, Italy. He has participated in numerous research activities and he is author of several international and national publications. His main research interests are rural and local development, agrarian and food policies, agricultural extension services, innovation and knowledge in agriculture, and cooperation in food supply chain. He is a member of the main Scientific Societies of the Agricultural Economics and Local Development fields.

Marco Lauteri is a senior researcher at CNR-IBAF, Italy, with research expertise in plant physiology and ecology. His main topics of concern are acclimation mechanisms to Mediterranean environmental constraints: seasonal drought, salinity, late frost, energy excess. He has expertise on gas exchange techniques and on stable isotope methodologies (IRMS) in the study of plant carbon-water economy.

Endro Martini is environmental geologist—an expert in planning and integrated planning of river basins/sub-basins, river contracts and participatory processes. He has experience in land defense service, quarries, mining, and mineral waters; hydrogeological risk, quarries, and environmental assessments; land valorization; and landscape protection.

Jan Masner is Assistant Professor and researcher in the Department of Information Technologies, FEM CULS, Prague. He is specialized in areas of web technologies, web design, programing, usability and user experience, data visualization, and data analysis, and is member of the Czech Society for Information Technology in Agriculture and other expert organizations.

Athanasios G. Mavromatis is Assistant Professor in Genetics and Plant Breeding in the School of Agriculture, Aristotle University of Thessaloniki. He works with cotton, tomato, chamomile, and wheat field and in vitro breeding. He is also a tissue and anther culture expert, focused on cotton.

Yvonilde Dantas Pinto Medeiros is a civil engineer with a PhD in Hydrology. Since 1984, she is lecturer in the MSc and PhD courses of the Polytechnic School. She coordinates the Resource Management Research Group, with research on integrated water resources management and semiarid hydrology.

Rosa Misso is Associate Professor in the Department of Economic and Law Studies at the University of Naples "Parthenope." Currently, she teaches "Sustainable governance of rural systems," "Quality management of agri-business and data base," and "Shared social responsibility and agri-food businesses." She has produced several publications on the topic of well-being sustainability.

Odysseas Moschidis is Professor in the Department of Marketing and Operation Management, University of Macedonia, Greece. He has participated in several research projects working on the areas of evaluating business data and market research and multidimensional statistical analysis. His research interests include scales evaluation, multidimensional statistical analysis, and evaluation of business data and market research.

Aliki D. Muradova is specialized in the fields of applied mathematics, computational mechanics, and mathematical modeling. She is currently an independent scientist and a research collaborator in the Institute of Computational Mechanics and Optimization, School of Production Engineering and Management, Technical University of Crete.

Michalis Neofytou is a production management engineer currently employed as a Quality Manager at the Mill of Crete SA. His research fields include the development of quality management and food safety systems, and business sustainability.

Vladimír Očenášek is Assistant Professor in the Department of Information Technologies, FEM CULS, Prague. Vladimír teaches courses such as Information and Communication Technologies, Internet Technologies, and Information and Communication. He is professionally involved in regional development and participates in several projects mainly in the education and pedagogy area.

Stefano Orsini is a researcher currently employed at the Organic Research Centre. His research fields include rural development agricultural and environmental policy evaluation and socioeconomic analysis of the agricultural sector.

Susanne Padel is senior program manager at the Organic Research Centre and coeditor of the *Organic Farm Management Handbook*, which summarizes business information for farmers and consultants. She has conducted research for more than 25 years on many aspects of organic farming in Europe, including planning and business analysis of organic farms, organic regulation, values and standards, in addition to qualitative social research.

Dimitra Papaefthimiou is currently employed as teaching and research staff (EDIP) at the Laboratory of Botany, Department of Biological Applications and Technology, University of Ioannina. Her academic qualifications include a PhD in Plant Physiology and a Licentiate of Philosophy in Plant Physiology, an MSc (by research) in Biochemistry, and a BSc (Special Honors) in Molecular Biology and Biotechnology.

Konstantinos Papageorgiou is a veterinarian microbiologist—army officer, working for the Greek army, and a research fellow in the Laboratory of Microbiology and Infectious Diseases, Faculty of Veterinary Medicine, School of Veterinary Medicine, Aristotle University of Thessaloniki, Macedonia, Greece. His research fields include microbial pathogenesis, epidemiology, and molecular virology, with special interest in next-generation sequencing.

Pierluigi Paris is a researcher at CNR-IBAF, Italy. He is an agroforestry specialist, with multiannual field experience on ecophysiology of agroforestry systems for valuable timber and bioenergy production. He is Project Manager for CNR-IBAF of several European projects, Coordinator of the Agroforestry Group of Italian Scientific Society S.i.s.e.f., and author/coauthor of approximately 90 scientific publications.

Evanthia Petridou is Associate Professor of Microbiology and Infectious Diseases, working in the School of Veterinary Medicine, Aristotle University of Thessaloniki, Macedonia, Greece. Her research fields include microbial pathogenesis, epidemiology and prevention of infectious diseases (vaccines), as well as the resistance of bacteria to antibiotics.

Dimitrios D. Piromalis is Assistant Professor in the University of West Attica. He holds an Automation Engineering degree and a PhD in the field of wireless sensor networks and embedded systems. He has participated in several research and development projects and has authored numerous publications. His current interests include among others Internet of Things (IoT), Smart Cities, Cyber-Physical Systems and Smart Grid, etc.

Andrea Pisanelli is a researcher at CNR-IBAF, Italy. His main research interests relate to rural development, focusing the attention on the needs and perspectives of stakeholders dealing with land use systems (farmers, entrepreneurs, institutional bodies, policy makers), and use of participative approaches and tools assessment of agroforestry practices according to farmers' expectations and willingness.

Shahin Rahimifard is Founder and Director of the Centre for Sustainable Manufacturing and Recycling Technologies (SMART) at Loughborough University. His research interests are sustainable product design, low carbon manufacturing, environmentally focused production planning and supply chain management, and product recovery, reuse, and recycling technologies.

Ian Rose is animal scientist currently employed as researcher at the Aristotle University of Thessaloniki, School of Veterinary Medicine and Laboratory of Animal Husbandry. His research interests include breeding programs, quantitative genetics, and bio-economic modeling, specializing in sheep.

Dimitrios Roustemis is animal scientist with a PhD in animal genetic improvement. He works for the Greek Center of Animal Genetic Improvement, and his main fields of expertise concern setting up breeding programs and defining the breeding goal for autochthonous breeds of sheep, goats, and bovine.

Giuseppe Russo is a research fellow at CNR-IBAF, Italy. He holds a PhD in Forest Ecology and has research interests in palaeoecology, ecophysiology of stable isotopes, and biocultural identity.

Alessandro Scuderi has a degree in Agricultural Sciences and a PhD in Agricultural Economics, with specialization in marketing of food products, and is a researcher at the University of Catania Di3A. Currently, he is Professor of Applied Economics at the University of Catania. He is the author of about 90 publications in national and international journals and books in the field of agricultural economics, marketing, ICT, and sociology. He has held and currently holds institutional roles within the agricultural food production nationally and internationally.

Pavel Šimek is a lecturer and researcher at FEM CULS, Prague. His research field is ICT in agriculture, data sharing, data visualization, and SW architecture and design. Pavel serves as the president of CSITA and he was board member of the EFITA during 2009–2013. He is a member of the Centre of Precision Agriculture CULS, Prague.

Michalis Skordoulis is a PhD candidate at the Democritus University of Thrace. He is the author of several articles published in scientific journals and the Proceedings of National and International Conferences. His research interests focus on total quality management, customer satisfaction, business management, marketing, corporate social responsibility, environmental management, management in education, banking, and finance.

Igor Sozinov is plant pathologist, employed as senior researcher at the Institute of Plant Protection, National Academy of Sciences of Ukraine. He is currently working on the biochemical analysis of quality in wheat (bread and durum), hulled wheat and triticale, and the molecular detection of biotic resistance in the aforementioned species.

Oksana Sozinova is plant pathologist, employed as researcher at the Institute of Plant Protection, National Academy of Sciences of Ukraine. She is currently working on the biochemical analysis of quality in wheat (bread and durum), hulled wheat and triticale, and the molecular detection of biotic resistance in the aforementioned species.

Bojan Srdjevic is Emeritus Professor with a PhD in Technical Sciences. His research interests are mainly in water resources systems analysis, multicriteria optimization, heuristics, and decision-making. His international publishing record includes more than 200 peer-reviewed papers.

Zorica Srdjevic is Full Professor with a PhD in Technical Sciences. Her research interests include multicriteria analysis, decision-making methodologies and supporting tools, and natural resources management. She has published papers mainly in peer-reviewed international and national journals.

Georgios E. Stavroulakis is a civil engineer specialized in the fields of computational mechanics, optimization, and control. He is currently employed as Professor at the School of Production Engineering and Management of the Technical University of Crete. His current research is focused on the modeling control and optimization of smart structures, including piezocomposites and neuro-fuzzy control. More details can be found in the web page of the Institute of Computational Mechanics and Optimization.

Michal Stočes is Assistant Professor in the Department of Information Technologies CULS, Prague. He has participated in many European and national research projects and published a number of papers in the field of information system in agriculture. Professionally, he has specialized in web technologies, metadata, and IoT in agriculture.

Luisa Sturiale is Full Professor in Economy Applied to Engineering and Economics and Environmental Evaluation at the University of Catania. She has authored about 100 publications in national and international journals and books in the field of marketing, web marketing, territorial marketing, economic analyses of production processes of the agricultural and food systems; application of MCDA in territorial and environmental issues; environmental economics; economic evaluation of natural resources; and green infrastructures in urban areas.

Eleni G. Symeonaki is a PhD candidate at Agricultural University of Athens. She studied Automation Engineering at the BA level and obtained her master's in Industrial Automation. She is employed as Laboratory Teaching Staff in the Department of Industrial Design and Production Engineering at the University of West Attica (Egaleo, Greece). Her research field is Internet of Things and Cloud Computing Technologies for the benefit of precision agriculture.

Georgios K. Tairidis is a production and management engineer specialized in dynamics and control. He is employed as a researcher at the Technical University of Crete (Institute of Computational Mechanics and Optimization) and as teaching fellow at the Technological Educational Institute of Crete. His research is focused on optimal fuzzy and neuro-fuzzy control of smart structures. Other fields of his research include energy harvesting, shunt piezoelectric damping, and optimal design.

Stavros Valsamidis is Associate Professor in the Eastern Macedonia and Thrace Institute of Technology, Greece. He has published more than 120 papers in international journals and conferences. His research interests include data analysis and data mining, digitalization and business models, and sustainability.

Jiří Vaněk is Head of the Department of Information Technologies FEM CULS, Prague. He teaches several courses, such as operation systems, computer networks, and information and communication technologies. He has been involved in ICT development for more than 30 years and specializes in agrarian sector and rural areas. He cooperates on several national and international projects.

Monica Varlese is a research assistant at the University of Naples "Parthenope." She collaborates with the Simone Cesaretti Foundation on the topic of quality management of agribusiness. She has produced publications on well-being sustainability, marketing, and ICTs.

Ioannis N. Xynias is Professor in Genetics and Plant Breeding, in the School of Agricultural Technology and Food Technology and Nutrition, at the Western Macedonia University of Applied Sciences. He is currently working on wheat, triticale, and rye field and in vitro (anther culture and wide hybridization) breeding and cotton and chamomile cytogenetics.

David Yáñez-Ruiz is Principal Research Scientist at the Spanish National Research Council. His research interests include developing new feeding strategies in ruminants using agro-industrial by-products and interventions applied in early life of the animal.

Rafaelle Zanoli is Professor at the Università Politecnica delle Marche. His research interests include economic and market research in the agro-food sector, with special experience in foresight analysis for policy-decision-making and consumer and supply chain research.

Konstantinos Zaralis is Assistant Professor at the Technological Educational Institute of Western Macedonia in Greece with research expertise in the area of livestock nutrition, feeding behavior and interactions between genotype and environment on animal welfare and productivity and on holistic sustainability assessments of livestock farms. He has been participating in a number of EU projects and has essential experience in EU-funded research.

Claudio Zaza graduated in Plant Medicine in 2013 and earned a PhD in Healthy Food Innovation and Management. He has worked on integration of ICT tools in agriculture, since he began PhD studies. His research project is titled "ICT tools for data management and analysis to support decisional process oriented to sustainable agri-food chains." He is very interested in precision agriculture tools and technics, such as unmanned aerial system and remote sensing, for agricultural inputs optimization.

Editors

Alexandros Theodoridis is Assistant Professor in Animal Production Economics and Applied Statistics at the School of Veterinary Medicine, Aristotle University of Thessaloniki, Greece. His main research interests include agricultural economics, agricultural policy, animal production economics, statistical analyses and evaluation of experimental data, and design of farm management surveys. Dr. Theodoridis has participated in several scientific projects funded by Greek governmental bodies, the European Union, and private companies and organizations. He has participated in 40 peer-reviewed journal articles, 5 book chapters, 18 conference papers, and 17 conference announcements in various fields of agricultural economics and applied statistics.

Athanasios Ragkos is an agricultural economist currently employed as a researcher at the Agricultural Economics Research Institute of the Greek Agricultural Organization "Demeter." He has taught several courses in Greek Universities and Technological Institutions. He has participated in 22 research projects (one of which as Coordinator) and has published more than 55 papers in peer-reviewed journals and conference proceedings. His research fields include the economics of agricultural and animal production, environmental economics, the multifunctional character of agriculture and extensive farming systems, project appraisal, and sustainable rural development. He is experienced in designing interdisciplinary approaches with regard to the above-mentioned research fields by maintaining close collaboration with scientists of relevant fields.

Michail Salampasis is Professor of Internet Technologies in the Department of Informatics at the TEI of Thessaloniki. He was also a Marie Curie Fellow at the Institute of Software Technology and Interactive Systems, Vienna University of Technology. His main research interests are in applied and interdisciplinary studies in information science, information seeking in large professional search systems, patent search, distributed information retrieval, and search systems evaluation. He is author or coauthor of over 80 publications in refereed journals and international conferences.

Part I
Contemporary Views of Rural Development, Sustainability and Cooperation

A New Development Paradigm for Local Animal Breeds and the Role of Information and Communication Technologies

Athanasios Ragkos, Georgia Koutouzidou, Stavriani Koutsou, and Dimitrios Roustemis

Abstract Local breeds are endowed with numerous advantages in terms of adaptability to specific conditions and the sustainable utilization of resources. Nonetheless, their role is neglected when their multiple societal, economic, and environmental contributions are not properly acknowledged and interest is focused only on productivity issues. This paper explores the role of local breeds in local development as an efficient and effective strategy to ensure their long-term sustainability. The paper focuses on the reasons behind the underestimation of the values with which local breeds are endowed and discusses how economic development, policies, and market competition have rendered many of these breeds in danger of extinction. The chapter proposes an integrated development framework combining the notion of territorial development and social innovation and focusing on multifunctionality considerations and resilience and vulnerability aspects. Based on these, the potential role of ICT, new technologies, and innovations is assessed.

Keywords Multifunctionality · Territorial development · Social innovation · Resilience · Vulnerability

---

A. Ragkos (✉)
Agricultural Economics Research Institute, HAO Demeter, Athens, Greece
e-mail: ragkos@agreri.gr

G. Koutouzidou
Department of Applied Informatics, University of Macedonia, Thessaloniki, Greece

S. Koutsou
Department of Agricultural Technology, Alexander Technological Educational Institute of Thessaloniki, Thessaloniki, Greece
e-mail: skoutsou@farm.teithe.gr

D. Roustemis
Center of Genetic Improvement of Animals, Nea Mesimvria, Thessaloniki, Greece

A. Theodoridis et al. (eds.), *Innovative Approaches and Applications for Sustainable Rural Development*, Springer Earth System Sciences,
https://doi.org/10.1007/978-3-030-02312-6_1

1 Introduction

Local farm animal breeds constitute important assets for European territories. It has been established in literature that local breeds are better adapted to local conditions (Legarra et al. 2007; Ragkos and Lagka 2014) and exhibit higher resistance to disease (Mpizelis 2013). Local breeds, depending on the production system, have relatively low requirements for purchased inputs (mainly feedstuff) (Roustemis 2012; Ragkos et al. 2014), especially when they utilize rangelands. Despite the fact that their milk yields are relatively low, the quality characteristics of their products are very satisfactory (high fat and protein content) (Vafiadakis 2013) (Table 1). Nonetheless, in Greece they are not always remunerated to their true potential because, in many cases, the pricing policy of dairy companies is based only on quantity (Roustemis 2012).

The economic advantages of the Greek farm animal breeds, in particular, include clearly better product quality (meat and milk), lower needs for fixed and variable capital, adaptability to grazing (which results in significant savings in feeding costs), resistance to disease (leading to reduction of veterinary costs), and more lactation periods per ewe (higher longevity). With these characteristics, local breeds are suitable for traditional farms that formulate multifunctional production systems throughout Greece, as in numerous European settings alike (OECD 2001; Lankoski

Table 1 Production characteristics and/or populations of selected Greek local breeds

Breed	Milk yield (kg/ewe/year)	Milk quality (fat%/protein%)	Population
Sheep[a]			
Kalaritiko	100–110	8–9%/6%	6.104
Katsika	90–170	N/A	1.633
Sarakatsani	50	N/A	1.205
Chios	260–380	5,9-6,8%/5,5%[b]	86.000
Lacaune[c]	270	7,1%/5,3%	–
Friesland[c]	500-700 (in Germany)	6/N/A	–
Assaf[c]	448	7,5%[d]	–
Goats			
Skopelou[a]	>350	5,5%/4,6%	8.400
Alpine[e]	500–600	3,4%/3%	–
Bovine			
Vrachykeratiki[f]	–	–	3.100
Greek Buffalo[g]	Only scattered milk production	N/A	4.789

[a]Unless specified otherwise, all data about Greek sheep and goat breeds were found in http://www.minagric.gr
[b]Roustemis (2012)
[c]https://mediasrv.aua.gr
[d]http://www.gaiapedia.gr/
[e]https://oceclass.aua.gr/
[f]http://www.kgbzath.gr/
[g]Center of Animal Genetic Improvement of Nea Mesimvria

and Ollikainen 2003; Rege and Gibson 2003; Ragkos and Lagka 2014). These systems play multiple roles (social, economic environmental, developmental) beyond the production of market goods, thus contributing to sustainability (Ripoll-Bosch et al. 2013). Consequently, the loss of genetic material entails the loss of traditional farming and jeopardizes the viability of local production systems (Rege and Gibson 2003) where the prevalence of local/autochthonous breeds plays an important role in sustainable land management. Lacking local breeds, farmers do not always have access to animals well-adapted to the local climate conditions, which are often very harsh.

The past decades have witnessed significant losses of genetic diversity. This is an issue of particular importance at the country level but also at European level. In 1992, it was reported that 28% of livestock breeds had already disappeared or had become threatened or rare during the last 100 years (World Conservation Monitoring Centre 1992). Since then, the recognition of the problem has been reflected in scientific and political discussions. In the context of the Interlaken Declaration in 2007, the main objective was the development of a World Plan of Action on Genetic Resources of Farm Animals (conservation, sustainable use, and development of genetic resources of farm animals for food and agriculture, global food security, improvement of human nutrition, and rural development). This Declaration called for registration, monitoring, characterization, sustainable use, development, and conservation of the genetic resources of farm animals. The European Association for Animal Production (EAAP) has developed a genetic material bank of animal breeds, by encouraging, *inter alia*, the use of breeds that contribute to the protection of genetic diversity.

In Greece, many local breeds are actually in danger of extinction – despite their advantages which were outlined above – and are supported through the Common Agricultural Policy (CAP). Under the most current policy framework, 5 bovine, 17 sheep, 1 goat, 1 swine, and 6 horse breeds are eligible for support. Also, there are other local breeds which are not endangered and have partially capitalized on their potential, but still have important roles to play in order to become mainstream in Greek livestock production. For sheep and goats, such examples include Chios breed, which shows important economic advantages (Theodoridis et al. 2012, 2013); Lesvos breed, which is very important for the island of Lesvos; and Frizarta sheep breed, which actually achieves high milk yields.

The preservation of local farm animal breeds is an important issue with two equally important dimensions (Ligda and Casabianca 2013). The first is to protect and rescue them in the short run, especially those under imminent danger of extinction or the vulnerable ones. The tools to achieve this goal include – but are not restricted to – the provision of income support in order to compensate farmers for the economic losses that they incur by choosing to maintain them. In Greece, relevant legislation has been introduced in order to implement European Rural Development legislation. Table 2 shows the evolution of such measures. The economic incentive and the support provided to farmers have been significant ever since the 20 years of implementation, although the actual amount paid to farmers has been changing according to the revisions of European and Greek legislation. The

Table 2 Policy measures supporting local animal breeds in Greece

European legislation	Greek legislation	Brief description
C (97) 551/ 28–4-97 Decision of the European Commission E(97) 3098 /24–11-97	Program 3.3.1.4 for the conservation of threatened indigenous breeds of farm animals (since 1998)	Aimed to increase populations: bovine (10–25%); sheep and goats (10–20%); equids (steady)
Reg. EC/1257/1999 Article 24, par. 2	Document of Rural Development Programming Axis 3: Measure 3.7 "Program for Threatened Indigenous Animal Breeds" Also Measure 5.1 permitted keeping herdbooks	2.220 farms (148% of target), 66.706 animals were supported. Total expenditure 5.050.000 €[a]
Reg. EC/1698/2005 Article 36 (a) (iv) and 39	Rural Development Program of Greece 2007–2013, Axis 2, Measure 214 Action 3.1 (Conservation of threatened indigenous breeds of farm animals) Action 3.4. (Conservation of genetic resources in livestock production)	
Reg. EC/1305/2013	Rural Development Program of Greece 2014–2020	Sub-measures 10.1 and 10.2 are the continuation of the previous Actions 3.1 and 3.2, respectively

[a]Source: Ex post assessment of the Document of Rural Development Programming of Greece 2000–2006

second dimension targets the long-term sustainability of production systems that rear local farm animal breeds. It refers to the ability of these breeds to support farms in competition in terms of productivity and economic performance against other types of farms which rear improved animal breeds. Interventions of this type aim at securing benefits and at strengthening the linkages of the former farms to markets, existing value chains, and other actors (see, for instance, Verrier et al. 2005). In other words, they seek to provide incentives and other motives to keep these breeds not as part of a campaign to maintain tradition or for environmental conservation, but as tools for development.

This chapter focuses mainly on the second dimension, as it endeavors to move beyond the "traditional" viewing of local breeds as "rustic" and "folklore". Its main purpose is to discuss and propose measures and remedies to protect local animal breeds from their declining trends. A framework is described by means of which a shift of interest is facilitated, from a policy-driven rationale to more market-oriented approach, with long-term strategic development effects. Firstly, the chapter endeavors a discussion of the reasons behind this reduction and seeks to find whether this trend is justified or whether there are factors which are disregarded, thus leading to the current situation. It continues with a presentation of the main external and internal barriers hindering the operation of production systems rearing local breeds. Afterward, the chapter presents a brief outline of the basic concepts of a territorial development pattern for local breeds, which builds upon social innovation and multifunctionality and aims to increase the resilience and to reduce the vulnerability of these systems. Then, potential policies and strategies to support the protection of these valuable resources are discussed, especially based on the use of modern technologies, ICT, and novel socioeconomic approaches.

2 The Situation So Far: The Need for a New Paradigm

In Greece, genetic diversity of local farm animal breeds is threatened by massive imports of foreign breeds which are reared either as purebreds or as reproduction animals for crossbreeding in order to improve the milk and meat productivity of farms. The adverse situation that many local animal breeds face nowadays is mainly due to the underestimation of their real values, as they have not been properly acknowledged as public goods (Ligda and Zjalic 2011). This is directly related to the economic development process, the agricultural policies in force for decades but also to market competition.

The economic growth process in the agricultural sector of most European Union (EU) countries, including Greece, was characterized by a transition from extensive labor-intensive to modern capital-intensive systems (Karanikolas and Martinos 2012). New modern farms started to emerge, which adopted novel technologies and developed an entrepreneurial organization model (Ragkos et al. 2014; Ragkos et al. 2015). Farms stopped to pursue subsistence and geared toward strategies to reduce production costs, including the achievement of economies of scale (De Rancourt et al. 2006). The average farm size increased and this was accompanied by specialization in the production of one or a few livestock products, signaling the end of the era of "economies of scope." In order to improve their economic performance, livestock farms were in need of animals which would be pertinent to a holistic management approach, so as to facilitate workers to include a new lifestyle, with less work on the farm and more free time to enjoy other amenities (Roustemis 2012). At the local level, the economic development pattern led to specialization in non-farming activities as a result of rural diversification. Under these circumstances, livestock systems were oriented to these market requirements, where the demand for products of animal origin has been ever-increasing. The increased productivity of the improved breeds was thus a particularly desirable feature for farmers, who further increased the import of such animals into their livestock farms. For all these reasons, local breeds were gradually substituted by others, fewer in number, of foreign origin and of lower resilience, as will be shown in subsequent sections.

Apart from economic growth, the policy framework favored a productivist model (Bye 2009) through the provision of subsidies (Ragkos et al. 2016), which has contributed substantially to the devaluation of animal genetic resources. EU price policies, especially in livestock production, encouraged high milk yields, thus leading to an increase of dairy breeds at the expense of dual-purpose local breeds. Policy incentives for safeguarding local breeds have not been proven adequate to inverse this trend (Canali and Ecogene Consortium 2006). At the same time, high subsidies to crops altered the competitive relationships between crop and livestock production substantially: the former became more profitable, and thus several livestock farmers abandoned livestock production. In addition, the alternative cost of grazing increased due to the increased profitability of crops, and the cultivation of fodder plants was substituted by other highly subsidized crops (Karanikolas and Martinos 2012), burdening the production costs of livestock farms and further reducing its competitiveness in relation to agriculture. The livestock farms that ceased to operate as a result of these issues were mainly those rearing local breeds. Policies also encouraged increased productivity at the national level also for political reasons, related to the consolidation of national sovereignty and the treatment of market uncertainty, noncyclical market fluctuations, and unexpected extreme conditions. Surprisingly, in livestock farming, productivity growth was pursued through the introduction of improved genetic material from abroad, ultimately increasing the reliance on imported genetic material and partly counterbalancing the advantages of self-sufficiency in domestic markets.

Market conditions combined with the characteristics of modern livestock farming intensified threats to local animal breeds. Companies often supported intensive production systems that require improved genetic material at the expense of genetic diversity. In this context, private consultants in rural areas, without the operation of a central extension service, often favored the expansion of improved breeds, without reassuring first that they are adapted to local conditions or are suitable for semi-extensive traditional production systems. So, although these breeds were inappropriate in numerous contexts, they displaced local ones. In addition, this trend limited the focus of breeding programs on only a small number of primary traits, with a direct and observable effect on farm economic performance (dairy production, prolificacy), thus increasing the demand for breeds with such characteristics and shifting interest from local breeds with desirable secondary traits (e.g., resilience against diseases).

3 New Conditions Call for New Actions

As is obvious from the above description of the situation in previous periods, the development model of livestock production so far has not been proved successful in the Greek European/Mediterranean setting. In addition, livestock production systems nowadays operate within a changing context, which sets specific barriers for development but also provides opportunities for future transformations. This context is defined by external factors – i.e., those attributable to the external environment, which are not under the control of farms – but also internal ones, which characterize production systems and farms.

3.1 External Factors

The Common Agricultural Policy

In the European Union, the CAP has been a major factor influencing the evolution of production systems in general. Actually, European livestock farms can benefit from payments and funding from both Pillars (Reg. EC/1307/2013 and Reg. EC/1305/2013). In particular, Pillar I payments are allocated to farmers either in the form of income support or as coupled payments for sectors of particular importance to member states. For instance, in Greece sheep and goat farms and cattle for meat are entitled to this latter type of payments. For Pillar II, a long list of measures supporting rural development in general is available to achieve general policy objectives such as the generational renewal in the livestock farming sector; new investments in infrastructure; protection of fragile, remote, and mountainous areas; etc. In all cases, all types of livestock farms are approached under a common framework by the CAP (Ragkos and Nori 2016). An exception is the payments that farms rearing rare local breeds receive (see Table 2). Nonetheless, these payments are the only compensating measure available, and in fact they sometimes isolate farms from markets instead of helping them to integrate in market conditions and existing value chains. Additional barriers are imposed to farms rearing local breeds by legislation regarding quality standards, which also poses restrictions to the production of local specific products (e.g., cheese from raw milk).

The Economic Crisis

Livestock production systems are now obliged to operate within globalized markets where agricultural goods are commoditized, experiencing the "agricultural squeeze." And while the optimal performance within this environment is the goal for large commercial farms, small-sized extensive or semi-extensive farms do not share this objective. On top of that, the economic crisis, already evident in several EU settings, puts additional pressure in the operation of the primary sector. Such crises are not infrequent, either as a result of changes in international conditions and the overall macroeconomic framework or for reasons associated with its own particularities (uncertainty in food production, volatility of farm income). Under the pressure of the crisis, the economic performance of farms is heavily burdened by excessive production costs (Ragkos et al. 2016), due to increases in input prices (Karanikolas and Martinos, 2012). Reduced liquidity deprives from opportunities for investment and innovation and potentially leads to lower levels of farm and agribusiness operation, decreased productivity, and efficiency losses. It should be stressed that empirical evidence has shown that extensive farms, the ones which usually rear local breeds, have demonstrated generally better behavior during the crisis (Ragkos et al. 2014, 2018). Nonetheless, late payments of income support to farms rearing local breeds have put many of these farms in a very unsound condition, which reveals their vulnerability to policy changes.

Changing Trends in Food Consumption and Food Marketing

There is strong evidence that an increasing proportion of consumers is very interested in products that cover for societal demands for food safety, animal welfare, and sustainable use of natural resources. On the other hand, the consumption trends of industrial products are increasing, also due to population dynamics and increased demand for food in general. Within this framework, food produced by local breeds, which is endowed with specific characteristics such as clear origin, close relationship with the territory, safety, etc., could find their way in markets, but with alternative marketing strategies. For instance, focusing on niche markets where consumers are particularly aware for food quality and safety would be more efficient that channeling these products to conventional markets, as they are not price-competitive. And while there are relevant examples in other EU countries (France – Roquefort – Lacaune sheep – Roquefort cheese; Spain – Basque country – Latxa and Carranzana sheep – Idiazabal cheese; etc.) in Greece, there is not even one example of cheese explicitly linked not only to a specific place of origin but also to a specific local breed. Similarly, there is much room for better marketing of meat from local breeds, as some of them have highly appreciated names (e.g., the Greek breed Kalaritiko is known for its good quality) but cannot cover for demand due to their small populations. For cattle and buffalo meat from local breeds, there is little scientific knowledge regarding its quality.

Climate Change

Climate change is thought to have a significant impact in the future development and adaptation of livestock production systems. Extensive systems are seemingly more exposed to these dynamics, which also affect the availability of forage and the grazing capacity of grasslands. However, these systems benefit from rearing local breeds, as they are better adapted to local conditions and resistant to diseases and adverse weather conditions. These breeds should be better examined and included in breeding programs in order to increase the endurance of animals reared within all types of systems, under the imminent effects of climate change. Hence, the protection and conservation of local breeds is an asset for the sustainability of livestock systems in general.

3.2 *Internal Factors*

Competition with Other Economic Activities

Because of competition from intensive/industrial/systems, farms rearing local breeds face increasing difficulties. As these local breeds have not been improved, their productivity is generally low, and so respective farms cannot compete with intensive farms in mainstream markets. Table 1 shows the average yields of selected local breeds compared to imported improved breeds. Extensive farms rearing local breeds also face competition from other more profitable economic activities (alternative energy sources in the rangelands they use, intensive crop production, tourism, industry, etc.). Under this pressure, farms are obliged to intensify, which is not always compatible with their culture, values, and lifestyles. Farmers decide to replace local breeds with improved ones, or, like the example of Greece, they decide to introduce males from improved breeds for crossbreeding in order to ameliorate their productivity. Uncontrolled matings of this sort are the cause of genetic diversity losses.

Low Profitability

The products of local breeds (raw or manufactured) nowadays face difficulties in capturing the market shares they deserve, as already outlined in Sect. 3.1.3. Some of the reasons behind this situation include small volumes of production, which burden industries with significant transportation costs; difficulties to maintain stable quality, due to the traditional management practices commonly applied; low levels of market information, as the breeds are often reared in remote territories; lack of or inadequate standardization (especially of meat); and high transportation costs even for transformed products, due to high distances from large consumption centers. Combined with their low productivity, these farms usually achieve low economic performance compared to intensive systems. This situation is further burdened by structural barriers (e.g., small average size).

Despite these external and internal barriers, the animal production systems typically rearing local breeds are endowed with significant benefits (Belibasaki et al. 2012). As it will be shown in what follows, these systems are able to cope with unexpected situations and market outcomes. Also, as mentioned above, policy measures and other interventions are available to support the continuation of rearing of these breeds. Efficient or not, these measures may contribute to safeguard local genetic resources by providing a strong foundation on which to build viable production systems for local breeds.

4 Toward an Integrated Development Framework: Basic Concepts

4.1 Multifunctionality

The weaknesses described in Sect. 3 are partially due to the low recognition of the various roles that local breeds play in their respective territories and within the systems they are reared. Indeed, they are multifunctional; as a working definition, this corresponds to the fact that these breeds – and their corresponding production systems – produce a wide range of goods and services jointly with food (milk, meat, dairy products, etc.). The former outputs are non-traded, which means that they have public good characteristics or are externalities, affecting various aspects of human activity (OECD 2001). The effects of these local collective goods reach society through complex chains with multiple stakeholders. Consequently, the result of rearing local breeds is the generation of products, services, and other outputs with particular social, economic, and environmental features, which are not properly priced through market mechanisms. As a result, their true contribution to societal objectives is not fully accounted for (Vermersch 2001; Allaire and Dupeuble 2002; OECD 2003) due to market failures (Ingersent and Rayner 1999).

The socioeconomic role of local breeds, as mentioned above, is of utmost importance, for mountainous/marginal/inland areas where the economic activity is not sufficiently diversified. Livestock production has been the main, if only, economic activity for centuries, and these breeds are the best-adapted ones (if not the only) to local conditions which are sometimes harsh. Nowadays, even in communities in which other sectors of the economy have emerged, rearing these breeds is still a significant source of income, always under the dangers and pressures of Sect. 3. Local breeds are therefore important for rural livelihoods and particularly responsible for keeping these areas alive and inhabited.

Farms rearing local breeds play a very important *environmental role* contributing to mitigating environmental degradation through the provision of *ecosystem services*. In a great variety of European agroecological settings – ranging from islands to highland territories and remote areas – rearing local breeds has contributed to maintaining diversity of flora and fauna species and has shaped important natural parks and reserve areas. Grazing in mountainous rangelands formulated particular landscapes and pseudo-alpine ecosystems, particularly affecting plant coverage and diversity, the soil, and bird species (Ispikoudis et al. 2002; Koocheki and Gliessman 2005; Ganatsou et al. 2008). The important element of grazing by local breeds is that they are well-adapted to the morphology and vegetation of specific territories, so they do not pose threats related to overgrazing, as local vegetation is able to accommodate their dietary needs.

The cultural heritage related to local breeds is important and encompasses heterogeneous sociocultural contributions. Some relevant elements include tacit knowledge concerning the functioning of local ecosystems; the specific needs and characteristics of animals, weather, and nature; farm management based on the specific needs of these animals; habits and customs related to the production cycle of local breeds (e.g., local festivals during the summer or at the sheering period); traditions, norms, and tacit rules related to land uses; and processing skills, including the manufacturing of typical products. An important element is their intangible cultural heritage (www.unesco.org/).

4.2 Resilience and Vulnerability

Resilience is based on the fundamental argument that economic units may survive only if they are able to change on time when they face changes in the economic, natural, or social environment (Polman et al. 2011). Under volatile conditions, uncertain outcomes, and turbulent environments, resilience examines the various ways by which farms respond to such changes and their respective defense mechanisms in order to maintain their essential structure, function, and feedbacks (Gunderson 2003). This also extends to future turbulences, as resilience also refers to the utilization of accumulated knowledge from the past in order to face future turbulences effectively and to reach new levels of balance (Berkes and Folke 2002; Chapin III et al. 2009). Resilience extends toward many more dimensions than ecology, also embracing social and economic considerations. In this context, an intriguing question is why local breeds have not been proven as resilient as the family agropastoral farms which rear them, which is reflected in their declining populations.

The lack of economic resilience is linked to vulnerability, because exogenous changes in the economic environment can increase the vulnerability of an area or system (Schouten et al. 2009). Vulnerability is originally an ecology-based concept, and, while there are several papers discussing resilience to climate change and environmental factors (Dong et al. 2011), little work has been done in discussing the "social" or "economic" vulnerability of farms and systems (López-i-Gelats et al. 2016). Eakin (2005) provided an interesting discussion of "social vulnerability" as opposed to "ecological" vulnerability. Her approach recognized complex relations of power, resource distribution, knowledge, and technological development in outlining the notion of vulnerability of rural communities. As a multidisciplinary issue, vulnerability examines a broad array of parameters (Dong et al. 2011).

4.3 Territorial Development

Most countries agree that the diversification of rural economies is a prerequisite for lively rural areas; however, in remote regions such as in EU's less favored areas, the role of agriculture is predominant in the local economy and in employment. Thus, taking into account the issues of competitiveness and sustainability, a strategic goal of rural economy diversification emerges toward a pattern of less reliance on agriculture and of introduction of new and novel economic activities. This pattern is highly pertinent to *territorial development*. A definition has been provided by the ECDPM (www.ecdpm.org) "Territorial development refers to integrated multi-sector development across a specific portion of territory, guided by a spatial vision of the desirable future and supported by strategic investments in physical infrastructure and environmental management. This definition makes no reference to scales (local, regional, national or transnational) and applies equally to any of them."

This type of development occurs through two pillars, i.e., the social capital in an area and the production of specific products characterizing territories. *Social capital*, which represents the ability of actors in a society to cooperate and act collectively (Koutsou et al. 2014), is an important asset in this process. It is formulated through trustworthy relationships and nourishes networking and collective action. The innovative element of the "territorial development" model is that it allows to build development socially through cooperation and networking among stakeholders of a specific area (e.g., producers, public services, local groups, associations, etc.), all of which aim at the sustainable use of resources. Territorial development requires to detect and focus on territorial-specific characteristics which can be used as development resources. Section 5 provides more insights regarding the potential role of local breeds as such resources, which could trigger a development process. In territorial development in rural areas, agriculture usually provides for all sectors of the economy, generating multiplicative effects.

4.4 Social Innovation

Social innovation refers to new strategies, initiatives, processes, or organizations that meet pressing social needs and profoundly change the basic routines, resource and authority flows, or beliefs of the social system in which they arise (Mulgan et al. 2007; Westley et al. 2013). It involves new institutions and governance systems, including networking or new partnerships among stakeholders – individual, private, or public. As it is not limited to technology, social innovation is particularly relevant to rural space in general, as it describes a type of innovation which is not limited to a particular group but can rather be developed by the vast majority or even by all members of a rural society and contributes to empowerment, especially of marginal actors, across the community. By definition, social innovation incorporates networking among stakeholders – individual, private, or public – in order to rearrange the agro-food system, to protect the environment and natural resources, and to restructure the relationships between rural and urban regions in fairer ways.

5 A New Framework for Rural Areas: Local Breeds as Development Resources

The consequences of the unrealistic pursuits of the formerly applied development patterns (see Sects. 2 and 3) have made the introduction of a new development pattern almost imperative. Its basic underpinnings were presented in Sect. 4. In what follows, a concept is outlined, where local breeds are viewed as development resources. It basically aims to increase benefits not only in the primary production system or across the value chain but also at a broader scale to other related actors, in order to generate more opportunities for employment and income in rural territories. Such a development pattern could induce the emergence of models where farms, dairies, manufacturers, tourism professionals, female entrepreneurs, and other stakeholders run their businesses efficiently and communicate directly.

Figure 1 illustrates the proposed development framework, by depicting the situation before and after its implementation. The left-hand side of the figure represents the actual situation. The territory is not clearly defined, and this is why it is represented by a dashed line. This is due to the fact that territorial products (i.e., resources for development) have not yet been clearly defined. The local economy is mainly based on non-livestock activities or on farms not rearing local breeds. Actors involved in these activities are in the center of the economic environment. On the other hand, local breeds are of low importance for the territory, and, most of all, they are interconnected to multifunctional characteristics (e.g., local products, culture, landscape, etc.) but not clearly. In other words, these characteristics are there, but are not properly recognized and attributed to rearing local breeds.

The right-hand side of the figure represents the effects of territorial development. Local breeds are now in the center, which is clearly defined. This is due to the fact that "territorial products" have been recognized and clearly connected to local breeds. Multifunctional characteristics, existing as a result of livestock production and rearing of local breeds in the area, have been "transformed to innovative products and services." Other stakeholders are still present in the territory, but

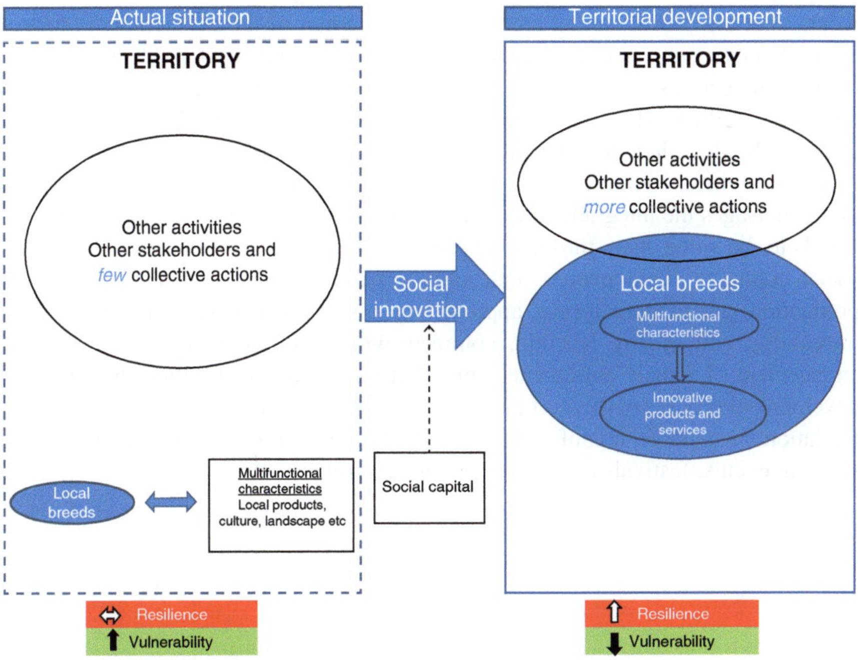

Fig. 1 The transition toward territorial development

now they maintain closer relationships with those rearing local breeds. They also undertake activities connected to innovative services and products. The transition to this type of development is based on social innovation. Local actors collectively (re) discover resources and put them into action. This process is facilitated by social capital: trustworthy relationships and collective actions. Although there are few of the latter in the territory in the actual situation, they are enough to initiate the process, acting as internal triggers. In the future situation, however, local actors are better organized in collective actions in order to maintain the benefits of the development process. Breeding associations, although they play a trivial role in the actual situation, could operate as facilitators in this process.

Under this situation, it is interesting to comment on the resilience and vulnerability of the system. Indeed, the resilience of the whole system is moderate in the actual situation, as systems rearing local breeds have generally demonstrated high resilience, which is not reflected, however, in the deteriorating populations of these breeds. On the other hand, what is perhaps more important is that the system is highly vulnerable to external shocks. Since there are no well-defined activities specific to these breeds, the system is dependent on external funding through CAP measures. Potential changes in this framework would be shocking and would radically threaten its viability. The development process brings about improvements in both these characteristics. The system becomes more resilient and less vulnerable to external shocks, as now market-based activities strengthen the role of local breeds, while collective actions are there to safeguard their role in the territory.

Within this novel pattern, the benefits, specific characteristics, and unique elements of rearing local animal breeds are in the center of the development process. Such characteristics would include traditional dairy and meat products, arts and crafts, and customs and festivals related to the production cycles of these animals, territory-specific production practices, particular landscapes, etc. They have all been collectively identified, and they all constitute "externalities" of rearing local breeds, i.e., their multifunctional characteristics (attributes). In other words, this new paradigm envisages the integration of multifunctional characteristics (attributes) of local breeds to the value chain through the process of social innovation. As a result, new social contracts could evolve such as new partnerships among artisanal cheese production and manufacturing, approaching new market niches, enhancement of traceability and quality labeling, contracts with hotels and other agencies, and engagement of female entrepreneurship. Activities generated within the territorial development process of this sort may include alternative marketing methods, accommodation, restaurants, manufacturing by developing tacit knowledge, naturalistic or sporting events, festivals associated with local products, etc.

6 Novel Approaches and New Technologies for the Protection and Promotion of Local Breeds

The overall objective of protection and conservation strategies should be to save, protect, and valorize existing populations of local breeds by employing interdisciplinary approaches and novel tools addressing all aspects of these multifunctional systems. These goals have been pursued through public genetic improvement programs but also through private initiatives by cooperatives, businesses, and nongovernment organizations. Whatever their form, the success of these actions is highly dependent on the mobilization of all actors through multi-actor approaches. Wider stakeholder involvement is a key to successful conservation programs, as they should be designed with a bottom-up approach, which should address the real problems of all actors. The genetic improvement process, which will enhance the overall state of the local breeds populations in order to support the achievement of the strategic goals of the territorial development approach, will need to utilize modern tools, methodologies, and technologies in order to achieve rapid, sound, and sustainable results. Considering that ICT continuously penetrate rural and urban societies, they are some of the basic tools to be incorporated in effective strategies of the sort. In what follows, some of these tools are briefly presented.

- Introduction of holistic farm management in farms rearing local breeds, especially in nucleus farms. This approach is rather a philosophy, which will allow farmers to make the best of the animals they rear and adopt innovation, without, however, altering their predominantly traditional production practices.
- An integrated reproduction strategy in farms. This includes the use of ICT tools for data recording, which will assist controlled mating and choice of animals, combined with artificial insemination and other modern technologies such as embryo transfer.
- Integrated databases with population statistics and historical, socioeconomic, and on-site measurement data, including productivity and quality characteristics. These data should be complemented with genotyping results, using cutting-edge genomic approaches. Such databases could inform current and future conservation and improvement efforts.
- Multimedia campaigns for the general public, in order to get them closer to the benefits that local breeds may bring to society. The Internet and social media constitute low-cost alternatives to contribute toward this way.
- Information and training are necessary also for farmers. More and more ICT techniques are included in curricula for farmer education, and they are useful also for this purpose. E-learning and the use of multimedia are important for providing aware farmers detailed information about the benefits of local breeds.
- Such methods are also important for establishing meaningful networks of farmers rearing such breeds. There are examples of organizations supporting local breeds at the local level (there are actually more than ten in Greece), many of which work on the genetic improvement of specific breeds. There also regional or even

national organizations (e.g., the CORAM (http://www.races-montagnes.com/) in France), and efforts are currently in force to support even supranational organizations (such as the European Shepherds Network – ESN). E-networking provides an ideal means of constant communication among them, in order to shape concrete linkages, also considering that these organizations usually operate with minimum funding and other means. E-networking can also be induced at the vertical level establishing food value chains (e.g., farmers – dairies – other SMEs – retailers – special interest tourism businesses (restaurants, visitable dairies, etc.) – consumers, etc.).
- The implementation of these strategies could be further enforced by a branding scheme and/or the certification of products (e.g., Protected Designation of Origin). This way, local breeds will favor the local short value chains, thus promoting sustainability and providing significant opportunities for innovation and development. All these could be contextualized by ICT applications such as e-commerce and e-sales platforms.

The strategies whose principles were described above can also be imprinted in policy measures. The new Rural Development Program of Greece (RDP) 2014–2020 is an example of the sort. Apart from measures tackling directly the issue of genetic resource losses (M10.1 and 10.2) (see Table 2), the protection of local breeds and genetic diversity is also in accordance with other RDP measures. For instance, reference should be made to M01 concerning cooperation for training, M03 about quality certification, M07 about cultural identities, M14 about animal welfare, M16.4 about short supply chains, etc. Strategies of this sort are in line with Europe 2020 strategy for SMART growth: smart (investments in education and innovation), sustainable (a low-carbon economy), and inclusive (emphasis on job creation and poverty reduction in rural areas).

7 Conclusions and Future Directions

This chapter outlined a development framework for Greek local breeds. This framework aspires to showcase their dynamics and to render them effective tools for rural and regional development. This framework can be equally applied to declining local breeds – like those presented in Table 1 – but also to local breeds with higher populations which are not endangered. Toward this direction, a comprehensive strategy is needed with a series of actions that will benefit the rural economy of targeted territories but also the country as a whole. The required interventions are basically long-term and are summarized here:

- Integrated genetic improvement programs and effective use of CAP incentives
- Combination of livestock farming with land use: improvement, rational distribution, integrated management of rangelands, and improved access to them

- Payment of milk and meat in relation to their quality characteristics, market research, increased public awareness, and focus on market niches interested in quality products and functional foods
- Clear diversification of production and marketing strategies for products of local breeds from conventional, industrial products
- Cooperation between producers, better communication with research centers, universities, genetic improvement agencies, and public services

Livestock farmers are in the heart of this process, as their real needs have to be recorded and taken into account (Ragkos and Abas 2015). Parallel to this, they should be involved in training and information activities concerning animal traits, the economic benefits of rearing local breeds, and development strategies for livestock production in general. Rural development programs (national and regional) combined with local/regional initiatives provide support for such actions, thus generating added value for local societies. Such activities should definitely be incorporated in the development framework presented here, as they will generate all the necessary prerequisites to stabilize networks, in order to consolidate favorable conditions for cooperation and collective actions. The participation of stakeholders and synergies with existing networks and food value chains will also be beneficial. Here, a particularly important role could be played by existing breeding associations, which are actually not operating in a satisfactory manner, but they could easily be boosted through specific interventions. Hence, the (re)discovery of territorial resources should also take into account other elements that shape the identity in these areas, and the whole framework should be built in a way that will enable and encourage synergies among different actors.

References

Allaire G, Dupeuble T (2002) De le multifonctionnalite de l' activite agricole a lamulti-evaluation de la production agricole: vers l' emergence d' un nouveau systeme de competences. Colloque SFER, La Multifonctionnalite de l' Activite Agricole, 21–22 mars 2002, Paris

Belibasaki S, Sossidou E, Gavojdian D (2012) Local breeds: can they be a competitive solution for regional development in the world of 'globalization'? The cases of Greek and Romanian local breeds. Scientific Papers Animal Sci Biotechnol 45(2):278–284

Berkes F, Folke C (2002) Back to the future: ecosystem dynamics and local knowledge. In: Gunderson L, Holling CS (eds) Panarchy: understanding transformations in human and natural systems. Island Press, Washington, DC, pp 121–146

Bye LM (2009) 'How to be a rural man': young men's performances and negotiations of rural masculinities. J Rural Stud 25:278–288

Canali G, Ecogene Consortium (2006) Common agricultural policy reform and its effects on sheep and goat market and rare breeds conservation. Small Rumin Res 62(3):207–213

Chapin FS III, Kofinas GP, Folke C (eds) (2009) Principles of ecosystem stewardship: resilience-based natural resource management in a changing world. Springer, New York

De Rancourt M, Fois N, Lavín MP, Tchakérian E, Vallerand F (2006) Mediterranean sheep and goats production: An uncertain future. Small Rumin Res 62(3):167–179

Dong S, Wen L, Liu S, Zhang X, Lassoie JP, Yi S, Li X, Li J, Li Y (2011) Vulnerability of worldwide pastoralism to global changes and interdisciplinary strategies for sustainable pastoralism. Ecol Soc 16(2):10 [online] URL: http://www.ecologyandsociety.org/vol16/iss2/art10/
Eakin H (2005) Institutional change, climate risk, and rural vulnerability: cases from Central Mexico. World Dev 33(11):1923–1938
Ganatsou E, Rapti D, Ispikoudis I (2008) Contribution of nomadic livestock husbandry on the creation of alpine and sub alpine formations «NATURA 2000» Areas. In: Proceedings of the 6th Greek Rangeland Conference, Leonidio, 2–4 October 2008, pp. 221–226
Gunderson LH (2003) Adaptive dancing: interactions between social resilience and ecological crises. In: Berkes F, Colding J, Folke C (eds) Navigating Social Ecological Systems: building resilience for complexity and change. Cambridge University Press, pp 33–52
Ispikoudis I, Soliou MK, Papanastasis VP (2002) Transhumance in Greece: past, present and future prospects. In: Bunce RGH, Pérez-Soba M, Jongman RHG, Gómez Sal A, Herzog F, Austad I (eds). Transhumance and Biodiversity in European mountains, pp. 211–226
Ingersent, K.A. and Rayner, A.J. (1999). "Agricultural policy in Western Europe and the United States", Edward Elgar, Cheltenham
Karanikolas P, Martinos N (2012) Greek agriculture facing crisis: problems and prospects. Available online at: http://ardinrixi.gr/archives/3811
Koocheki A, Gliessman SR (2005) Pastoral nomadism, a sustainable system for grazing land management in arid areas. J Sustain Agric 25:113–131
Koutsou S, Partalidou M, Ragkos A (2014) Young farmers' social capital in Greece. Trust levels and collective actions. J Rural Stud 34:204–211
Lankoski J, Ollikainen M (2003) Agri-Environmental externalities: a framework for designing targeted policies. Eur Rev Agric Econ 30:51–75
Legarra A, Ramon M, Ugarte E, Perez-Guzman MD (2007) Economic weights of fertility, prolificacy, milk yield and longevity in dairy sheep. Animal 1:193–203
Ligda C, Casabianca F (2013) Adding value to local breeds: challenges, strategies and key factors. Animal Genetic Resources/Resources génétiques animales/Recursos genéticos animales 53:107–116
Ligda C, Zjalic M (2011) Conservation of animal genetic resources in Europe: overview of the policies, activities, funding and expected benefits of conservation activities. Animal Genetic Resources/Resources génétiques animales/Recursos genéticos animales 49:75–86
López-i-Gelats F, Fraser ED, Morton JF, Rivera-Ferre MG (2016) What drives the vulnerability of pastoralists to global environmental change? A qualitative meta-analysis. Glob Environ Chang 39:258–274
Mpizelis I (2013) Evaluation and use of Greek sheep and goat breeds. Animal Sci Rev (Special Issue) 39:40–43
Mulgan G, Tucker S, Ali R, Sanders B (2007) Social innovation: what it is, why it matters and how it can be accelerated. Working paper, Skoll Centre for social entrepreneurship
OECD (2001) Multifunctionality: towards an analytical framework. OECD, Paris
OECD (2003) Multifunctionality: the policy implications. OECD, Paris
Polman N, Peerlings J, Slangen L (2011) Resilience of European farms under different CAP scenarios. EAAE 2011 Congress "Change and Uncertainty: Challenges for Agriculture, Food and Natural Resources", August 30 to September 2, 2011, ETH Zurich, Zurich, Switzerland
Ragkos A, Lagka V (2014) The multifunctional character of sheep and goat transhumance in Greece. 8th Panhellenic Rangeland Science Conference, 1–3 October 2014, Thessaloniki. (In Greek)
Ragkos A, Siasiou A, Galanopoulos K, Lagka V (2014) Mountainous grasslands sustaining traditional livestock systems: The economic performance of sheep and goat transhumance in Greece. Options Mediterraneennes 109:575–579
Ragkos A, Abas Z (2015) Using the choice experiment method in the design of breeding goals in dairy sheep. Animal 9(2):208–217
Ragkos A, Theodoridis A, Fachouridis A, Batzios C (2015) Dairy farmers' strategies against the crisis and the economic performance of farms. Procedia Economics and Finance 33:518–527

Ragkos A, Koutsou Σ, Manousidis T (2016) In search of strategies to face the economic crisis: evidence from Greek farms. South European Society and Politics 21(3):319–337. https://doi.org/10.1080/13608746.2016.1164916

Ragkos A, Koutsou S, Theodoridis A, Manousidis T, Lagka V (2018) Labor management strategies in facing the economic crisis. Evidence form Greek livestock farms. New Medit:2018–2011

Ragkos A, Nori M (2016) The multifunctional pastoral systems in the Mediterranean EU and impact on the workforce. Options Mediterraneennes, Series A: Mediterranean Seminars No. 114, pp. 325–328

Rege JEO, Gibson JP (2003) Animal genetic resources and economic development: issues in relation to economic valuation. Ecol Econ 45:319–330

Ripoll-Bosch R, Joy M, Bernués A (2013) Role of self-sufficiency, productivity and diversification on the economic sustainability of farming systems with autochthonous sheep breeds in less favoured areas in Southern Europe. Animal, available on CJO 2013. https://doi.org/10.1017/S1751731113000529

Roustemis D (2012) Improvement of Chios sheep breed – design of the breeding goal. Doctoral dissertation, Democritus University of Thrace, Orestiada, Greece

Schouten M, van der Heide M, Heijman W (2009) Resilience of social-ecological systems in European rural areas: theory and prospects. Paper presented in the 113th EAAE Seminar 'The Role of Knowledge, Innovation and Human Capital in Multifunctional Agricultural and Territorial Rural Development', Belgrade, Republic of Serbia

Theodoridis A, Ragkos A, Roustemis D, Galanopoulos K, Abas Z, Sinapis E (2012) Assessing technical efficiency of Chios sheep farms with data envelopment analysis. Small Ruminant Research, 107, pp. 85–91

Theodoridis A, Ragkos A, Roustemis D, Arsenos G, Abas Z, Sinapis E (2013) Technical indicators of economic performance in dairy sheep farming. Animal, 8(1), pp. 133–140

Vafiadakis T (2013) Evaluation and development of Greek sheep and goat breeds. Animal Sci Rev (Special Issue) 39:52–55

Verrier E, Tixier-Boichard M, Bernigaud R, Naves M (2005) Conservation and value of local livestock breeds: usefulness of niche products and/or adaptation to specific environments. Animal Genetic Resources/Resources génétiques animales/Recursos genéticos animales 36:21–31

Westley FR, Tjornbo O, Schultz L, Olsson P, Folke C, Crona B, Bodin Ö (2013) A theory of transformative agency in linked social-ecological systems. Ecol Soc 18(3):27

World Conservation Monitoring Centre (1992) Global biodiversity: status of the Earth's living resources. Chapman and Hall, London

Vermersch D (2001) Multifunctionality: applying the OECD framework. A review of literature in France. In: Directorate for Food, Agriculture and Fisheries. OECD

Internet Sources (Accessed in 30/8/2018)

http://ecdpm.org/great-insights/territorial-development-2/what-is-territorial-development/

https://mediasrv.aua.gr/eclass/modules/document/file.php/EZPY104/%CE%B1%CE%B9%CE%B3%CE%BF%CF%80%CF%81%CE%BF%CE%B2%CE%B1%CF%84%CE%BF%CF%84%CF%81%CE%BF%CF%86%CE%AF%CE%B1/B1/B1_breeds_foreign.pptx

http://www.minagric.gr/index.php/el/for-farmer-2/animal-production/sheep-and-goats/1075-autofilesaigoprobaton

http://www.gaiapedia.gr/gaiapedia/index.php/%CE%A6%CF%85%CE%BB%CE%AD%CF%82_%CE%B1%CE%B9%CE%B3%CE%BF%CF%80%CF%81%CE%BF%CE%B2%CE%AC%CF%84%CF%89%CE%BD

https://oceclass.aua.gr/modules/document/file.php/OCDASA102/zp_th_2995_010a.pdf

http://www.kgbzath.gr/index.php/breed/2011-10-03-05-10-54/2011-10-03-05-12-10

Restructuring the Greek Forest Sector in Order to Facilitate Rural Development in Greece

Panagiotis Koulelis

Abstract In Greece, priorities such as knowledge transfer and innovation in agriculture, forestry, and rural areas, but also the competitiveness of the agricultural sector and sustainable forestry, are now under negotiation. The ultimate goal for forestry is to improve life quality of the urban population and preserve the environment through the sustainable exploitation of forest resources, with an anticipated contribution of 1% to the national GDP. Overcoming important constraints, the Greek forest sector must be developed across all sub-sectors. The production of wood and non-wood forest products and services must be improved. Following a descriptive approach, representative new scopes and actions are suggested for the implementation of such a new forest strategy. Some new scopes, such as strengthening forest exploitation rather than protection, the valorization of forest resources, the active participation of the local population, the focus on small-scale and medium-scale forestry, and the promotion of advantages for certain regions, need to be followed by specific actions such as the sustaining of existing raw material resources from small-scale industrial plantations, the improvement of production and transportation infrastructures, the development of a qualified and cost-competitive labor force, the creating of an attractive investment climate, and paying more attention to NWFPs and services. The need for rural and national economic development in Greece after the severe negative impacts of the financial crisis is now more crucial than ever.

Keywords Forest sector development · Rural development constraints · Mediterranean forest products · NWFPs · Rural communities

P. Koulelis (✉)
Institute of Mediterranean Forest Ecosystems, Athens, Greece
e-mail: pkoulelis@fria.gr

A. Theodoridis et al. (eds.), *Innovative Approaches and Applications for Sustainable Rural Development*, Springer Earth System Sciences,
https://doi.org/10.1007/978-3-030-02312-6_2

1 Introduction

Greece, like other Mediterranean countries, is a net importer of wood and wood products. In 2010, according to the Food and Agriculture Organization, Mediterranean countries collectively imported wood and wood products of a value of more than US $40 billion, of which 80%, equivalent to US $32 billion, originated from non-Mediterranean countries (FAO 2013). Mediterranean forest products and services have represented a fundamental factor in the life of local inhabitants for centuries, playing an important role in rural development and/or in general economic and social development. Traditionally, the forest sector, including all economic activities that mostly depend on the production of goods and services from forests, makes a significant contribution to necessary social changes and sustainable economic development of the rural community. The ultimate goal is to improve the life quality of the urban population and preserve the environment through the sustainable exploitation of forest resources. According to the literature, Mediterranean forest ecosystems provide multiple wood and non-wood forest goods and services, which are crucial for the socio-economic development of rural areas as well as for the welfare of the urban populations of the Mediterranean region (Palahi et al. 2008). They play a key role in the welfare of urban and rural Mediterranean societies, by providing highly appreciated marketed goods (e.g., firewood, cork, pine kernels, and mushrooms) (EFI-Efimed 2010). Other environmental services which are complex to estimate, such as erosion control, water production services, landscape quality, and carbon sequestration, are provided continuously by Mediterranean forests. In general, forest ecosystems also provide a wide range of opportunities for leisure, mental development, and spiritual enrichment (DeGroot et al. 2002). The products and services that being produced by the Mediterranean forests have represented a fundamental factor in the life of local inhabitants for centuries, but not many studies have focused on them on a large scale (UNEP/MAP-Plan Bleu 2009; FAO 2013). Although, Mediterranean forests require special attention and research agenda (EFI-Efimed 2010).

At national level, the forest sector is facing important constraints. The modern problems for the Greek forest sector can be placed in three major categories according to Koulelis (2011).

1. A main characteristic of Greece is that most of the forest area is located in areas with high mountains and slopes (harvesting is extremely difficult). The production and often the quality of the produced wood for many managerial and ecological reasons are limited (e.g., a lot of knots). Koulelis (2009) placed Greece among EU countries with low productivity. Greek forests mainly produce (i) logs for sawn wood production mostly used by the state power company or telecommunication companies or as mining timber and (ii) wood chips for particleboard, MDF, paper, and pellet production.
2. The high cost and the outdated methods of harvesting and production. The highland terrain of the Greek mountains and the low level of automation in harvesting increase production and transportation costs. On the other hand, the

timber sales system seems not to fit efficiently in the Greek sector, because it does not ensure stable prices and quantities for timber industries. Greece has no capability to cover its domestic needs in wood and forest products from its own production, so its dependence on importing wood and wood products is very high (Arabatzis and Klonaris 2009). The importing character of the country changed during the last years and especially after 2008, during the economic crisis. According to Koulelis (2011), Greek wood product imports have increased rapidly over recent years, which have both economic and ecological implications. The levels of forest consumption per capita declined in all cases, mainly due to the collapse of the construction industry during the crisis (Koulelis 2016). He observed that the decrease of the deficit in forest products, mainly due to export increase, had a positive impact on the deficit. In addition, according to FAOSTAT (2016), the imports of forest products that were measured in m^3 declined 54% (2.011.810 m^3 in 2008 to 916.246 m^3 in 2014), and the imports in tons saw a slight increase of 0.06% (847.380 tons in 2008 to 897.287 tons in 2014).

3. The third major category of constraints concerns stemming from forest policy implementation in Greece. However, Greece has a long history of contributing through its policies to adopting plans regarding rural development policy. The National Strategic Rural Development Plan (NSRDP) 2007–2013 determined the priorities of Greece for the period 2007–2013, in accordance with Article 11 of Regulation (EC) 1698/2005 on the support of rural development by the European Agricultural Fund for Rural Development (EAFRD). In particular, they focused on three basic axes: improving competitiveness in the agricultural and forestry sectors, enhancing the environment and the countryside, and finally ameliorating life quality in rural areas and differentiation of rural economy. The rural development plan that was adopted by the European Commission on 11 December 2015 and last modified on 16 December 2017 outlined Greece's priorities for using the € 5.7 billion of public funds that is available for the period 2014–2020 (€ 4.7 billion from the EU budget and € 1 billion of national co-founding). Among those priorities were also the knowledge transfer and innovation in agriculture, forestry, and rural areas but also the priority of competitiveness of agricultural sector and sustainable forestry. During the last years, in European Mediterranean environments, public funds from the European Union have been available to encourage farmers to turn their cropland into forest plantations (Rey-Benayas 2005). Those well-planned actions could promote new investments in ecosystem goods and services and address the issue of forest land abandonment.

Nowadays, the new forest strategy is under public negotiation in Greece (Ministry of Environment and Energy 2018). This strategy will adopt a new model for forestry that will enhance between others: (a) the contribution of the forest sector in rural development and employment, (b) the further exploitation of public and private forests, and (c) the adaptation to climate change considering the Mediterranean characteristics of the forest resources based on the existing expertise (Hellenic Ministry of Environment, Energy and Climate Change 2018).

The new forest strategy is focused on a more effective participation of the forest sector to the Greek GDP. The unofficial and quite ambitious target of the Ministry is 1% of the GDP. In order to achieve this target, the participation of rural communities and investments in forestry are necessary. Given the particular characteristics of the sector and modern challenges such as climate change and the fragility of Mediterranean forests, the implementation of this modern strategy will not be a simple case.

The development of communities based on the forest sector is a multifaceted socioeconomic process (Tykkyläinen et al. 1997). On a global scale, forestry and the small-scale forest industry are of increasing importance for rural regions, often representing the major employer. From a theoretical perspective, regional economic theories and the multidisciplinary field of rural studies cover the subject of economic development in rural regions (Terluin 2001). Regional economic theories have provided a solid theoretical framework (Niskanen 2005). For example, Hyttinen et al. (2000) have indicated the importance of development theories in the late twentieth century. Furthermore, they have studied the possibilities for employment and income generation along European borders, including Greece. Another classification has been provided by Terluin (2001), where the competitiveness of companies is considered a key element for rural development. In the COST Action E30 (2006), (Niskanen 2005), it has been reported that it is common for many regional economic development theories to emphasize innovations by explaining economic development. Furthermore, Porter's theory on the comparative advance of nations and regions is used to explain economic development (Porter 1998, 2000). The applied model (see more at COST Action E30) was based on the competitiveness of enterprises (or a region) and has been widely used to illustrate the international competitiveness of countries or regions (Hazley 2000). However, a number of details are required to approach and explain economic rural development based on those theories. Such details, as reproduction conditions, company and organization/institution structures and interactions, economic frame conditions in terms of factor inputs, as well as markets and competition, are required to examine and explain the economic development of a particular region. As expected, these methodologies based on the above theoretical framework require in-depth research, resources, and mainly time. In this sense, the present study adopts a descriptive approach and not a specific methodology.

The main goal of this study is to recognize and promote the importance of the Greek forest sector to supporting rural development and, more specifically, the development of the communities where forest exploitation can play an important role in the regional economy. Taking into account some of the hypotheses that were examined by Hyttinen et al. (2000) to support rural development, as well as the development constraints of the national forest sector, as presented by Koulelis (2011), some representative actions are proposed for the successful implementation of new strategies.

2 Forestry and Connected Secondary Activities in Greece

Compared with Italy, Spain, Portugal, and Cyprus, Greece as a forest product supplier holds a low position, as its production is higher only than relevant Cyprus. The output of forestry and connected secondary activities amounts to approximately 100 million Euro, while other countries reach 1.2 to 1.6 billion Euros, without considering France due to its large-scale production (Fig. 1). Italy has an extremely high production because of its furniture industry and its global importance as a huge exporter. The low production in Greece, therefore, requires restructuring of the forest sector.

At the national level, the illustration of gross value added (GVA = output-intermediate consumption) and the measure of the values of goods and services produced in this sector benefited from the introduction of Euro, followed by a severe decrease during the financial crisis mostly after 2009 (Fig. 2). In particular, in 2014, the levels of the Greek GDP were below 22.000 $ per capita, after a serious decrease of 32.6% (Koulelis 2016). During the last years, the Greek construction industry was severely struck by the financial crisis (Sfakianaki et al. 2015). Greek households cut back on their consumption expenditure by a commensurate percentage of almost 30% (Kaplanoglou and Rapanos 2015). As it was expected, the demand for wood products declined. In general, the forest sector became a side issue and was hardly considered by decision-makers.

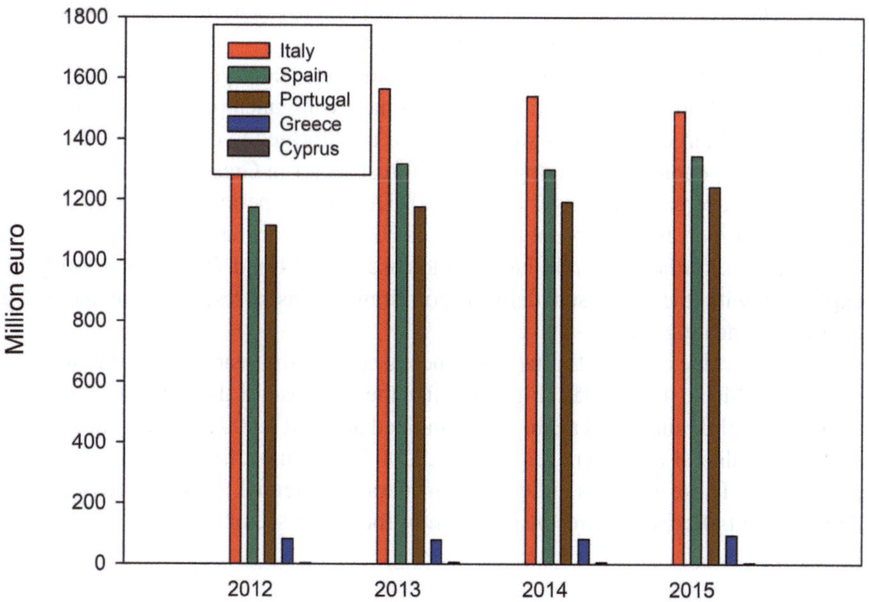

Fig. 1 The output of forestry and connected secondary activities for Mediterranean countries (Eurostat 2018)

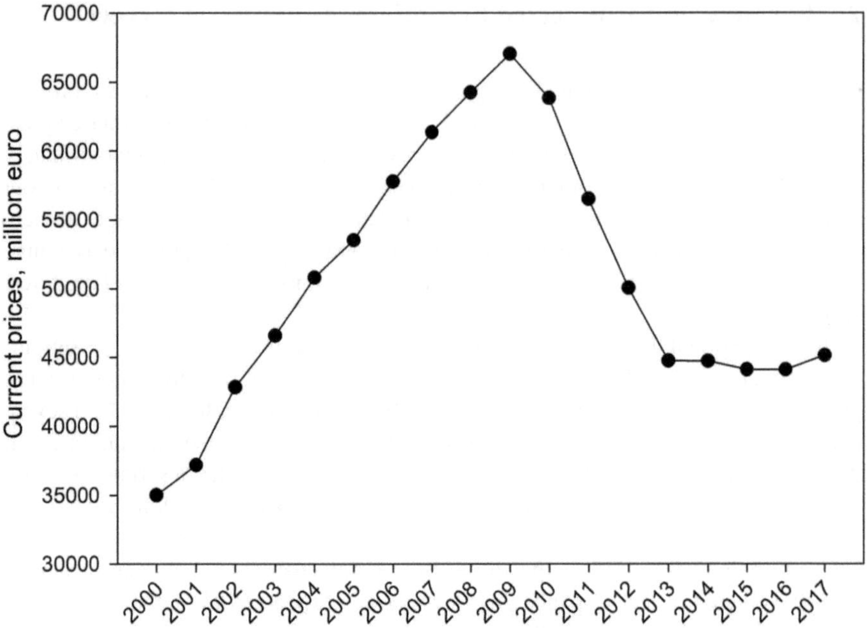

Fig. 2 Gross value added by agriculture forestry and fishing in Greece (OECD 2018)

Along with the value added, wages and salaries followed the same trend (Figs. 3 and 4). The crisis resulted in lower incomes and employment in rural areas, almost reaching the same levels of about 2000 and before.

Employment in forestry has decreased rapidly in recent years, especially from 2008 to 2013. Following the national trend of increasing unemployment, small-scale industries and large enterprises were closed or transformed to commercial. In 2008, around 7.000 people were employed in the forest sector in Greece, whereas in 2011, this number decreased to fewer than 4.000. During the period of 2014 and 2015, employment increased almost at the level of 2009, but declined again in 2016. This issue requires further research, but the increase in 2014 and 2015 can possibly be explained with the increased demand for firewood as a result of the increased taxation on heating oil.

The total value of imports after 2002 increased until the year 2007, just before the crisis (800 M$ to 1.400 M$) (Fig. 5). After the decrease of the demand, the value declined, slightly increased again, and stabilized at about 1.000 M $. During all these years, the value of exports never exceeded 200 M $, almost five times less than the import cost. These numbers indicate a considerable potential for the decrease of this deficit, albeit after serious reconsideration of the entire sector.

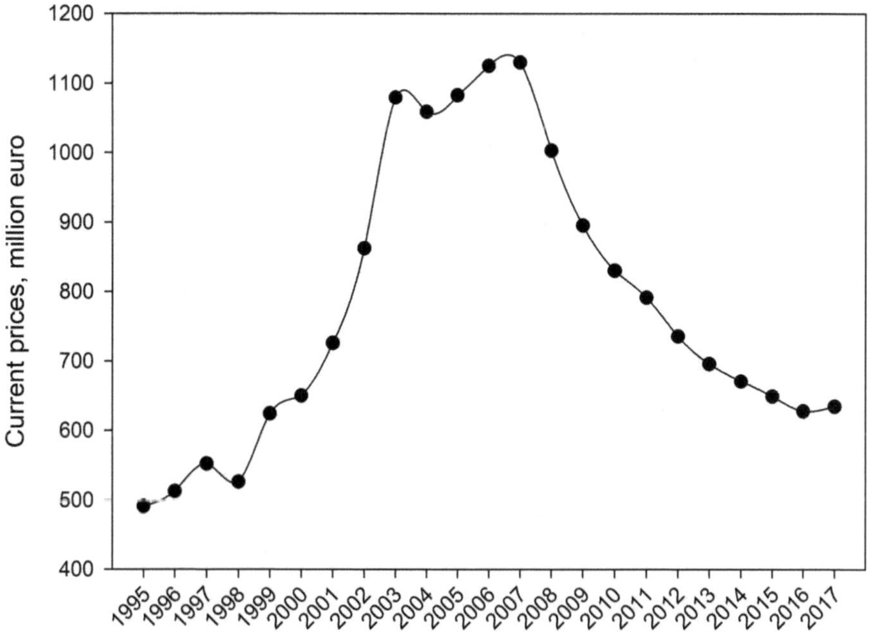

Fig. 3 Wages and salaries in agriculture forestry and fishing in Greece (OECD 2018)

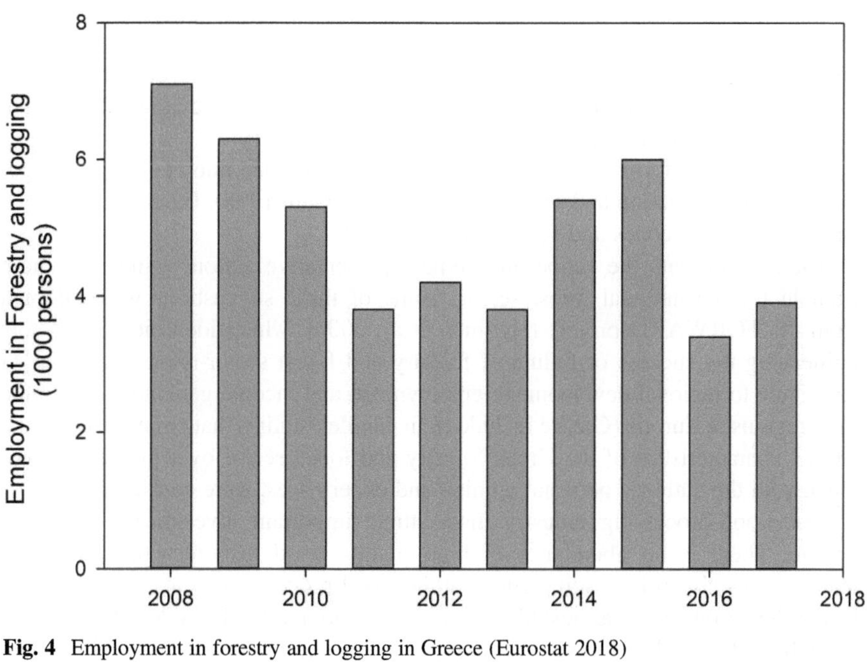

Fig. 4 Employment in forestry and logging in Greece (Eurostat 2018)

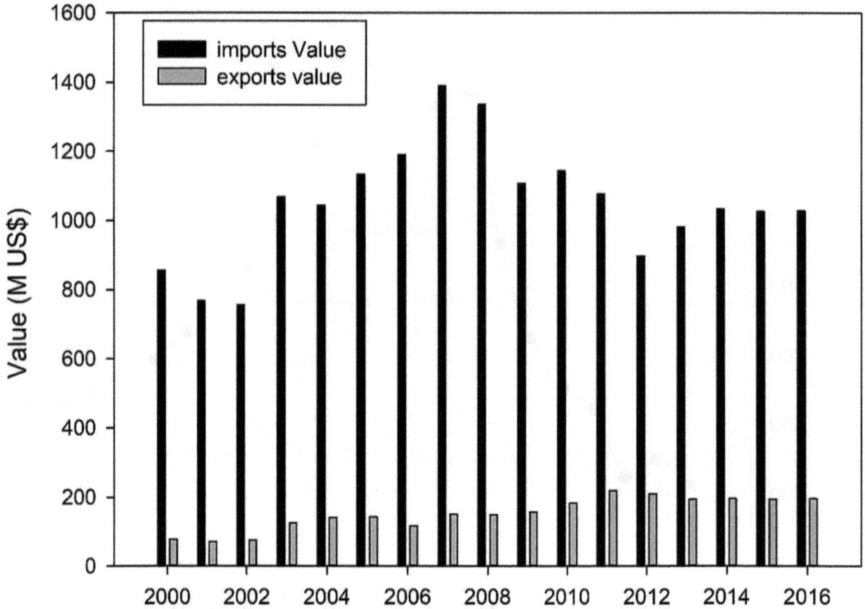

Fig. 5 Import and export value in Greece (FAOSTAT 2018)

3 New Scopes and Suggested Actions

The structural problems of the Greek forest sector cannot be resolved quickly. There are several possible solutions with positive outcomes for rural development. However, such solutions require the reassessment of a number of policies, a wider vision, and the inclusion of new parties such as the energy industry.

The objectives for forestry development can be classified into two broad categories those of production and service functions (Wiersum 1998). Based on those two categories, some scopes and actions are discussed.

Table 1 presents the scope and some representative actions which are more suitable for the national forest sector. Some of these suggestions were adopted from the FORWARD project (Hyttinen et al. 2000), which identified the factors influencing the success or failure of forestry and forest sector-related activities to contribute to regional development, employment, and income generation in remote rural regions in Europe (Greece included). In parallel, studies that emphasized on the special characteristics of the Greek forestry and forest sector over the years, combined with the author's personal opinion and experiences, were considered.

The wood-processing industry must attract important investments, public or private. There is no abundance of high-quality wood resources providing raw materials at stable prices, something that the wood-processing sector highly desires. Under the rationale that almost 80% of the forests are public (FRA 2015), one target could be that the public forests will supply wood processors/investors with important

Table 1 National forest resources under a new perspective (partially based on Hyttinen et al. 2000 examined hypotheses)

Scope	Representative actions
From traditional protection to higher exploitation	Limit the storage of forest natural capital Pay more attention to alternative opportunities of exploitation Increase logging activities based on SFM Adopting or create a national certification scheme Fund industrial plantations Ensure stable prices and quantities of raw material from the national forest resources for long-term periods based on management plans Focusing on illegal logging
Valorization of forest resources in order to encourage sustainable development and also ensure that international commitments in this area are met	Collection of specific data regarding timber and non-timber products value, possible fishing and hunting values, and recreation and tourism values
Enhance the role of the local population for the marketing of forests products and services	Awareness of local populations Promote new opportunities for investments Motivation Regional forest programs Subsidies increase
The potential for income and employment in rural areas from recreation activities should increase	More attention to alternative tourism Awareness of local populations Motivation
The potential for high benefits from NWFP and services	Investments Mushrooms, cork, pine nuts, chestnut, honey, and truffles are among the most important NWFPs in the Mediterranean region (FAO 2013) Regional forest programs Subsidies increase
Development of export markets	Investments to innovative technology and education Investments in infrastructures (in order to reduce transportation cost, develop communications, etc.) Labor continuing training Follow EU specifications for forest products More competitive forest products (i.e., bio-based products)
Creation of employment from small- and medium-scale forestry business	Development of the existing public state industries Attract new investors
Promotion of the advantages of a region in the production of a particular product	Investments in specific sub-sectors meaning different technology and market behavior
Well-informed investors, industry owners, policy, and decision-makers	Communication network between Greece and countries with more developed rural areas based on forest resources. Adoption of good practices from abroad

quantities of their total wood raw material. The creation of a large number of small firms and a relatively small number of large companies, depending on regional forests or even industrial plantations, seems to be more suitable for Greece. The potential production based on the existing forest resources, the cost of production, and the parallel protection of ecosystems sensitive to climate change are not conducive to large-scale production and the establishment of large companies. Timber and timber products require large sites for loading and transshipment and generate substantial demands for transportation by road, rail, or water; such demands are difficult to meet in Greece. However, logging activities produce quantities of residues/biomass, which, after careful examination of the local soil conditions, can be used in bio-based production. This issue requires more investigation and research focused on the Greek forest ecosystems. Milling and sawing produce large quantities of sawn dust, which can potentially be used as a source of fuel in pelletized form.

Regarding illegal logging and trade (in both state-owned and private forests), such activities only account for less than 1% of the total amount of legal logging (UNFF 2003), which is a result of efficient surveys, monitoring, and control. However, the National Forest Service is considerably understaffed, consequently facing several problems due to the lack of resources.

In Greece, the total forest area under certification is zero. According to FAO/FRA (2015), the area of forest under an independently verified forest certification scheme between 2005 and 2009 was around 35.000 ha. Before and after these years, there were no certified forest areas. A possible participation of Greece in a forest certification scheme (external such as the FSC or PEFC or even national), which will cover all types of forests, will have positive impacts on rural development processes. Georgiadis and Cooper (2007) have reported that this action will increase the recognition of the contribution of forestry to the economic development of the country, based on the broad participation associated with such a certification.

The new strategy must also target export markets. The certification scheme will ensure some production specifications. For example, investments in innovative technology will facilitate the production of innovative and more competitive products, such as products to support the bio-economy in Europe. In addition, later stages of processing, including the production of veneers, particle boards, and furniture, require an appropriate waste treatment technology and new investments.

The local population should be involved in the maintenance, protection, and development of forest resources. Its contribution to fire protection, timber production, tourism, recreation, and the NWFP (non-wood forest products) should also be crucial. On the other hand, local communities could play an important role in the marketing of forest products and services. The potential of rural areas as high-quality living environments is poorly recognized. The culture of alternative recreation activities in the forest and the consumption of traditional foods are important advantages that must be promoted. Some actions, such as promoting new opportunities for investments and jobs, providing motivation through regional forest programs, and increasing subsidies, are therefore suggested.

Furthermore, it is further necessary to invest in the ongoing education of all staff in public and private structures related to the forest sector. Investments in infrastructures will manage to reduce transportation cost and develop more efficient communication between rural and urban regions in Greece. Hyttinen et al. (2000) have proposed that stronger links between the urban demand and the rural supply should be created. The cooperation between regional development administrators and entrepreneurs needs to be strengthened, mainly because general bureaucrats' views seem to be distanced from the entrepreneurial spirit. Every region has specific characteristics and different potentials for development, depending on, e.g., the suitability of forests for the production of quality timber or the local protection priorities. However, investments in specific sub-sectors result in different technologies and market behavior and must be carefully evaluated by local decision-makers.

The establishment of industrial plantations is increasingly important as the primary source of timber, with considerable environmental, social, and local economic impacts. According to FAO data from 2015, Greece contains 3.84 Mha of forest, accounting for 29% of the total area. Naturally regenerated forests cover an area of 3.73 Mha, while plantations only account for 115.000 ha (Global Forest Watch 2018). Not enough has been done in the past to promote afforestation (afforestation programs) or even reforestation to anticipate the effects of large-scale wildfires that had occurred in Greece during the last years. Larger productions could be achieved only with new plantations (exploiting the advantages of short rotations; *Populus* spp. and *Cupressus* spp. have been proposed for this reason) in parallel with the development of the current forest ecosystems. To facilitate such development, future-targeted research on the testing and use of valuable native trees and shrubs is crucial (Koulelis and Daskalakou 2017).

In addition, new opportunities for non-wood forest products and services might be an interesting approach. The rich biodiversity, which characterizes the Mediterranean region, offers the potential for the production of a wide variety of NWFPs; most of them are referred in Table 1. Some countries in the Mediterranean region have specific legislation, facilitating further exploitation and overcoming bureaucratic obstacles. One example is the production of forest fruits. There is an ongoing interest in traditional forest fruits due to their nutritional value and health effects. All these fruits and plants are routinely integrated with the pharmaceutical properties beyond their nutritive value. Traditional forest fruits contain a number of compounds such as essential oils, phenols, anthocyanins, and flavonoids (collectively called "nutraceuticals" or "food additives"). Forest fruits could potentially strengthen the regional/national economy, preserve the forest genetic resources for forest fruit species, positively contribute to the Mediterranean diet, and support the sustainable development of rural areas.

Without a doubt, the adoption of methods and practices already applied in other EU countries will be a helpful approach. Although the potential is different, such an adoption will result in well-informed investors, industry owners, as well as policy and decision-makers regarding the possibilities and capabilities of the forest sector in Greece.

4 Conclusions

The structural problems of the Greek forest sector cannot be resolved quickly, but require rather thorough reassessment of policies, a more wide-ranging vision, and the inclusion of new parties, such as the energy industry based on bio-economy, in the policy planning process. Regarding traditional production, domestic production must be improved. Currently, much of the existing industrial wood-processing capacity is not fully used. An increased domestic production of all categories of products and marketing would gradually reduce imports, and this improvement, combined with the establishment of a new highly organized structure of small-scale industrial plantations, could make domestic roundwood products more competitive. Actions such as sustaining existing raw material resources, improving production infrastructures, developing a qualified and cost-competitive labor force, creating an attractive investment climate, and providing economic stimulus measures in the primary wood-processing and furniture industry sectors are underlying requirements for development. The suggested actions target the restructuring, strengthening, and connection of forest businesses with partners from urban areas and, tentatively, from abroad.

On the other hand, NWFP and services could provide opportunities for rural development under specific conditions and the determination of local and national authorities to complete these actions. The role of the state in promoting the competitiveness of a region is crucial. Regardless of what kind of new scopes the new forest strategy is going to serve, it should take into account region-specific, sectoral, and policy-related factors to enhance the role of forest businesses in the rural and national economy.

References

Arabatzis G, Klonaris S (2009) An analysis of Greek wood and wood product imports: evidence from the linear quadratic aids. Forest Policy Econ 11:266–270

Cost Action E30 (2006) Economic integration of urban consumers' demand and rural forestry production. University of Joensuu, Finland. http://www.joensuu.fi/coste30

DeGroot R, Wilson MA, Roelof MJB (2002) A typology for the classification, description and valuation of ecosystem functions, goods and services. Ecol Econ 41:393–405

EFI-Efimed (2010) A Mediterranean Forest Research Agenda – MFRA 2010–2020, 32

European Commission (2018) Eurostat. Retrieved from http://ec.europa.eu/eurostat/data/database. Accessed on May 2018

FAO (2013) State of Mediterranean Forests (2013). Retrieved from http://www.fao.org/docrep/017/i3226e/i3226e.pdf

FAO/FRA (2015) Global Forest resources assessment, country report, Greece Rome 2014

FAOSTAT (2016) Food and Agricultural Organization of the United Nations. Retrieved from http://www.fao.org/faostat/en/#data/FO. Accessed on May 2018

Georgiadis NM, Cooper RJ (2007) Development of a forest certification standard compatible with PEFC and FSC's management requirements. A case study from Greece. Forestry, Oxford Journals 80(2):113–135

Global Forest Watch (2018) Tree covers loss in Greece. Accessed on March 30th 2018 from www.globalforestwatch.org

Hazley CJ (2000) Forest-based and related industries of the European Union Industries districts, clusters and agglomerations. ETLA (The Research Institute of the Finnish Economy). Series B. No. 113, 230 p

Hellenic Ministry of Environment, Energy and Climate Change (2018) National Forest Strategy. http://www.ypeka.gr/Default.aspx?tabid=389&sni[524]=5606&locale=enUS&language=el-GR

Hyttinen P, Niskanen A, Ottitsch A (2000) New challenges for the forest sector to contribute to rural development in Europe. Land Use Policy 17(3):221–232

Kaplanoglou G, Rapanos VT (2015) Evolutions in consumption inequality and poverty in Greece: the impact of the crisis and austerity policies. Economics Discussion Reports No 2015/9 Economics Department, University of Athens

Koulelis P (2009) Cluster analysis in primary Roundwood production of 25 countries of European Union. Ann For Res 52:163–168

Koulelis P (2011) Greek timber industries and wood product markets over the last century: development constraints and future directions. Ann For Res 54(2):229–240

Koulelis P (2016) Forest products consumption and trade deficit in Greece during the financial crisis: a quantitative statistical analysis. OJBM April:258–265

Koulelis PP, Daskalakou EN (2017) Modeling per capita demand of wood products in Greece: variables with impact to demand and Forest management implications. Environ Manag Sustainab Develop 6(2):347. https://doi.org/10.5296/emsd.v6i2.11735

Niskanen A (2005) Forest sector entrepreneurship in Europe – summary of country studies of COST Action E30. Acta Silv. Ling. Hung. Special Edition 2005, 7–15.

OECD (2018) Value added by activity (indicator). https://doi.org/10.1787/a8b2bd2b-en. Accessed on 30 April 2018

Palahi M, Mavsar R, Gracia C, Birot Y (2008) Mediterranean forests under focus. Int For Rev 10 (4):676–688

Porter ME (1998) Clusters and the new economics of competition. Harvard Business Review November–December:77–90

Porter M (2000) Location, clusters, and company strategy. In: Clark G, Gertler M, Feldman F (eds) Oxford handbook of economic geography. Oxford University Press, Oxford 742 p

Rey-Benayas JM (2005) Restoring forests after land\abandonment. Forest restoration in landscapes: beyond planting trees, pp 356–360, Springer, New York, NY

Sfakianaki E, Iliadis T, Zafeiris E (2015) Crisis management under an economic recession in construction: the Greek case. Int J Manag Decis Mak 14:373–389 https://doi.org/10.1504/ijmdm.2015.074015

Terluin IJ (2001) Rural regions in the EU: exploring differences in economic development. Rijksuniversiteit Groningen, Utrecht

The FAO's Forest Product Statistics (FAOSTAT) (2018). Forestry production and trade. Available online: http://www.fao.org/faostat/en/#data/FO. Accessed on May 2018

Tykkyläinen M, Hyttinen P, Mononen A (1997) Theories on regional development and their relevance to forest sector. Silva Fennica 31(4):447–459

UNEP/MAP-Plan Bleu (2009) State of the environment and development in the Mediterranean – 2009. UNEP/MAP-Plan Bleu, Athens

United Nations Forum on Forests (UNFF) (2003) National Report for UNFF- Greece

Wiersum F (1998) The role of forestry in rural development – regional variations. In: Forestry in the context of rural development. Final report of COST action E3. Danish Forest and Landscape Research Institute, Horsholm, 11 p

Best Practices for Frugal and Sustainable Innovation

Stavros Valsamidis

Abstract While innovation is defined as the successful exploitation of new ideas incorporating new technologies, design, and best practice, the frugal innovation involves the use of local resources to come up with affordable, functional products that provide value for money. Frugal innovation is about creating efficient products to provide desirable user experience, at low cost by using resources frugally, while maintaining the safety and quality standards. Sustainable innovation could be defined as innovation that improves sustainability performance, where such performance includes ecological, economic, and social criteria. Sustainable innovation is a process where sustainability considerations are integrated into company systems from idea generation through to research, development, and commercialization. This applies not only to products and services, as well as to new business, technologies, and organizational models. Businesses nowadays cannot be competitive anymore just because of their industry or labor cost advantages. They can only remain competitive if they are highly focused on sustainability. On the other hand, in frugal innovation you must have the ability to do "more with less" by focusing on the creation of significant more business and social value while minimizing the use of resources. This study explores the best practices for frugal and sustainable innovation by investigating internal and external stakeholders in the top European countries in terms of sustainability and innovation. We discuss the implications of our findings aiming to build robust influential efforts for policy makers.

Keywords Frugality · Sustainability · Resources · Best practices · Business model canvas · Stakeholders

S. Valsamidis (✉)
Eastern Macedonia and Thrace Institute of Technology, Kavala, Greece
e-mail: svalsam@teiemt.gr

A. Theodoridis et al. (eds.), *Innovative Approaches and Applications for Sustainable Rural Development*, Springer Earth System Sciences,
https://doi.org/10.1007/978-3-030-02312-6_3

1 Introduction

Businesses nowadays cannot be competitive anymore just because of their industry or labor cost advantages. They can remain competitive if they are highly focused on *sustainability* (Radjou and Prabhu 2014). *FrugalPlease check if edit to sentence starting "Frugal innovation must..." is okay. innovation* must have the ability to do "more with less" by focusing on the creation of significant more business and social value while minimizing the use of resources (Bound and Thornton 2012; Fraunhofer ISI 2016). Resource constraint is considered an opportunity, not a liability. The target group of frugal innovations are the middle-class people in developing and emerging countries (Seyfang and Smith 2007). Frugal innovation is all about innovations for the 90% of the world's population (European Commission 2016).

Western philosophy is founded on an individual-based rights system, utilitarianism, and the commoditization of natural resources (Bijoy 2007). In contrast, frugal innovation combats with different belief systems, motivations, resource scarcity, and conflicting interests (Joshi and Chelliah 2013; Erika and Watu 2010). Frugal innovation requires a deep understanding of the customer's unique requirements, the specific conditions in which the product will be used, and the kind of trade-offs that can be made to keep the costs low without compromising on safety and quality (Krishnan 2010). Frugal solutions are not easily achieved and often created by those who are outside the formal sectors of society. People in the informal sector face resource shortages, hardships, and challenges of subsistence that many times do not leave them any other possibility than to draw on their knowledge and skills and come up with creative, low-cost solutions (Prahalad and Mashelkar 2010).

Innovation is defined as "the successful exploitation of new ideas –incorporating new technologies, design and best practice which is the key business process that enables businesses to compete effectively in the global environment" (DTI 2005). The following definition captures the essence of sustainable innovation as we see it (Charter et al. 2008; Charter and Clark 2007): "Sustainable innovation is a process where sustainability considerations (environmental, social, and financial) are integrated into company systems from idea generation through to research and development (R&D) and commercialization. This applies to products, services and technologies, as well as to new business and organizational models" (Boons et al. 2013). Innovation is an interesting focus of study in this respect, as technology and society come together, as do culture, economy, and politics, to create hybrid "socio-technical networks" (Latour 1993). Innovation is the fruit of the emergence of a hybrid socio-technical system, which is a set of complementary mechanisms and of diverse and multi-positioned actors, who evolve together, developing a common identity and a common language (Yalçın-Riollet et al. 2014). Sustainable innovation could be defined as "innovation that improves sustainability performance," including ecological, economic, and social criteria (Boons et al. 2013).

Sustainable development has been defined as a development progress which meets the needs of the present without jeopardizing the needs of future generations at the World Commission on Environment and Development (Brundtland

Commission) of the United Nations (DeSimone and Popoff 2000). In the context of sustainability, each value is considered not only from the perspective of economic value but also from the perspectives of environmental and social value (Yang et al. 2017). Indeed, the quest for sustainability is already starting to transform the competitive landscape, which will force companies to change the way they think about products, technologies, processes, and business models. The key to progress, particularly in times of economic crisis, is innovation (Nidumolu et al. 2015).

A sustainable organization expresses its purpose, vision, and/or mission in terms of social, environmental, and economic outcomes. Profits are a "means" to achieve sustainable outcomes—sustainable organizations must make a profit to exist, but they do not just exist to make a profit (Stubbs and Cocklin 2008). Organizations can make significant progress toward achieving sustainability through their own internal capabilities, but ultimately organizations can only be sustainable when the whole system of which they are part is sustainable (Jennings and Zandbergen 1995).

Companies started to give more attention to environmental concerns, as they aim to secure future existence of their resources and business (Karagülle 2012). A sustainable business is able to manage its financial, social, and environmental risks, obligations, and opportunities. These three impacts are referred to as profits, people, and planet. But more importantly, a sustainable business is able to survive shocks, because they are closely connected to healthy economic, social, and environmental systems. These businesses generate economic value and contribute to healthy ecosystems and strong communities (Bocken et al. 2014).

The term *ecopreneurship* consists out of "eco" and "entrepreneur." An ecopreneur creates and sells environmentally friendly products or services. Instead of focusing on valid management styles or technical procedures, an ecopreneur is driven by his/her own motivation, skills, passion, and fulfillment. Ecopreneurs work for a return on the environment while saving and restoring natural resources (Schaltegger and Wagner 2017). This also implies that they act in a social responsible way. All human beings and future generations are considered stakeholders. There exists a great focus on cooperation and collaboration (Schaltegger and Petersen 2000).

For business sustainability a number of best practices can be described:

1. *Stakeholder commitment*: businesses are able to learn from their employees, customers, and surrounding community.
2. *Environmental management systems*: these systems provide a structure and specific processes in order to help integrating environmental efficiency into the business' culture. There exist numerous standards, such as ISO 14001.
3. *Reporting and disclosure*: measurement and control are highly important in sustainable practices. A sustainable business can be entirely transparent with outsiders.
4. *Life cycle analysis*: a systematic analysis of the environmental and social impact of the products the business is using and producing will help the business to have a competitive advantage and be sustainable in the long run (Bocken et al. 2014).

The businesses are distinguished between internal and external stakeholders and therefore can be applied to two practices. The first practice is concerned with managers and staffs (internal stakeholders) and includes top-level support, staff empowerment, incentive and reward, collaboration, and effective communication. In terms of the second practice, the primary focus is on supply chain and clients (external stakeholders) and includes understanding clients' need, being proactive, interacting early with clients, building relationships, communicating effectively, incentivizing, partnering, and matching or facilitating different services (Abuzeinab and Arif 2014). Comparative analyses of sustainable innovation systems in different countries can reveal good practices (Boons et al. 2013).

Since all the positive factors of picking ideas from different practices of European companies and sustainable innovators are necessary in order to propose a superset, an analysis of existing business practices from three European countries (France, Finland, and the United Kingdom) is performed. Those findings are combined into viable business practices.

2 Baseline

The analysis is based on criteria out of three different sections, which are *innovation*, *entrepreneurship*, and *sustainability*. Those three were chosen because they are considered prerequisites for countries to perform sustainable innovations (Baldsarre et al. 2017).

The first section Innovation consists out of six different indicators, which are described in Table 1.

The second section Entrepreneurship consists out of four different criteria. Out of those, three were based on the EIP (Eurostat-OECD entrepreneurship indicator program). This program provides a basis on how to measure entrepreneurship. No single indicator can capture the different facets of entrepreneurship. Due to that reason, three main fields were defined: enterprises, employment, and wealth (Eurostat 2007). For this analysis indicators of each field were taken into consideration (Table 2).

The third section Sustainability consists out of five indicators, which are described in Table 3.

3 Approach

As a basis, existing business practices are evaluated. As a starting point, a quantitative analysis of all 28 European countries will be performed. The goal of this analysis is to determine the top three countries to choose models from. The criteria chosen for the analysis were explained in detail in the previous section.

Table 1 Indicators for innovation

Innovation	
ECO Innovation Index	This index measures the eco-innovation performance across the EU member states, compared to the EU average (which is equated with 100). The goal of the index is to capture different aspects of eco-innovation by applying 16 indicators, clustered into the following five dimensions: eco-innovation inputs, eco-innovation outputs, eco-innovation activities, eco-innovation resource efficiency, and socioeconomic outcomes (Eurostat 2016a, b)
Innovation activities: enterprises engaged in market introduction of innovations	This indicator shows the number of enterprises engaged in market introduction of innovation. The market introduction of innovation encompasses a new or significantly improved good or service, including market research and launch advertising (Community Innovation Survey 2014)
Innovation activities: enterprises engaged in external R&D	This indicator shows the number of enterprises engaged in external R&D. This means the company contracted-out R&D to other enterprises or to public/private research organizations. As a result, it shows how willing companies are to corporate concerning R&D (Community Innovation Survey 2014)
Innovation activities: enterprises engaged in internal R&D	This indicator shows the number of enterprises engaged in internal R&D. Internal R&D consists out of all development activities undertaken by the company to create new knowledge, with which scientific or technical problems can be solved (Community Innovation Survey 2014)
Innovation activities: enterprises engaged in training for innovation activities	This indicator shows the number of enterprises engaged in training for innovation activities. It describes in-house or contracted-out training for personnel specifically for the development and/or introduction of new or significantly improved products and processes (Community Innovation Survey 2014)
Innovation activities: gross domestic expenditures on R&D (R&D intensity)	The gross domestic expenditure on R&D is also called the R&D intensity. It is defined as the R&D expenditure as a percentage of the gross domestic product. This indicator is a key indicator to monitor resources dedicated to science and technology worldwide (Eurostat Statistics Explained 2017a)

In order to conduct this quantitative analysis, the three sections have to be split up in comparable key performance indicators for all countries of the European Union. These key performance indicators (as described in Tables 1, 2, and 3) will be evaluated individually to create a ranking from 1 (best) to 28 (worst) per section. This means there will be independent rankings for the sections "innovation," "entrepreneurship," and "sustainability." As to develop an expressive ranking, the indicators will be weighted depending on their importance for their superior section. Each section will be at hundred percent. As mentioned before, depending on the indicators' importance, these hundred percent will be spread over the number of indicators. The weighted percentage will show the contribution of each indicator to

Table 2 Indicators for entrepreneurship

Entrepreneurship	
Net business population growth	Growth rate of active companies within a certain amount of time is seen as the net business population growth. This growth rate considers both births and deaths of active enterprises. In order to be considered in this rate as an active enterprise, the companies have to achieve a certain amount of employment and/or turnover in the predefined period of time (Innovationdata 2014)
Birth rate of enterprises	This indicator belongs to the field of enterprises. The birth rate is calculated by the number of enterprises born in the reference period (t) divided by the number of enterprises active in t. This indicator is seen as one of the key determinants of job creation and economic growth. Every birth of a new enterprise increases the competitiveness of a country's enterprise population. Due to new competitors on the market, innovation and the adoption of new technologies are stimulated. This increases the overall productivity within an economy (Eurostat Statistics Explained 2017b)
Number of high-growth enterprises	This indicator belongs to the field of employment. A high-growth enterprise (growth by 10% or more) is defined as an enterprise with an average annualized growth in number of employees greater than 10% per year over a 3-year period. Also the enterprise must have at least ten employees at the beginning of growth. Obviously, high-growth enterprises contribute a lot to the economic growth and the creation of jobs in a country (Eurostat 2007)
Value added at factor cost (total business economy)	This indicator belongs to the field of wealth. The value added at factor costs describes the gross income from operating activities after adjusting for operating subsidies and indirect taxes (Eurostat Statistics Explained 2017c)

the overall ranking of the section. For example, in the section innovation, the indicator "ECO Innovation Index" is evaluated with 27%, whereas "Enterprises engaged in internal R&D" is evaluated only with 9%. This shows that the "ECO Innovation Index" is more expressive than the indicator "Enterprises engaged in internal R&D." After the evaluation of each indicator, a sum will be drawn for each section, and a ranking for this section will be determined. At the end, three weighted rankings will be established, and they will be combined to an overall ranking.

The approach is depicted in Fig. 1.

4 Results

This secondary research was realized in 2017, and the sources are files derived from Eurostat and OECD. The top three European countries with the best score are shown in Table 4. The detailed analysis can be seen in the appendixes. The three sections had to be split up in comparable key performance indicators for all countries of the European Union. These key performance indicators were evaluated individually to

Table 3 Indicators for sustainability

Sustainability	
Environmental benefits due to innovation in the enterprises	This indicator measures the amount of companies that creates environmental benefits through innovation of products. An innovation is a new or a significantly improved product. This product can either be a good, a service, a process, an organizational change, or a marketing method that creates benefits for the environment. This benefits can either occur through the production or the consumption of the product. The environmental benefit can also be the main goal of the innovation or just a positive side effect of the overall objective (Community Innovation Survey 2014)
Development of environment-related technologies, % of all technologies	This indicator describes the number of environment-related invention expressed as a percentage of all domestic inventions in all technologies. It is constructed by measuring inventive activity using patent data across a wide range of environment-related technological ranges. The development of environment-related technologies is seen as an important indicator of development and delivery of green growth (OECD Green Growth Indicators 2016)
Environmentally related government R&D budget, % total government R&D	The total government R&D budget is also referred to as GBAORD (Government Budget Appropriations or Outlays for Research and Development). This indicator expresses a percentage of environment-related government R&D budget of all purpose-GBOARD. The environment-related government R&D budget focuses on research directed at the control of pollution and on developing monitoring facilities to measure, eliminate, and prevent pollution. As a result, it indicates the effectiveness of policies to deliver green growth in a country (OECD Green Growth Indicators 2016)
Resource productivity	Resource productivity is measured as gross domestic product over domestic material consumption. This shows how much revenue is created by using 1 kg of a specific resource. The higher the value of the output per kilogram, the more sustainable is the production. Therefore, the companies and the country they are located in are becoming more sustainable by trying to increase the value per kilogram (Eurostat Statistic Explained 2016)
Production value of cleaner and resource-efficient products	This indicator provides an estimated production value of products that were produced with additional market activities. These additional activities were made to create cleaner and resource-efficient products. These products are part of a sector that tries to generate environmental protection (restoring the environment after it has been degraded) and resource management (preservation, maintenance, and enhancement of the stock of natural resources) (Eurostat 2016a, b)

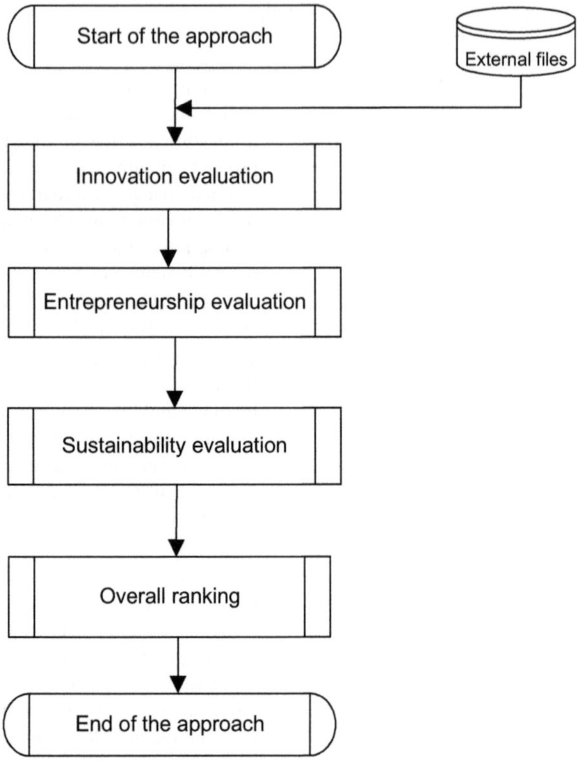

Fig. 1 The four-step approach

create a ranking from 1 (best) to 28 (worst) per section. This means there were independent rankings for the sections "innovation," "entrepreneurship," and "sustainability." After the evaluation of each indicator, a sum was drawn for each section, and a ranking for this section was determined. At the end three weighted rankings were established, and they were combined to an overall ranking (OECD Green Growth Indicators 2016; The World Bank, 2016). The ranking of the European countries is shown in Table 4.

The top three European countries with the best score are France, Finland, and the United Kingdom. There are countries which are at top positions regarding one index, but they are in low positions regarding the other indices. Greece is a good example of this situation. Greece is leading in sustainability but is not in the first ten countries in innovation and entrepreneurship.

In the next subsections, there are three characteristic examples of frugal and sustainable innovation in the top three countries. It is hard to consider that it's because only of these innovations that these countries are in the top. On the other hand, the selected examples are emblematic and different and, more importantly, match each other for the proposed model in the best practices.

Table 4 Ranking of the European countries regarding innovation, entrepreneurship, and sustainability

Country	Final position	Average total	Ranking innovation	Ranking entrepreneurship	Ranking sustainability
France	1	4,00	1	2	9
Finland	2	6,67	6	12	2
United Kingdom	3	6,67	12	1	7
Portugal	4	9,00	8	9	10
Sweden	4	9,00	2	14	11
Cyprus	6	9,67	10	16	3
Malta	7	10,00	14	3	13
Greece	8	10,33	11	19	1
Poland	8	10,33	16	7	8
Austria	10	11,33	3	17	14
Netherlands	10	11,33	5	25	4
Croatia	12	11,67	17	13	5
Spain	13	13,33	4	21	15
Italy	14	14,33	13	5	25
Bulgaria	15	15,33	18	6	22
Germany	16	15,67	9	26	12
Luxembourg	17	16,67	25	8	17
Latvia	18	17,33	26	20	6
Belgium	19	18,00	23	10	21
Slovenia	19	18,00	24	4	26
Czech Republic	21	18,33	15	24	16
Denmark	22	19,00	7	27	23
Ireland	23	19,33	20	11	27
Slovakia	24	20,00	19	22	19
Estonia	25	21,33	22	18	24
Lithuania	25	21,33	21	15	28
Romania	27	23,00	28	23	18
Hungary	28	25,00	27	28	20

4.1 Finland

In order to support the government's key goals, Finland is launching a digital platform called Kokeilun Paikka. This platform provides small and local innovators with an opportunity to present their ideas and initiatives to develop public services. This also directly links the innovators with reformers and sources of funding. Furthermore, this enables the country to divide complicated challenges into smaller more manageable parts (OECD 2017).

4.2 France

For a long period, the production of energy was centralized in France. The market was dominated by two big companies EDF (Electricité de France) and GDF (Gaz de France). Due to this dominance in the energy sector, more and more regional problems occurred for local farmers in different areas in France, especially in Le Mené. Not only agricultural aspects but also challenges in other sectors led to the emergence of the grass root project of Le Mené. Le Mené tried to combine the challenges of all sectors to unify them and encourage them to develop innovative ideas for an independent energy production. This collaboration of a lot of different characters created a unified and strong identity, which finally led to an independent production of fuel oil and wind energy (Yalçın-Riollet et al. 2014).

4.3 The United Kingdom

In the process of developing frugal innovations, a very important part is the so-called community leadership. A community can be created through a shared vision, similar problems, or nationality. In the United Kingdom, communities tried to do community energy projects. These projects are done by civil society actors to develop either or both energy saving and renewable energy generation measures. By offering incentives for community leaders, the development of frugal innovation can be increased (Seyfang et al. 2014).

4.4 Analysis of Three European Business Models of Frugal and Sustainable Innovation

According to the results from the quantitative analysis, business models from Finland, France, and the United Kingdom were analyzed. The detailed analysis is based on the business model canvas (BMC) which consists out of the following parts: key activities, value proposition, customer relationship, customer segment, revenue stream, channels, key resources, costs, and key partners (Table 5).

4.5 The Proposed Practices

In order to establish this business model, a combination of the most important aspects of the three models mentioned above has to be done. The main service of the business model will be a platform like the one of the Finnish business model. This aspect allows us to shorten the distance between innovators and companies, by

Table 5 BMC of three business models

	Finland	France	United Kingdom
	Digital platform Kokeilun Paikka	French case: Le Mené	Community innovation for sustainable energy
Value proposition	Kokeilun Paikka provides the citizen of Finland with an opportunity not only to present their own innovative ideas but also with the chance to help another innovator by funding their ideas. This also offers a huge collection of collective knowledge for new innovators	The governmental representatives of Le Mené offered their citizens from different sectors a central way to combine their innovative ideas, in order to develop new projects and policies	Since the context of innovations comes from the bottom-up, it has consequences for the viability and resourcing of putative actor and for the policy context in which they operate. The nature of the protection benefits from the challenge of niche development. The diffusion and influence of wider energy systems
Customer relationships	The platform allows users to browse content, obtain ideas for their own project, and communicate with each other to help market and share their innovations. Since the platform was launched only recently, it is too early to gauge impacts or assess results	All people that live in the French region of Le Mené are customers that could possibly benefit from an independent production of energy People that moved to Le Mené due to the creation of jobs for the independent production of energy	The actors who work within the community energy sector including dedicated energy intermediaries, policy actors such as local and national government, and private sector organizations such as energy utilities and independent consultants
Customers segment	The goal at this level is to identify and support the best results from local and regional experiments and ultimately to test them on a larger scale. Finally, municipalities, academics, civil society organizations, and citizens can use the application to promote their innovation, with each actor individually monitoring their own activities	The customer segments are the people that live in Le Mené but also people that are just working in this region, since they contribute to the production of energy.	The community energy practitioners aiming to build robust influential niches and for policy makers.
Revenue stream	The users will be joined and stay involved, so there will be indirect revenues. However, it is hard to estimate the exact revenues, since there will be eventually many partners	Since this project is rather a social approach to be more economic and independent than creating revenue, the revenue stream must be seen differently. Other than money this project	Local household, community, and school efficiency

(continued)

Table 5 (continued)

	Finland	France	United Kingdom
		provided Le Mené with way and tool to push further innovative ideas of local people	
Channels	The platform will connect innovators, and this is the means for the communication. It is believed that users will reach into the thousands and that the platform will enable innovators to establish links with support and funding networks	Actors, which were at the origins of pioneering projects, are the references in this area. Social innovation networks help to provide encouragement for citizens who come across similar obstacles. The municipalities which harbor these civic projects	Voluntary organizations, cooperatives, informal associations, and partnerships with social enterprises, schools, businesses, faith groups, local government, or utility companies
Key resources	The huge amount of possible users provides all the necessary resources, such as funding, connections, other ideas, and most importantly direct feedback	The willingness of the local actors (e.g., farmers, municipalities, citizens) to participate in the project. The integration into the central system and the critical relationship with national authorities	Sharing learning is an important activity for this case. The engagement in networking activities in a variety of ways, with a diverse set of partners, to gain support and information and share their experiences. The development of shared expectations and visions is considered a prerequisite for the development of local projects
Costs	The cost is relatively low. It is hoped that the replicability in other countries will reduce the final cost	The cost was reduced since the farmers limited their purchases and replaced them with their own production. Beneficial cost reduction derived from the local development with the creation of jobs, the fight against exclusion, protection of the environment, etc.	Grant funding needed to carry out the project, to buy equipment, and to pay for key staff
Key activities	The main activity is to offer a central platform for innovators. Furthermore, the finish government offers expertise and guidance for innovators	The government of Le Mené acted as link between different sectors in order to push forward innovation. Another huge activity was the building	The most important activity in this case was the introduction of incentives. This led to a rise in motivation for people to become a

(continued)

Table 5 (continued)

	Finland	France	United Kingdom
		of a network for further partnerships	community leader and bring forward ideas
Key partners	Apart of nongovernmental users such as organizations, municipalities, academics, civil society, and citizens will contribute individually by monitoring their own activities	Native leaders who declared themselves to be mainly motivated by their love for the region People who did not come from Le Mené but who had chosen to live and/or work there, bringing their technical, economic, managerial, and social skills to projects	Intermediary organizations which influence and support in framing and coordinating the project

using the Internet for a fast exchange of information. The second aspect will be provided by the French business model. This facet bears the idea of building a central basis for communication not only between companies and innovators but also between two innovators by offering a huge list of former innovators. Such a central basis allows innovators to cooperate and create even better innovations such as the farmers did in the business model of France. The last but very important part for our business model is provided by the case study of the United Kingdom.

The business model will include two evaluation stages for the idea to pass. It is also important to note that the submitted ideas will not be publically displayed. The first stage will have three members of an evaluation committee to check if the submitted innovation is worth investing in. After the idea was evaluated worthy, a second committee will evaluate the submitted innovation. After this final stage, the innovation is ready to go into further work and development. In order to get rid of the innovator's fear, all of the mentioned members of the committee will be bound to contracts similar to nondisclosure agreements. Additionally, as a further motivator for innovators to share their ideas, the business model will include monetary incentives for the ideas that successfully make it through the first stage to the second committee. The following paragraphs describe each section in detail.

Value Proposition

The business model provides fast access not only to the company that owns the platform. By using the offered platform, the innovator can get access to funding, a team for further development, and the needed capacity in order to produce the innovation. This aspect is highly sensible, in consideration of the short product life cycle of innovations. The platform also offers access to previous innovators, if wanted. On voluntary basis, the innovators can ask for immediate feedback from the company and other innovators, which could provide the innovators new information.

Target Customers

The target customers for companies, applying the business model, apart from the ordinary customers are also the innovators. These people are an excellent customer segment to provide new ideas to companies, because they have a better understanding of the problems, as well as the resource constraints in their everyday life. In a broader sense, also nongovernmental organizations (NGOs) can be a customer segment, which may help to get into contact with innovators.

Channels

As innovators can get into contact with the company via a platform, obviously, this online platform will be the first and most important channel. To promote the platform, buzz marketing will be of high importance.

Customer Relationship

The innovators are the most important sources of ideas for the companies. Hence, a good relationship with them is much needed. The platform helps to establish a first contact and relationship. Once the idea was picked up from the company, a personal face-to-face relationship will be established.

Key Activities

The activities of the companies can be guided in four phases. Companies should start with an observational phase, to see which ideas are submitted to the platform. Companies can learn about certain problems, customer needs, and requirements. This phase also provides the basis for the next one: conceptualization. During this phase the potential of the idea is evaluated, and the companies will experience the advantage of introducing the innovation. This enables them to set market boundaries, rules, and regulation for potential customers. Thus, a competitive advantage on a certain market can be achieved. In the third phase, the development of the product starts. After the development, the product is ready for the production. As the companies want to sell in the future, also market preparation, like marketing, is a key activity. Besides that, the constant support, maintenance, and review of the platform and the submitted ideas are another key activity.

Key Partners

A very important part of the proposed BMC is the key partners. These are people or organizations that are related to the companies' activities to develop innovative ideas. Identified partners are of course the innovators themselves, since they are the source of any idea. As already mentioned also organizations and companies can be seen as key partners. The last group of identified partners are NGOs that can recommend new innovators to the companies.

Key Resources

As per definition, key resources are all necessary resources to find, develop, and sell innovations. In the case of our best practice, these are of course through the Internet, since it is necessary to run the central basis and manage the innovators and their know-how as well. Another resource is the committee that is needed to evaluate the submitted ideas. Very important for the development of the ideas are free capacities within the company or at manufacturers. In order to provide the innovators incentives, money is seen as a central resource. Lastly, the network of previous innovators can be seen as a very valuable key resource.

Revenue Stream

Since this business model is established to improve the collaboration between companies and innovators, no source of revenue in the sense of money can be determined. Only ideas and knowledge can be identified as a "revenue." The revenue in monetary value of the innovated products is generated later on during selling the product on the target market.

Cost Structure

Concerning the cost structure, various factors have to be taken into account. Obviously, the company will have expenditures caused by the support of the platform and the intensive support of ideas through the innovation committee. Also, research and production costs need to be considered. Those costs can be considered normal costs, which every company has to cope with. But working with innovators, more costs occur. Depending on which model of collaboration is chosen, different expenses have to be covered. One option is a one-time compensation: It is possible that the company buys all intellectual property rights related to the idea and patent by paying one time a certain amount of money. Another option would be a royalty: With this option, the company has monthly or annually expenses, because the innovator is paid a certain amount regularly. The third option would be an investment into the innovator, to enable him/her to develop and produce the product himself/herself. What also has to be considered as an expense are the incentives for the best innovation ideas.

5 Discussion and Conclusions

This study proposes best practices for European companies to successfully pick up and implement frugal and sustainable innovations. A quantitative analysis based on criteria of entrepreneurship, sustainability, and innovation for all European countries was carried out to determine the top three countries namely Finland, France, and the United Kingdom. The proposed business model for European companies is based on existing ones from the aforementioned countries.

It was observed that European companies need to pay special attention to the short life cycle of products as well as to a good relationship with the innovators. The way of collaborating and compensation is a highly sensitive process in picking up ideas. The research also revealed that a collaboration between European companies and innovators is highly required. The proposed business model considers the observed factors and is therefore providing a solid basis for European companies to successfully pick up innovations.

During the literature research, it was observed that picking up ideas for frugal innovations from emerging economies is now starting to get more popular in the West (Fraunhofer 2016). Frugal innovations are a possibility for companies to create more value from less resources for more people. European companies can enable innovators with limited resources to develop and produce their ideas. The collaboration with frugal innovators will help European companies to reach new markets with minimized effort in research. By scouting and investing in such ideas from every corner of the earth, large companies can reduce the cost of sales without compromising the quality and environmental integrity. As a result, the benefit of science, technology, and social innovation can be reached to the lowest level of the society and farthest corner of the globe. Due to those reasons, the creation of a viable business practices is of high importance.

The findings show that a model is very much required by innovators as well as organizations. Both the parties could be part of such collaborations if their respective interests and rights are protected. Apart from financial and logistical advantages, such projects will also place companies at high moral ground. When people are questioning and debating over ethical practices of the organizations, connecting themselves with innovators, people will give the organization high moral advantage and social satisfaction. It can help them to improve their image in the market and can also create a sense of loyalty among customers. Innovators must understand the pressure on companies to make profit on their investment as well as companies should understand the desires of innovators to be financially independent.

It is hardly possible to conduct a study that does not contain weaknesses or an element of bias. Therefore, there are some limitations in this study as well. This study did not include more examples of the top three countries. Emblematic examples of specific countries which are top in a specific index, e.g., Greece, in sustainability were not discussed. Nevertheless, all the practices raised in this research can be applied to companies expanding from the specific ones of this study. As the concept is quite new, only the future will tell if picking up frugal and sustainable innovation will become a common and successful business model for European companies.

Appendices

	INNOVATION						ENTREPRENEURSHIP				SUSTAINABILITY				
	Innovation activities						Entrepreneureship activities				Sustainability activities				
	ECO Innovation Index	enterprises engaged in market introduction of innovation	enterprises engaged in external R&D	enterprises engaged in internal R&D	enterprises engaged in training for innovation activities	Gross Domestic Expenditure on R&D (R&D intensity)	Net business population growth	Birth rate=number of enterprise births in the reference period (t) divided by the number of enterprises active in t	number of high-growth enterprises	value added at factor cost (total business economy)	Environmental benefits due to innovation in the enterprises	Development of environment-related technologies, % all technologies	Environmentally related government R&D budget, % total government R&D	Ressource productivity	production value of cleaner and resource-efficient products
Source	Eurostat	Eurostat	Eurostat	Eurostat	Eurostat	Eurostat	OECD	OECD	OECD	Eurostat	Eurostat	OECD	OECD	Eurostat	Eurostat
Unit	Number	Number of Enterprises	Number of Enterprises	Number of Enterprises	Number of Enterprises	% of GDP	%	%	Number of Enterprises	Value	Number of Enterprises	%	%	Euro per kilogram	Million Euros
Year	2015	2014	2014	2014	2014	2014	2014	2014	2014	2014	2014	2012	2012	2015	2013
		2012			2012		2012								2014
Austria	108	3.038	1.900	3.709	4.124	3,06	-0,98	7,13	2.962	169.625,5	5.841	13,51	2,36	1,65	19.377,64
Belgium	97	2.217	2.531	4.313	3.304	2,46	2,52	4,37	2.364	194.938,2	:	8,83	2,29	2,45	10.032,52
Bulgaria	49	583	190	374	855	0,79	1,62	11,82	2.823	19.887,6	733	18,13	:	0,28	799,11
Croatia	67	670	442	811	1.090	0,79	0,66	7,77	1.239	20.426,7	1.344	14,02	:	1,10	:
Cyprus	60	216	116	212	360	0,48	-0,49	6,87	:	6.996,1	246	:	:	1,85	:
Czech Republic	99	2.665	1.913	3.929	3.454	1,97	5,52	9,13	4.659	83.864,3	4.780	7,61	1,83	1,03	1.650,07
Denmark	167	:	447	981	894	3,02	0,77	11,14	2.166	127.418,5	1.135	22,87	1,88	2,26	7.024,18
Estonia	80	301	301	501	323	1,45	1,46	10,73	550	10.485,8	360	23,95	3,62	0,50	:
Finland	140	1.885	2.094	3.107	1.650	3,17	1,06	7,88	2.058	87.139,0	2.746	12,11	1,49	1,12	:
France	115	11.135	8.974	18.934	14.876	2,24	7,23	9,93	14.860	888.564,0	:	13,19	1,77	2,82	28.554,83
Germany	129	24.541	12.559	34.068	42.254	2,89	-5,17	7,20	32.298	1.531.691,5	57.355	13,42	2,86	2,12	:

	INNOVATION						ENTREPRENEURSHIP				SUSTAINABILITY				
	Innovation activities						Entrepreneureship activities				Sustainability activities				
	ECO Innovation Index	enterprises engaged in market introduction of innovation	enterprises engaged in external R&D	enterprises engaged in internal R&D	enterprises engaged in training for innovation activities	Gross Domestic Expenditure on R&D (R&D intensity)	Net business population growth	Birth rate=number of enterprise births in the reference period (t) divided by the number of enterprises active in t	number of high-growth enterprises	value added at factor cost (total business economy)	Environmental benefits due to innovation in the enterprises	Development of environment-related technologies, % all technologies	Environmentally related government R&D budget, % total government R&D	Ressource productivity	production value of cleaner and resource-efficient products
Greece	72	1.846	1.063	2.154	1.978	0,84	:	:	:	50.454,2	3.341	6,91	1,36	1,35	:
Hungary	81	801	369	1.310	1.009	1,36	1,19	9,98	3.536	51.089,1	1.340	5,61	1,65	0,88	:
Ireland	134	:	912	1.977	:	1,51	-1,86	6,82	1.852	:	:	7,12	1,29	2,19	910,73
Italy	106	12.284	4.679	16.193	13.034	1,38	-1,47	7,14	11.397	643.135,2	25.156	10,65	3,36	3,04	:
Latvia	75	229	124	398	238	0,69	4,08	15,05	1.130	10.086,7	505	11,89	:	0,50	1.331,69
Lithuania	73	623	289	712	1.001	1,03	12,37	24,50	1.921	14.557,6	1.620	10,52	:	0,80	819,18
Luxembourg	124	264	183	296	445	1,28	3,38	10,03	421	21.640,3	547	16,38	3,85	3,39	:
Malta	64	47	12	81	60	0,75	4,42	11,30	:	4.168,7	121	14,29	:	1,36	:
Netherlands	98	3.794	4.902	8.494	5.195	2	2,29	10,11	5.631	314.920,3	:	9,83	0,89	3,44	:
Poland	59	3.199	1.986	3.082	4.656	0,94	0,50	12,50	9.378	181.284,8	4.183	15,64	6,12	0,64	11.286,97
Portugal	102	2.258	1.738	3.112	3.983	1,29	0,41	14,77	3.106	68.977,5	6.237	12,52	3,48	1,10	:
Romania	82	155	45	797	372	0,38	0,89	10,19	1.058	55.859,7	827	14,28	8,33	0,31	5.861,06
Slovakia	72	422	236	741	577	0,88	9,82	19,79	1.509	30.934,4	949	7,75	3	1,04	:
Slovenia	96	546	586	1.015	538	2,38	2,11	11,03	532	18.562,8	1.135	10,59	2,98	1,35	1.125,97
Spain	106	2.592	2.913	6.596	3.246	1,24	-0,27	9,77	10.768	430.070,4	:	15,63	3,77	2,77	:
Sweden	124	2.680	2.122	5.011	2.043	3,15	1,08	7,23	4.659	214.126,2	5.018	10,5	1,94	1,74	:
United Kingdom	106	:	:	:	:	1,68	4,33	14,27	26.705	1.212.506,6	:	12,14	2,8	3,4627	:

	INNOVATION													
	ECO Innovation Index	Weighted Points	enterprises engaged in market introduction of innovation	Weighted Points	enterprises engaged in external R&D	Weighted Points	enterprises engaged in internal R&D	Weighted Points	enterprises engaged in training for innovation activities	Weighted Points	Gross Domestic Expenditure on R&D (R&D intensity)	Weighted Points	Summe	Ranking
Weighting		**0,27**		**0,17**		**0,15**		**0,09**		**0,10**		**0,22**		
Austria	8	2,16	6	1,02	6	0,90	7	0,63	9	0,90	6	1,32	**6,93**	3
Belgium	15	4,05	11	1,87	22	3,30	24	2,16	18	1,80	23	5,06	**18,24**	23
Bulgaria	28	7,56	17	2,89	10	1,50	8	0,72	8	0,80	10	2,2	**15,67**	18
Croatia	24	6,48	15	2,55	16	2,40	17	1,53	17	1,70	4	0,88	**15,54**	17
Cyprus	26	7,02	23	3,91	1	0,15	1	0,09	1	0,10	5	1,1	**12,37**	10
Czech Republic	13	3,51	8	1,36	19	2,85	22	1,98	24	2,40	13	2,86	**14,96**	15
Denmark	1	0,27	13	2,21	14	2,10	14	1,26	14	1,40	12	2,64	**9,88**	7
Estonia	19	5,13	20	3,40	13	1,95	13	1,17	12	1,20	22	4,84	**17,69**	22
Finland	2	0,54	12	2,04	5	0,75	5	0,45	10	1,00	18	3,96	**8,74**	6
France	7	1,89	3	0,51	2	0,30	2	0,18	2	0,20	8	1,76	**4,84**	1
Germany	4	1,08	1	0,17	17	2,55	18	1,62	14	1,40	24	5,28	**12,10**	9
Greece	22	5,94	13	2,21	4	0,60	3	0,27	3	0,30	14	3,08	**12,40**	11
Hungary	18	4,86	14	2,38	25	3,75	26	2,34	23	2,30	27	5,94	**21,57**	27
Ireland	3	0,81	13	2,21	24	3,60	23	2,07	25	2,50	26	5,72	**16,91**	20
Italy	10	2,7	2	0,34	20	3,00	21	1,89	16	1,60	19	4,18	**13,71**	13
Latvia	20	5,4	22	3,74	23	3,45	25	2,25	21	2,10	17	3,74	**20,68**	26
Lithuania	21	5,67	16	2,72	18	2,70	15	1,35	15	1,50	15	3,3	**17,24**	21
Luxembourg	5	1,35	21	3,57	27	4,05	27	2,43	26	2,60	25	5,5	**19,50**	25
Malta	25	6,75	25	4,25	3	0,45	4	0,36	4	0,40	9	1,98	**14,19**	14
Netherlands	14	3,78	4	0,68	11	1,65	9	0,81	6	0,60	3	0,66	**8,18**	5
Poland	27	7,29	5	0,85	9	1,35	12	1,08	5	0,50	20	4,4	**15,47**	16
Portugal	12	3,24	10	1,70	12	1,80	10	0,90	7	0,70	16	3,52	**11,86**	8
Romania	17	4,59	24	4,08	26	3,90	19	1,71	22	2,20	28	6,16	**22,64**	28
Slovakia	23	6,21	19	3,23	15	2,25	16	1,44	20	2,00	7	1,54	**16,67**	19
Slovenia	16	4,32	18	3,06	21	3,15	20	1,80	19	1,90	21	4,62	**18,85**	24
Spain	9	2,43	9	1,53	8	1,20	11	0,99	13	1,30	1	0,22	**7,67**	4
Sweden	6	1,62	7	1,19	7	1,05	6	0,54	11	1,10	2	0,44	**5,94**	2
United Kingdom	11	2,97	13	2,21	14,5	2,18	14,5	1,31	14	1,40	11	2,42	**12,48**	12

ENTREPRENEURSHIP										
	Net business population growth	Weighted Points	Birth rate=number of enterprise births in the reference period (t) divided by the number of enterprises active in t	Weighted Points	number of high-growth enterprises	Weighted Points	value added at factor cost (total business economy)	Weighted Points	Summe	Ranking
Weighting		**0,3**		**0,26**		**0,22**		**0,22**		
Austria	9	2,70	27	7,02	14	3,08	8	1,76	**14,56**	17
Belgium	12	3,60	7	1,82	13	2,86	21	4,62	**12,90**	10
Bulgaria	4	1,20	18	4,68	8	1,76	13	2,86	**10,50**	6
Croatia	18	5,40	9	2,34	15	3,30	11	2,42	**13,46**	13
Cyprus	27	8,10	22	5,72	1	0,22	1	0,22	**14,26**	16
Czech Republic	13	3,90	11	2,86	23	5,06	24	5,28	**17,10**	24
Denmark	26	7,80	26	6,76	18	3,96	14,5	3,19	**21,71**	27
Estonia	14,5	4,35	14,5	3,77	13	2,86	17	3,74	**14,72**	18
Finland	22	6,60	17	4,42	5	1,10	5	1,10	**13,22**	12
France	3	0,90	16	4,16	3	0,66	3	0,66	**6,38**	2
Germany	19	5,70	20	5,20	20	4,40	20	4,40	**19,70**	26
Greece	25	7,50	23	5,98	4	0,88	4	0,88	**15,24**	19
Hungary	23	6,90	25	6,50	13	2,86	26	5,72	**21,98**	28
Ireland	7	2,10	3	0,78	21	4,62	25	5,50	**13,00**	11
Italy	1	0,30	1	0,26	17	3,74	23	5,06	**9,36**	5
Latvia	8	2,40	14	3,64	25	5,50	19	4,18	**15,72**	20
Lithuania	14	4,20	15	3,90	10	2,20	16	3,52	**13,82**	15
Luxembourg	5	1,50	8	2,08	13	2,86	27	5,94	**12,38**	8
Malta	10	3,00	13	3,38	7	1,54	6	1,32	**9,24**	3
Netherlands	24	7,20	24	6,24	12	2,64	10	2,20	**18,28**	25
Poland	20	6,00	6	1,56	6	1,32	9	1,98	**10,86**	7
Portugal	21	6,30	4	1,04	11	2,42	14	3,08	**12,84**	9
Romania	17	5,10	12	3,12	22	4,84	15	3,30	**16,36**	23
Slovakia	11	3,30	10	2,60	24	5,28	22	4,84	**16,02**	22
Slovenia	2	0,60	2	0,52	19	4,18	18	3,96	**9,26**	4
Spain	16	4,80	19	4,94	16	3,52	12	2,64	**15,90**	21
Sweden	15	4,50	21	5,46	9	1,98	7	1,54	**13,48**	14
United Kingdom	6	1,80	5	1,30	2	0,44	2	0,44	**3,98**	1

SUSTAINABILITY										
	Environmental benefits due to innovation in the enterprises	Weighted Points	Development of environment-related technologies, % all technologies	Weighted Points	Environmentally related government R&D budget, % total government	Weighted Points	Ressource productivity	Weighted Points	production value of cleaner and resource-efficient products	Weighted Points	Summe	Ranking
Weighting		**0,31**		**0,26**		**0,09**		**0,22**		**0,12**		
Austria	11,5	3,57	22	5,72	13	1,17	7	1,54	4	0,48	**12,48**	14
Belgium	17	5,27	3	0,78	11,5	1,04	28	6,16	12	1,44	**14,69**	21
Bulgaria	6	1,86	24	6,24	16	1,44	21	4,62	7	0,84	**15,00**	22
Croatia	13	4,03	2	0,52	15	1,35	8	1,76	5	0,60	**8,26**	5
Cyprus	1	0,31	11	2,86	10	0,90	10	2,20	14	1,68	**7,95**	3
Czech Republic	20	6,20	1	0,26	5	0,45	26	5,72	7	0,84	**13,47**	16
Denmark	11,5	3,57	25	6,50	21	1,89	9	1,98	10	1,20	**15,14**	23
Estonia	8	2,48	26	6,76	20	1,80	16	3,52	7	0,84	**15,40**	24
Finland	11,5	3,57	6	1,56	4	0,36	6	1,32	7	0,84	**7,65**	2
France	11,5	3,57	12	3,12	17	1,53	5	1,10	1	0,12	**9,44**	9
Germany	11	3,41	9	2,34	11,5	1,04	18	3,96	7	0,84	**11,59**	12
Greece	2	0,62	17	4,42	7	0,63	4	0,88	7	0,84	**7,39**	1
Hungary	21	6,51	14,5	3,77	11,5	1,04	11	2,42	7	0,84	**14,58**	20
Ireland	19	5,89	16	4,16	11,5	1,04	25	5,50	8	0,96	**17,55**	27
Italy	10	3,10	19	4,94	11,5	1,04	23	5,06	11	1,32	**15,46**	25
Latvia	18	5,58	4	1,04	3	0,27	3	0,66	7	0,84	**8,39**	6
Lithuania	12	3,72	27	7,02	18	1,62	22	4,84	7	0,84	**18,04**	28
Luxembourg	22	6,82	7	1,82	11,5	1,04	14	3,08	7	0,84	**13,60**	17
Malta	11,5	3,57	21	5,46	22	1,98	2	0,44	7	0,84	**12,29**	13
Netherlands	4	1,24	10	2,60	12	1,08	13	2,86	2	0,24	**8,02**	4
Poland	7	2,17	5	1,30	2	0,18	24	5,28	3	0,36	**9,29**	8
Portugal	3	0,93	13	3,38	6	0,54	19	4,18	7	0,84	**9,87**	10
Romania	16	4,96	8	2,08	1	0,09	27	5,94	6	0,72	**13,79**	18
Slovakia	14	4,34	18	4,68	9	0,81	15	3,30	9	1,08	**14,21**	19
Slovenia	15	4,65	23	5,98	8	0,72	20	4,40	7	0,84	**16,59**	26
Spain	9	2,79	15	3,90	19	1,71	17	3,74	7	0,84	**12,98**	15
Sweden	5	1,55	20	5,20	14	1,26	12	2,64	7	0,84	**11,49**	11
United Kingdom	11,5	3,57	14	3,64	11	0,99	1	0,22	7	0,84	**9,26**	7

References

Abuzeinab A, Arif M (2014) Stakeholder engagement: a green business model indicator. Procedia Econ Financ 18:505–512
Baldsarre B, Calabretta G, Bocken NMP, Jaskiewicz T (2017) Bridging sustainable business model innovation and user-driven innovation: a process for sustainable value proposition design. J Clean Prod 147:175–186
Bijoy C (2007) Access and benefit-sharing from the indigenous peoples' perspective: the TBGRI-Kani 'model'. Law Environ Dev J 3(1):18
Bocken NMP, Short SW, Rana P, Evans S (2014) A literature and practice review to develop sustainable business model archetypes. J Clean Prod 65:42–56
Boons F, Montalvo C, Quist J, Wagner M (2013) Sustainable innovation, business models and economic performance: an overview. J Clean Prod 45:1–8
Bound K, Thornton I (2012) Our frugal future: lessons from India's innovation system. Nesta, London Available at: http://www.nesta.org.uk/publications/our-frugal-future-lessonsindias-innovation-system. Accessed on 22 April 2017
Charter M, Clark T (2007) Sustainable innovation. The Centre for Sustainable Design, Farnham
Charter M, Gray C, Clark T, Woolman T (2008) Review: the role of business in realising sustainable consumption and production. In: Tukker A, Charter M, Vezzoli C, Stø E, Andersen MM (eds) Perspectives on radical changes to sustainable consumption and production. System innovation for sustainability. Greenleaf, Sheffield pp. 46e69
Community Innovation Survey (2014) The harmonized Questionnaire. Retrieved March 11, 2017, from http://www.mof.gov.cy/mof/cystat/statistics.nsf/All/D7103836F6734F2BC2257ED200364104/$file/INNOVATION_SURVEY-2012_2014-EN-021015.pdf?OpenElement
DTI (2005) Innovation. Available at: www.innovation.gov.uk. Accessed 22 Oct 2016
DeSimone, L. D., & Popoff, F. (2000). Eco-efficiency: the business link to sustainable development. MIT press, Cambridge, MA.
Erika, K. M., & Watu, W. (Eds.). (2010). Innovation and the development agenda. IDRC.
European Commission (2016) A conceptual analysis of foundations, trends and relevant potentials in the field of frugal innovation (for Europe). Retrieved March 18, 2017 from http://www.isi.fraunhofer.de/isi-wAssets/docs/p/de/publikationen/DG-RTD-publication.pdf
Eurostat (2007) OECD manual on business demography statistics. Retrieved March 11, 2017, from http://ec.europa.eu/eurostat/documents/3859598/5901585/KS-RA-07-010-EN.PDF/290a71ec-7a71-43be-909b-08ea6bcdc521?version=1.0
Eurostat (2016a) ECO innovation index. Retrieved March 11, 2017, from http://ec.europa.eu/eurostat/tgm/table.do?tab=table&init=1&plugin=1&language=en&pcode=t2020_rt200
Eurostat (2016b) Production, value added, employment and exports in the environmental goods and services sector. Retrieved March 11, 2017, from http://ec.europa.eu/eurostat/cache/metadata/en/env_egs_esms.htm
Eurostat Statistic Explained (2016) Resource productivity statistics. Retrieved March 11, 2017, from http://ec.europa.eu/eurostat/statistics-explained/index.php/Resource_productivity_statistics
Eurostat Statistics Explained (2017a) R&D intensity. Retrieved March 11, 2017, from http://ec.europa.eu/eurostat/statistics-explained/index.php/Glossary:R_%26_D_intensity
Eurostat Statistics Explained. (2017b). Business demography statistics. Retrieved March 11, 2017, from http://ec.europa.eu/eurostat/statistics-explained/index.php/Business_demography_statistics#Birth_rate
Eurostat Statistics Explained (2017c) Value added at factor cost. Retrieved March 11, 2017, from http://ec.europa.eu/eurostat/statistics-explained/index.php/Glossary:Value_added_at_factor_cost
Fraunhofer ISI (2016) Cheaper, better, more relevant: is frugal innovation an opportunity for Europe?. Retrieved January 8, 2017, from http://www.isi.fraunhofer.de/isi-wAssets/docs/p/de/projektberichte/FrugalInnovationSummary_ISI_Nesta_mit-ISI.pdf

Innovationdata (2014) Net business population growth. Retrieved March 11, 2017, from http://www.innovationdata.be/i/BUS_ENT_4/Net-business-population-growth
Jennings PD, Zandbergen PA (1995) Ecologically sustainable organizations: an institutional approach. Acad Manag Rev 20:1015–1052
Joshi RG, Chelliah DJ (2013) Sharing the benefits of commercialisation of traditional knowledge: what are the key success factors? The Intellectual Property Forum, No. 93
Karagülle AÖ (2012) Technology and Innovation Management Green business for sustainable development and competitiveness: an overview of Turkish logistics industry, International Conference on Leadership. Procedia Soc Behav Sci 41:456–460
Krishnan R (2010) From Jugaad to systematic innovation. Utpreraka Foundation, Bangalore
Latour B (1993) We have never been modern. Harvard University Press, Cambridge, MA
Nidumolu R, Prahalad CK, Rangaswami MR (2015) Why sustainability is now the key driver of innovation. IEEE Eng Manag Rev 43(2):85–91
OECD (2017) Embracing innovation in government: global trends: case study: citizen as experts. Retrieved March 18, 2017 from https://www.oecd.org/gov/innovative-government/embracing-innovation-in-government.pdf
OECD Green Growth Indicators (2016) Database documentation. Retrieved March 11, 2017, from http://stats.oecd.org/
Prahalad CK, Mashelkar RA (2010) Innovation's holy grail. Harvard Business Review, July–August. Accessed through http://www.andersbraekken.com/RESOURCES/Innovations%20Holy%20Grail.pdf
Radjou N, Prabhu J (2014) Frugal innovations: how to do more with less. In: The Economist Books, New York
Schaltegger S, Petersen H (2000) Ecopreneurship – KonzeptundTypologie. R.I.O. Impuls, Luzern
Schaltegger S, Wagner M (2017) Managing the business case for sustainability: the integration of social, environmental and economic performance. Routledge, London
Seyfang G, Smith A (2007) Grassroots innovations for sustainable development: towards a new research and policy agenda. Env Polit 16:582–603
Seyfang G, Hielscher S, Hargreaves T, Martiskainen M, Smith A (2014) A Grassroot sustainable energy niche? Reflections on community energy in the UK. Environ Innov Soc Trans 13:21–44
Stubbs W, Cocklin C (2008) Conceptualizing a sustainability business model. Organiz Environ 21:103–127
The World Bank (2016) World Bank list of economies (July 2016). Retrieved May 19, 2017, from http://iccmoot.com/wp-content/uploads/2016/08/World-Bank-list-of-economies_July2016.pdf
Yalçın-Riollet M, Garabuau-Moussaoui I, Szuba M (2014) Energy autonomy in Le Mené: a French case of grassroots innovation. Energy Policy 69:347–355 Available at: http://www.sciencedirect.com/science/article/pii/S0301421514001050
Yang M, Evans S, Vladimirova D, Rana P (2017) Value uncaptured perspective for sustainable business model innovation. J Clean Prod 140:1794–1804. https://doi.org/10.1016/j.jclepro.2016.07.102

Group Evaluation of Water Management Plans with Analytic Hierarchy Process and Social Choice Methods

Bojan Srdjevic, Zorica Srdjevic, and Yvonilde Dantas Pinto Medeiros

Abstract Selection of a good water management plan for the river basin is a complex decision-making problem because interests of stakeholders are rarely in complete agreement. If water committee has to emulate interest and power of key parties, decision-making process can be organized in many different ways, depending on adopted methodology for deriving decisions and formalizing setup to implement solutions. Group context brings individuals with different background, attitude, and (in)consistency they will demonstrate when evaluating and/or judging options. This chapter shows how methodologically distinct tools can efficiently support group decision-making at a group and sub-group level within water committee. We propose to firstly use analytic hierarchy process (AHP) to rank management plans in strictly multi-criteria environment and, secondly, to use social choice (voting) methods Borda Count (BC) and Approval Voting (AV) for the final ranking of reduced set of top-ranked plans as identified in the AHP. Illustrative example from Brazil is used to show usefulness of combined approach.

Keywords Decision-making · AHP · Borda Count · Approval Voting · Water management plan

B. Srdjevic (✉) · Z. Srdjevic
Faculty of Agriculture, Department of Water Management, University of Novi Sad, Novi Sad, Serbia
e-mail: bojans@polj.uns.ac.rs

Y. D. P. Medeiros
Escola Politécnica da Universidade Federal da Bahia, Salvador, Brazil

A. Theodoridis et al. (eds.), *Innovative Approaches and Applications for Sustainable Rural Development*, Springer Earth System Sciences,
https://doi.org/10.1007/978-3-030-02312-6_4

1 Introduction

The MCDM (multi-criteria decision-making) method known as analytic hierarchy process (AHP) (Saaty 1980, 2003) and two SC (social choice) methods, known as Borda Count (BC) and Approval Voting (AV) (d'Angelo et al. 1998; Srdjevic 2007), are employed to manage a group decision-making process aimed at assessment of and selection of the best among five long-term water management plans across five criteria. During one of multiple workshops related to hierarchical decision-making processes held at the School of Polytechnic, Federal University of Bahia, Salvador, Brazil, a special session has been organized to analyze possibility of establishing decision-making framework related to water management at a catchment scale and within different group and sub-group contexts. A group of 21 professionals took part in two sessions that lasted for 2 h, with one 3-hour break.

AHP is used to support individuals' and sub-groups' cardinal assessments and prioritization of the decision elements and to rank management plans. Participants used evaluation sheet given in Annex III (sheet #1) and judged decision elements within given hierarchy by strictly following standard AHP procedure. Once evaluation sheets are collected, prioritization of criteria and plans is performed, followed by the final AHP synthesis of all local priority vectors. Values of weights obtained within sub-groups are aggregated by weighted geometric mean method (WGMM) to obtain the final weights and corresponding ranking of plans. The AHP results are finally used to reduce set of alternative plans to a new set of three top-ranked plans for voting by BC and AV in the session after break.

Note that in the second part of the first session, immediately after AHP, all participants voted plans within their interest sub-group by two social choice methods, BC and AV. These two methods are selected to be a part of the decision framework due to their simplicity, easiness to be explained to participants, and as less time and effort consuming than AHP. Evaluation sheet #2, also presented in Annex III, is distributed to individuals to set their preferences (by BC voting) and approvals (by AV voting) of plans considering all criteria as implicitly condensed into unique criterion; note that this is the case how voters usually do in real-life elections. Collected preferential and approval opinions of participants, as fully ordinal information (differently from cardinal information obtained by AHP), are summarized to rank alternative plans and indicate most desired as a group decision.

Worthy to mention is that when applying AHP, BC, and AV methods, awareness is required because individual "background and knowledge are generally very different, particularly if groups (and sub-groups) are large. Also important is that in large river-basin water committees, a decision-making process related to planning and overall water management will expectedly be performed with participation of 'oriented committee members' bringing particular background and mostly narrowed interest from social, political, or economic environment they are coming from. These facts have also been a part of our recent research, but are not discussed in details in this paper.

In strictly AHP context, reported approach recognizes importance of using equal weights of individuals and, at the later stage, different weights of sub-groups based on the number of members in the sub-groups (larger a sub-group – higher is its weight). At a committee level, two situations may occur: (1) sub-groups act via their delegates, actually as new (virtually single) individuals; and (2) individuals are acting independently (with inherent behavior of interest groups they represent). In either case, the final preferences are usually determined by weighted geometric aggregation (see, e.g., Forman and Peniwati 1998; Ramanathan and Ganesh 1994).

For the sake of completeness and to justify our methodological choices, we consulted rich literature around group decision-making in agriculture, ecology, etc. which treat problems with multiple criteria and multiple participants and multiple evaluation table settings, notably a lot of them applying the aggregation-disaggregation paradigm (e.g., Morais and de Almeida 2012; Zendehdel et al. 2010; Jonoski and Seid 2016; Jonoski and Seid 2016). More methodological sources were consulted to check our ideas related to consistency, consensus and aggregation schemes (Kadzinski et al. 2013; Cabrerizo et al. 2014), and especially recently published excellent book (Dong and Xu 2016) devoted completely to search for the consensus in group decision-making context.

Outline of the paper is as follows: after brief description of mathematical bases of AHP, BC, and AV, illustrative example section contains statement of the decision problem, description of decision elements, and final outcomes of AHP + BC + AV application. Concluding remarks are given at the end of paper followed by list of relevant references and three annexes.

2 AHP, Borda Count, and Approval Voting Mathematics

2.1 Analytic Hierarchy Process (AHP): Multi-criteria Method

A philosophy of the analytic hierarchy process (AHP) is easy to understand. Assume that hierarchy of the decision problem consists only of a goal (G), a set of criteria C_j ($j = 1, 2, M$), and a set of alternatives A_i ($i = 1, 2, N$). This hierarchy may be called three-level hierarchy, with levels counting from top to bottom (Fig. 1).

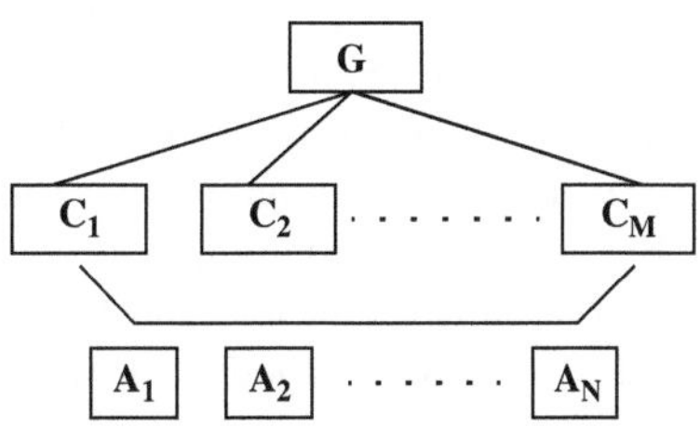

Fig. 1 Hierarchy of a decision-making problem

Table 1 Original Saaty's scale for pairwise comparisons

Numerical values	Judgment definition
1	Equal importance
3	Weak dominance
5	Strong dominance
7	Demonstrated dominance
9	Absolute dominance
2,4,6,8	Intermediate values

The AHP starts by performing a sequence of *M*x(*M*-1)/2 pairwise comparisons of criteria with respect to a goal by using the 9-point Saaty's scale, Table 1 (Saaty 1980).

This way a judgment matrix (1) of size *M*x*M* is created

$$A = \begin{bmatrix} a_{11} & a_{12} & \dots & a_{1M} \\ a_{21} & a_{22} & \dots & a_{2M} \\ \dots & \dots & \dots & \dots \\ a_{M1} & a_{M2} & \dots & a_{MM} \end{bmatrix} \tag{1}$$

with entries a_{ij} $(i,j = 1,2,\ldots,M)$ being numericals given in the first column of Table 1 as representation of preferences elicited from the individual as judgments defined in the right-hand side of the same table. Reciprocal property of matrix *A* means that $a_{ij} = 1$ for all $i = j$ $(i, j = 1, 2, M)$, and $a_{ij} = 1/a_{ji}$.

If we assume that entries of the vector $\boldsymbol{w} = (w_1, w_2, \ldots, w_M)^{\mathrm{T}}$, commonly called priority vector, are weights of criteria, then it is desired to determine these values so that matrix (2) is best approximate of judgment matrix (1).

$$\tilde{A} = \begin{bmatrix} 1 & w_1/w_2 & \dots & w_1/w_M \\ w_2/w_1 & 1 & \dots & w_2/w_M \\ \cdot & \cdot & \dots & \cdot \\ \cdot & \cdot & \dots & \cdot \\ \cdot & \cdot & \dots & \cdot \\ w_M/w_1 & w_M/w_2 & \dots & 1 \end{bmatrix} \tag{2}$$

In standard AHP, for matrix *A* the maximum eigenvalue $\lambda_{\max}$ is determined, and related eigenvector is adopted as vector **w**. This method is generally recognized as the eigenvector method (Saaty 1980). There are, however, more than 20 other methods described in scientific articles for deriving vector **w**, for instance, additive normalization (Saaty 1980), direct least squares (Chu et al. 1979), weighted least squares (Chu et al. 1979), logarithmic least squares (Crawford and Williams 1985), fuzzy preference programming (Mikhailov 2000), logarithmic goal programming (Bryson 1995), evolution strategy prioritization (Srdjevic and Srdjevic 2011), and most recently cosine maximization (Kou and Lin 2014). Useful information on these and many other methods can be found in rich scientific literature (e.g., Golany and

Kress 1993; Mikhailov and Singh 1999; Srdjevic 2005; Ishizaka and Labib 2011; Blagojevic et al. 2016a, b).

Next, *N*x(*N*-1)/2 pairwise comparisons of alternatives are performed at level 3 with respect to each criterion at level 2. This way a set of *M* matrices of size *N*x*N* is created. Local eigenvectors (for each matrix one vector) are computed as before, and a new matrix (3) of size *N*x*M*is is created. Computed local vectors represent columns of this new matrix *X*. Recall that elements of *j*th vector are partial ratings of alternatives with respect to the *j*th criterion and sum to 1.

$w_1\ w_1\ \ldots\ w_M$

$$X = \begin{bmatrix} x_{11} & x_{12} & \cdots & x_{1M} \\ x_{21} & x_{22} & \cdots & x_{2M} \\ \cdots & \cdots & \cdots & \cdots \\ x_{N1} & x_{N2} & \cdots & x_{NM} \end{bmatrix}. \tag{3}$$

Finally, local priority vectors are multiplied by the weights of related criterions to obtain matrix (4) which aggregates performance ratings of all alternatives with respect to all criteria.

$$Z = \begin{bmatrix} w_1 w_{11} & w_1 w_{12} & \cdots & w_M x_{1M} \\ w_1 w_{21} & w_1 w_{22} & \cdots & w_M x_{2M} \\ \cdots & \cdots & \cdots & \cdots \\ w_1 x_{N1} & w_2 x_{N2} & \cdots & w_M x_{NM} \end{bmatrix} = \begin{bmatrix} z_{11} & z_{12} & \cdots & z_{1M} \\ z_{21} & z_{22} & \cdots & z_{2M} \\ \cdots & \cdots & \cdots & \cdots \\ z_{N1} & z_{N2} & \cdots & z_{NM} \end{bmatrix} \tag{4}$$

Summing the elements in each row of the matrix *Z* gives the final result (5): weights for alternatives at fingertips of the hierarchy with respect to the goal at the top of hierarchy.

$$w_i = \sum_{j=1}^{M} z_{ij}, i = 1, 2, \ldots, N \tag{5}$$

The alternative with the highest weight coefficient value w_i should be considered as “the best alternative,” i.e., the best choice in the multi-criteria sense.

Measuring Consistency of Decision-Makers in AHP Applications

In either individual or group AHP applications, it is typical to perform checkup of decision-maker(s) consistency. Several indicators are proposed to measure quality of weights of decision elements which are extracted from the comparison matrix created in pairwise manner at given node of hierarchy (usually by using Saaty’s scale, Table 1). For instance, a consistency ratio (CR) is used if eigenvector method (Saaty 1980) performs prioritization of decision elements; fuzzy intersection measure (μ) is used when prioritization is performed by fuzzy preference programming method (Mikhailov 2000); and geometric consistency index (GCI) is used with logarithmic square method (Aguaron et al. 2003). There are more general consistency measures applicable to all prioritization methods such as a generalized L^2 Euclidean distance (ED) and order reversal indicator known as minimum violations (MV) criterion (Golany and Kress 1993; Mikhailov and Singh 1999; Srdjevic and Srdjevic 2014). All aforementioned indicators serve to measure both the accuracy of the solution and the ranking order properties and are widely accepted by researchers (Srdjevic 2005).

Measuring consistency of AHP solutions in group context receives continuous attention of researchers in this subject area. Different methods of prioritization and different methods for aggregating local priority vectors produced various measures of consistency. There are many reported examples of measuring consistency of a group (e.g., Blagojevic et al. 2016a, b; Moreno-Jimenez et al. 2008; Dong and Zhang 2014), but still there is no agreement about which measure is the best and widely applicable in AHP group contexts.

Aggregation of Individual Priorities of Alternatives Obtained by AHP into the Group Priorities

There are two commonly used methods for aggregating individual priorities of alternatives into the final group priorities of alternatives vs. goal: (1) the weighted arithmetic mean method (WAMM) and the weighted geometric mean method (WGMM). It is generally accepted that the later method gives more trustful results, especially in cases when priorities provided by individuals significantly differ. In our research decision made is to use this method for aggregation of AHP priorities within each sub-group and finally for aggregation of the sub-groups' priorities.

In short, WGMM performs as follows. Given are alternative A_i and its resulting priority weight z_{ik} for individual k. If weight w_k is assigned to an individual k in a group of G individuals, then the weighted geometric mean is

$$z_i^g = \sum_{k=1}^{G} z_{ik}^{w_k} \tag{6}$$

where z_i^g stands for final (composite) priority weight of the alternative A_i. The individuals' weights w_k are by assumption additively normalized.

Referring to a claim found in Ramanathan and Ganesh (1994) that only WAMM of aggregation does not violate the Pareto (agreement) principle, Forman and Peniwati (1998)argue that their research showed that either WGMM or WAMM may be used in most cases of group decision-making without violating the Pareto principle. At the final instance, they provide reasons for using the WGMM rather than WAMM.

2.2 Borda Count: Preferential Social Choice Method

Preferential voting methods from the SC theory exclusively use ordinal preference information contained in the preference table (Table 2), created by collecting ballots (in real elections). A constructed preference table usually has the following properties. The size of the table is *M*x*N*, where *M* is the number of individuals and *N* is the number of possible alternatives (choices). Each row represents the ranking of alternatives performed by one individual. If *j* is the best alternative for individual *i*, then the rank number is $r_{ij} = 1$; if *j* is the second-best alternative, then $r_{ij} = 2$, and so on; if alternative *j* is the worst one, then $r_{ij} = N$.

In the Borda Count method, each alternative gets 1 point for each last place vote received, 2 points for each next-to-last point vote, etc., all the way up to *N* points for each first-place vote. The alternative with the largest point total wins the election and is declared to be the social choice.

Following notation given in (d'Angelo et al. 1998; Srdjevic 2007) for each r_{ij} in the preference schedule, a number

$$q_{ij} = N - r_{ij} + 1 \tag{7}$$

is assigned by the above procedure, and the total score for alternative *j* is given as

$$Q_j = \sum_{i=1}^{M} q_{ij} = \sum_{i=1}^{M} (N - r_{ij} + 1) = M(N+1) \sum_{i=1}^{M} r_{ij}. \tag{8}$$

The alternative *j** with the highest *Q* value can be selected as the winner, i.e., social choice:

$$Q_{j*} = \max_{1 \le j \le N} Q_j. \tag{9}$$

Table 2 Preference table

	Alt. 1	Alt.2		Alt. *j*		Alt. N
Indiv. 1	r_{11}	r_{12}	...	r_{1j}	...	r_{1N}
Indiv. 2	r_{21}	r_{22}	...	r_{2j}	...	r_{2N}
...	...	...	...	...	...	...
Indiv. i	r_{i1}	r_{i2}	...	r_{ij}	...	r_{iN}
...	...	...	...	...	...	...
Indiv. M	r_{M1}	r_{M2}	...	r_{Mj}	...	r_{MN}

2.3 Approval Voting: Non-preferential Social Choice Method

Differently from Borda Count method, the Approval Voting (AV) method does not use information directly from a preference schedule and is therefore considered as non-preferential. In AV method, voters vote for as many candidates as they wish, and each approved candidate receives one vote. The candidate with the most votes wins. This method is considered simple for voters to understand and use (Bolloju 2001; Srdjevic 2007; Lakicevic et al. 2014). Its mathematical background can be formulated as follows:

Referring to preference Table 1, let's define

$$Q_j = h(r_{ij}) = \begin{matrix} 1 \;\; \text{if } r_{ij} \text{is ticked (here`} + \text{')} \\ 0 \;\; \text{if } r_{ij} \text{isn't ticked (left empty cell)} \end{matrix} \tag{10}$$

for all i (individuals) and j (alternatives). Then, for each alternative j, let

$$V_j = \sum_{i=1}^{M} h(r_{ij}) \tag{11}$$

The number V_j indicates how many times alternative j has been approved by all individuals in the sub-group. The alternative j^* with the largest V_j value is then selected as the social choice:

$$V_{j*} = \max_{1 \le j \le N} \{V_j\}. \tag{12}$$

3 Combined Use of Multi-criteria and Social Choice Theory Methods for Deriving BEST Water Management Plan: An Illustrative Example

3.1 Statement of the Decision Problem

The problem is stated as to select the most desired long-term water management plan for river basin by authorized institution such as the water committee (WC). The WC is considered to be a decision body (global group), and what is said hereafter to be “the group choice” should be understood as “the WC choice.” Assuming that certain decisions will be made by individuals in sub-groups, the final decision should certainly be made at the WC level in a democratic manner with respect to the preferences derived by participating sub-groups and/or their delegates.

3.2 *Hierarchy*

A decision problem is stated as a three-level hierarchy with (1) a goal at the top of hierarchy, (2) five evaluating criteria under goal, and (3) five alternative management plans under criteria level that is at the bottom of hierarchy, Fig. 2.

The hierarchy is adopted after each decision element is briefly described to all participants at a plenary part of a session. Main decision elements (goal, criteria set, and alternatives) are as follows:

- *Goal*

  Select the best (most desired) plan using given set of criteria.

- *Criteria set*

Political influence criterion is considered as the gradually exposed impact of various state and in-basin agencies and bodies, representatives of cities/villages, stakeholders, producers, and local leaders.

Economic criterion relates to real possibilities to implement the economical process; reliability of economical parameters; estimated costs of investment, operation, and maintenance; and expected direct and indirect benefits.

Social issues criterion relates to issues such as infrastructure, demographic changes (migration), health care, and working conditions.

Environmental protection criterion relates to specific environmental and ambient conditions such as the distribution of pleasant resorts, preservation of historical sites and cultural values, accessing the objects and facilities, protecting water quality, and particularly preserving acceptable sanitary conditions.

Technical criterion encapsulates interests in preserving proper spatial distribution of projects, technical conditions for project operations, technologies involved, and eligibility for technical improvements.

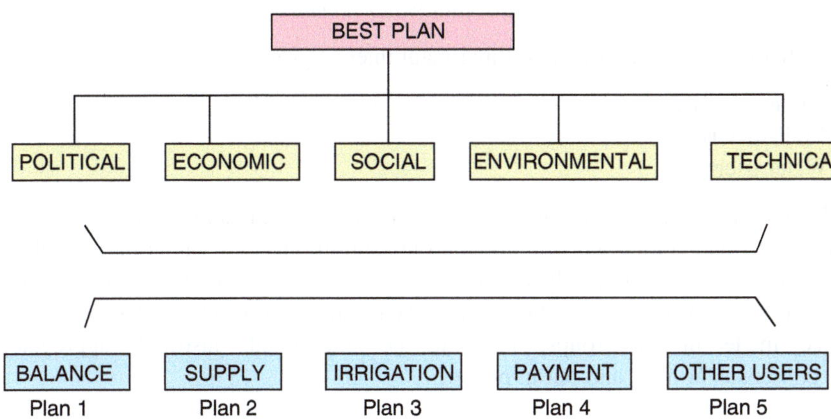

Fig. 2 Hierarchy of the problem

- *Decision alternatives (management plans)*

Plan 1 (Balance). High industrial developments are foreseen as well as intensive irrigation. Electric power production will increase by 20% after certain reconstructions of the existing hydroelectric objects and facilities. All users, including big users such as irrigation and hydroelectric production, will have approximately equal treatment. However, ecological and urban water requirements will receive top priority in water allocation.

Plan 2 (Supply). Water supply (municipal and rural, human and animal) will absolutely get an increased concern from the state agencies responsible for water management. It will be dominantly realized by means of reservoir management. Demographic movements from rural areas to cities and state capitals will continue with an actual increasing trend but will be significantly decreased by the middle of the planning period. Irrigation will rise to only 50% of that amount estimated as maximum by the end of the period.

Plan 3 (Irrigation). Irrigation will have a dominant role with respect to the other water uses throughout the basin. Priority will be given to large irrigators (development at a level higher than 80% of the estimated maximum). No water payments are expected until 2020; only irrigation and industrial uses will be charged afterward.

Plan 4 (Payment). Water payments will start progressively by 2020, with revisions of payment policy every 5 years (2025 and 2030). Pricing will be combined with an advanced system for obtaining the water rights. Other elements of the plan are the same as in Plan 3.

Plan 5 (Other users). This plan is a modification of Plan 1 in a way to emphasize importance of small irrigation users, tourism, ecotourism, and other small users (such as handmade manufacturers, ceramic industry). Intent is to enable that various users (other than large ones) will receive a higher priority by obtaining proper water rights and excluding payments; compensation for their uses of water will come from large consumers in irrigation, hydroelectric production, and industry by proper pricing policy.

- *Decision-makers, interest groups, and water committee as a global group*

Participants were divided into three main interest groups:

1. Public authorities
2. Civil society
3. Water users

Each participant is an individual decision-maker and fully autonomous. Note that within the water committee as a global group, sub-groups may gather individuals in different ways for differently organized decision-making processes. In adopted context, a group is the entire body of a water committee where "delegated" decisions, made in sub-groups, have to be interpreted, justified, aggregated (by consensus or not), and put in power.

3.3 Remarks

Several main remarks can be given to better describe the decision-making framework:

Remark 1 WC decides by applying scientifically sound multi-criteria (AHP) and election (Borda Count) methods, followed by common aggregating techniques.

Remark 2 WC recognizes panel meetings as principal mean of its work where mediating rules must be adopted by consensus and where the final decisions are to be made. WC also recognizes "decentralized part" of the decision process performed at separate meetings of each entity. Entities are by assumption authorized to make their own decisions and forward them to be aggregated at the WC level.

Remark 3 Each entity has "its own point of view" while evaluating possible decision alternatives and ranking them appropriately. An outcome of the decision process conducted through each entity is forwarded to the WC level (for aggregation) as it is. That means that no any changes, interpretations, or justifications are permitted.

Remark 4 As first part of the decision framework, the method used by each entity in assessing criteria and management plans is AHP. Consensus at sub-group level is assumed as logical and appropriate. Although AHP produces cardinal preferences of decision alternatives, represented by computed weights, only ordinal information is analyzed, i.e., ranking of the alternatives.

Remark 5 Each entity (individual or specific sub-group) assesses the same set of management plans across the same criteria set.

Remark 6 By applying voting method Borda Count and Approval Voting to the reduced set of best-ranked plans, it is possible to come up to the final decision: the most desired management plan or at least to additionally reduce set of alternative plans and enable voting and following majority principle to select just one plan as most desired.

3.4 Procedure and Results

A group of 21workshop participants is considered as a water committee (WC), and participants are divided into three sub-groups to represent the following interest groups (IGs) of stakeholders as follows: public authorities (7participants), civil society (5 participants), and water users (9 participants). Decision-making process is organized in two separate sessions (identified hereafter as *Session 1* and *Session 2*), each lasting in 1.5 hour, with a 3 h break. In each session individual opinions are expressed at delivered evaluation sheets (Annex III) for AHP, Borda Count, and Approval Voting. Evaluations (AHP), ranking (Borda Count), and approvals

(Approval Voting) are collected and synthesized for each interest group, and the sub-group decisions are forwarded to an upper level (the WC level) where the final aggregation and interpretation of results are performed.

The decision-making process and obtained results were as given below.

- **Application of AHP** (first part of the *Session 1*)

Delegates individually assessed hierarchy given in Fig. 1 by using AHP within each sub-group (IG). Each delegate had to fill in six pairwise comparison matrices with numbers from Saaty's 9-point fundamental scale (see Annex I for delivered evaluation sheet #1). Local weights of criteria vs. goal and alternatives versus criteria are computed by the eigenvector method, and standard AHP synthesis generated the final weights of alternative plans versus global goal (best plan) for each individual.

Based on the number of individuals in the sub-groups, participation weights of sub-groups in the WC are defined as 40% for interest sector of water users, 34% for public authorities, and the rest 26% for civil society. By applying these weights to the sub-group aggregated weights, as given in Annex I, the final aggregation is performed to obtain the final group decision corresponding to the WC level, Table 3. The best plan, as the WC final choice, is Plan 1 (Balance), second ranked is Plan 2 (Supply), and third one is Plan 5 (Other users). Least desired plan is Plan 3 (Irrigation).

It is easy to see that Plan No. 1 (Balance) is selected as the best by two IGs: public authorities and water users. Top ranked by civil society is Plan 5 (Other users), while Plan 1 is ranked as second. Worthy to notice is also that Plans 1, 2, and 5 are ranked as top 3 by all sub-groups. Notice also that final ranking mostly reflects preferences of the third interest group (water users).

- **Application of Borda Count and Approval Voting** (second part of *Session 1*)

In the second part of *Session 1*, the Borda Count is firstly employed, and within all three sub-groups, participants individually assessed set of alternative plans.

Different from AHP methodology, this time participants did not rank criteria within criteria set. Rather, all original criteria are considered as virtually "unique criterion" which describes general desire and implicitly contains "a flavor" of each

Table 3 Sub-groups' weights (ranks) of alternative plans obtained by AHP and final aggregation

		Plan (alternative)				
IG	IG weight	1 Balance	2 Supply	3 Irrigation	4 Payment	5 Other users
Public authorities (7)	$\alpha_1 = 0.34$	0.258 (1)	0.249 (2)	0.117 (5)	0.145 (4)	0.231 (3)
Civil society (5)	$\alpha_2 = 0.26$	0.314 (2)	0.178 (3)	0.037 (5)	0.056 (4)	0.415 (1)
Water users (9)	$\alpha_3 = 0.40$	0.309 (1)	0.289 (2)	0.139 (4)	0.119 (5)	0.144 (3)
Aggregated (WGMM)		**0.306**	**0.254**	**0.098**	**0.110**	**0.232**
Rank		**1**	**2**	**5**	**4**	**3**

criterion from the original criteria set. Appropriately briefed during this part of Session 1, each participant is expected to be aware that his/her ranking of five alternative plans will carry on information about his/her general desire and preference. Participants are asked to express their individual (ordinal) preference by filling in appropriate boxes with integers 1–5 in distributed evaluation sheet. Borda Count additions of points were straightforward afterward.

Individual and final ranking of alternative plans derived within each interest group by Borda Count is summarized in Table 4.Assuming that obtained three rankings in interest groups are additionally aggregated at the WC level (by following the same Borda Count procedure and associating equal weights to each sub-group ranking), the best alternative in a social choice context is Plan 2 (Supply), while Plan 5 (Other users) and Plan 1 (Balance) share the second and third place. The last two rows in Table 4 represent aggregated values: (1) sum of ranks and (in parenthesis) the total number of points received in all three sub-groups and (2) the final ranks of alternatives based on values in above row. Note that the final ranks of alternative management plans obtained by sub-groups and by total Borda Count points from all individuals coincide, that is, ranks of alternative plans are unique.

Session 1 is concluded by applying non-preferential Approval Voting method to rank five alternative plans. Again, participants are briefed about social choice context and instructed how to express their desire. By using the evaluation sheet #2 from Annex III, they individually approved one up to five plans (without giving them preferences; they have just to express with sign "+" that they accept certain plan to fulfill their expectations. All votes (approvals) are collected and presented in Annex I and Table 5.

Obviously, participants in both cases (Borda and Approval) identified Plans 1, 2, and 5 as the most desired for implementation. If preferential framework is considered with multiple criteria as in AHP and condensed criteria "all-as-one" as in case of Borda Count, preference structure is changed: in case of AHP, the first ranked is Plan 1 (Balance) and in case of Borda Count it is Plan 5 (Other users). Explanation for this could be that in case of Borda Count, three best-ranked plans (1,2, and 3) received similar number of points (53, 48, and 51, respectively; see Table 4). Condensing criteria into one (differently from AHP which is very strict in this regard) probably

Table 4 Final Borda Count assessment at the group (water committee) level

IG	IG weight	Borda Count points received for plans (alternatives) 1 Balance	2 Supply	3 Irrigation	4 Payment	5 Other users
Public authorities	$\alpha_1 = 0.33$	3 (21)	1 (13)	4 (23)	5 (31)	2 (17)
Civil society	$\alpha_2 = 0.33$	1 (6)	3 (14)	4 (21)	5 (22)	2 (12)
Water users	$\alpha_3 = 0.33$	3 (26)	1 (21)	4 (31)	5 (35)	2 (22)
Sum of ranks and total no. of points received		**7 (53)**	**5(48)**	**12 (75)**	**15 (88)**	**6 (51)**
Final ranks of management plans		**3**	**1**	**4**	**5**	**2**

Table 5 Final Approval Voting results at the group (water committee) level

		Number of approvals (out of 21) of alternative plans				
IG	IG weight	**1 Balance**	**2 Supply**	**3 Irrigation**	**4 Payment**	**5 Other users**
Public authorities	$\alpha_1 = 0.33$	6	6	4	3	6
Civil society	$\alpha_2 = 0.33$	5	0	0	0	4
Water users	$\alpha_3 = 0.33$	6	6	4	3	7
Total number of approvals		**17**	**12**	**8**	**6**	**17**
Final ranking		**1–2**	**3**	**4**	**5**	**1–2**

Table 6 Ranking of alternative water management plans obtained with AHP, BC, and AV methods (*Session 1: All Plans*)

Method	1 Balance	2 Supply	3 Irrigation	4 Payment	5 Other users
1. AHP	1	2	5	4	3
2. Borda Count	3	1	4	5	2
3. Approval Voting	1–2	3	4	5	1–2
Total	**5.5**	**6**	**13**	**14**	**6.5**
Final rank	**1**	**2**	**4**	**5**	**3**

led to smoothing effect and incapability of participants to sharply distinguish plans among each other. Regarding Plan 3 and Plan 4, there is no dilemma that these plans are inferior.

Table 6 summarizes the results obtained during Session 1 by all three methods. By adding obtained ranks of all five plans, it appears that the Plan 1 (Balance) is most desired by the complete group of participants, so that it would be the best choice of the water committee. Second ranked is Plan 2 (Supply) and the third is Plan 5 (Other users). Note that this result almost completely coincides with the final outcome of the AHP method; only the last two positioned plans change place, namely, Plan 3 (Irrigation) and Plan 4 (Payment).

- **Borda Count and Approval Voting for Assessing Top-Ranked Alternative Plans** (*Session 2*)

During *Session 2* which started after a break of 3 h, all participants are briefed about voting they will have to do as members of complete group, that is, not anymore within sub-groups like during *Session* 1. Each participant received evaluation sheet (Annex III, sheet #3) and performed individual voting for three top-ranked plans identified during previous session by AHP and confirmed by Borda Count and Approval Voting. The resulting ranking of plans in reduced set of plans is given in Table 7.

Differences occurred in final ranks for plans 1, 2, and 5 and all three methods (AHP, BC, and AV). Plan 1 was twice ranked as the best; interesting to note is that once it was by AHP as preferential method and secondly by AV as social non-preferential method. According to points received from BC preferential method,

Table 7 Ranking of top three water management plans 1, 2, and 5 obtained with AHP, BC, and AV(*Session 2: Three Plans*)

Method	Plan 1 Balance	2 Supply	5 Other users
1. AHP	1	2	3
2. Borda Count	3	1	2
3. Approval Voting	1	2	3
Total	**5**	**5**	**8**
Final rank	**1–2**	**1–2**	**3**

Plan 2 is ranked as the most desired by WC which in final summation of ranks equalizes that plan with the Plan 1 (cf. last two rows in Table 7). The last positioned Plan 5 was never selected by the WC as the best.

4 Conclusions

The paper presents group decision-making framework that could be applicable as a part of paradigm decision-making in any water committee responsible for water management on the river basin scale. The problem is stated as to select the most desired long-term management plan among several offered plans by assessing plans across more or less conflict criteria. We presented approach which is illustrated by using two different methodological options in decision-making and showed how they can be combined for establishing common professional, social, and political environment where people ought to make decisions by using advanced scientifically sound techniques.

The first part, organized as *Session 1*, included more complex, more detailed, and time- and effort-consuming evaluation of five management plans across five criteria. A group of 21 participants in one workshop dedicated to hierarchical decision-making is divided into 3 sub-groups to simulate different stakeholders' interest while searching for most desired plan. We believed that it is more convenient for participants to use two simple voting methods from social choice theory in the second part of that session to check agreement of voted solutions with solution obtained by the analytic hierarchy process as representative of multi-criteria class of methods. Although voting process performed by Borda Count (as preferential voting method) and Approval Voting (as non-preferential voting method) implied situation where participants considered all criteria as virtually one criterion, which is different from AHP methodology, an agreement was complete regarding three top-ranked plans. This provoked us to undertake additional *Session 2* after few hours and to repeat voting process about only those three top-ranked plans. A complete group, acting as water committee, reconfirmed results from the *Session 1* and indicated that even additional shrinking of set of plans is possible. That is, *Session 2* indicated that two plans can be considered as equally desired by the water committee and that it might be opportune in real-life situation to apply majority principle in simple voting for just two alternatives.

From the methodological point of view, the final ranking does not necessarily depend on the number of voters and on the size of sub-groups. If setup of voting processes is such that large sub-group has the same power as a small sub-group, this could yield to ranking which does not satisfy the conditions of social welfare functions and especially the principle of majority decision. An issue of number of members in groups or sub-groups, importance of individuals (experts) within sub-groups and across representatives on the group level, and related problems of preserving fairness, competence, and consistency, is always subjective to controversial discussion in practical applications. And of course, our approach is not immune of it.

In voting part of a methodology, we imply that all members of a group vote on equal importance base. This must not be a rule. For instance, within relatively large water committee, only representatives of different sub-groups (one person for one sub-group) may be allowed to vote. A possible new direction of research could be how to effectively and consistently avoid any early confrontation of individuals and sub-groups within water committee, i.e., how to define their different weights based on competences, i.e., expert knowledge, education, attitude, willingness, political impacts, etc.

It is important to mention that involved decision-makers found proposed methodology transparent, easy to understand and implement, and results trustful.

Acknowledgment Authors acknowledge the financial support from the Ministry of Education and Science of Serbia under the Fundamental scientific research program in Mathematics, Computer Science, and Mechanics; Grant No. 174003 (2011-2014): Theory and application of the analytic hierarchy process (AHP) in multi-criteria decision-making under conditions of risk and uncertainty (individual and group context). Support from the CNPq Federal Agency for science in Brazil is also acknowledged.

Annex I. Results of Evaluation (*Session 1*)

Final priorities of alternative plans derived by AHP, Borda Count, and Approval Voting in three interest sub-groups

AHP

PUBLIC AUTHORITIES No. of participants = 7

Plans vs. Goal

Participants	Plan					
	1 Balance	2 Supply	3 Irrigation	4 Payment	5 Other Users	
1	0.280	0.319	0.055	0.080	0.266	
2	0.208	0.181	0.106	0.395	0.111	
3	0.191	0.402	0.047	0.074	0.285	
4	0.306	0.162	0.287	0.087	0.158	
5	0.300	0.220	0.112	0.274	0.094	
6	0.068	0.175	0.120	0.116	0.522	
7	0.429	0.163	0.113	0.082	0.213	
WGMM aggregated	**0.258**	**0.249**	**0.117**	**0.145**	**0.231**	
Rank	**1**	**2**	**5**	**4**	**3**	

CIVIL SOCIETY No. of participants = 5

Plans vs. Goal

Participant	Plan					
	1 Balance	2 Supply	3 Irrigation	4 Payment	5 Other Users	
1	0.302	0.175	0.034	0.059	0.430	
2	0.305	0.259	0.035	0.061	0.340	
3	0.316	0.171	0.035	0.048	0.431	
4	0.319	0.149	0.041	0.055	0.436	
5	0.319	0.149	0.041	0.055	0.436	
WGMM aggregated	**0.314**	**0.178**	**0.037**	**0.056**	**0.415**	
Rank	**2**	**3**	**5**	**4**	**1**	

USERS No. of participants = 9

Plans vs. Goal

Participant	Plan					
	1 Balance	2 Supply	3 Irrigation	4 Payment	5 Other Users	
1	0.274	0.167	0.106	0.290	0.162	
2	0.272	0.456	0.072	0.068	0.132	
3	0.282	0.236	0.234	0.122	0.126	
4	0.320	0.156	0.283	0.141	0.099	
5	0.097	0.314	0.154	0.076	0.359	
6	0.371	0.259	0.054	0.079	0.236	
7	0.504	0.234	0.131	0.075	0.056	
8	0.201	0.356	0.115	0.172	0.158	
9	0.431	0.298	0.151	0.075	0.045	
WGMM aggregated	**0.309**	**0.289**	**0.142**	**0.119**	**0.141**	
Rank	**1**	**2**	**3**	**5**	**4**	

BORDA COUNT

PUBLIC AUTHORITIES No. of participants =7

Participant	Plan				
	1 Balance	2 Supply	3 Irrigation	4 Payment	5 Other Users
1	2	3	5	4	1
2	5	1	2	4	3
3	5	1	3	4	2
4	2	3	1	5	4
5	3	1	5	4	2
6	2	3	4	5	1
7	2	1	3	5	4
Total points	**21**	**13**	**23**	**31**	**17**
Rank	**3**	**1**	**4**	**5**	**2**

CIVIL SOCIETY No.of participants =5

Participant	Plan				
	1 Balance	2 Supply	3 Irrigation	4 Payment	5 Other Users
1	1	3	4	5	2
2	1	2	4	3	5
3	1	3	4	5	2
4	1	3	4	5	2
5	2	3	5	4	1
Total points	**6**	**14**	**21**	**22**	**12**
Rank	**1**	**3**	**4**	**5**	**2**

USERS No. of participants = 9

Participant	Plan				
	1 Balance	2 Supply	3 Irrigation	4 Payment	5 Other Users
1	4	1	3	5	2
2	3	1	4	5	2
3	3	2	5	4	1
4	1	2	3	5	4
5	3	4	5	2	1
6	4	3	2	1	5
7	3	1.5	4.5	4.5	1.5
8	3	2	5	4	1
9	2	5	1	3	4
Total Points	**26**	**21.5**	**32.5**	**33.5**	**21.5**
Rank	**3**	**1-2**	**4**	**5**	**1-2**

APPROVAL VOTING

PUBLIC AUTHORITIES No. of participants =7

Participant	Plan				
	1 Balance	2 Supply	3 Irrigation	4 Payment	5 Other Users
1	+				+
2	+	+			+
3		+	+	+	+
4	+	+	+	+	+
5	+	+	+	+	+
6	+	+	+		+
7	+	+			
No. of approvals	**6**	**6**	**4**	**3**	**6**
Rank	**1-3**	**1-3**	**4**	**5**	**1-3**

CIVIL SOCIETY No. of participants =5

Participant	Plan				
	1 Balance	2 Supply	3 Irrigation	4 Payment	5 Other Users
1	+				+
2	+				
3	+				+
4	+				+
5	+				+
No. of approvals	**5**	**0**	**0**	**0**	**4**
Rank	**1**	**3-5**	**3-5**	**3-5**	**2**

USERS No. of participants = 9

Participant	Plan				
	1 Balance	2 Supply	3 Irrigation	4 Payment	5 Other Users
1	+		+		
2	+	+			+
3	+	+			+
4	+	+			
5					+
6			+	+	+
7	+	+	+	+	+
8	+	+	+	+	+
9		+			+
No. of approvals	**6**	**6**	**4**	**3**	**7**
Rank	**2-3**	**2-3**	**4**	**5**	**1**

Annex II. Results of Social Choice Methods *(Session 2)*

Final priorities of alternative plans derived by Borda Count and Approval Voting methods in three interest groups for three top-ranked plans by AHP during *Session 1*

BORDA COUNT

PUBLIC AUTHORITIES No. of participants =7

Ranking alternative plans

Participant	1 Balance	2 Supply	3 Irrigation	4 Payment	5 Other Users
1	2	3	5	4	1
2	5	1	2	4	3
3	5	1	3	4	2
4	2	3	1	5	4
5	3	1	5	4	2
6	2	3	4	5	1
7	2	1	3	5	4
Points	21	13	23	31	17
Rank	**3**	**1**	**4**	**5**	**2**

CIVIL SOCIETY No. of participants =5

Ranking alternative plans

Participant	1 Balance	2 Supply	3 Irrigation	4 Payment	5 Other Users
1	1	3	4	5	2
2	1	2	4	3	5
3	1	3	4	5	2
4	1	3	4	5	2
5	2	3	5	4	1
Points	6	14	21	22	12
Rank	**1**	**3**	**4**	**5**	**2**

USERS No. of participants = 9

Ranking alternative plans

Participant	1 Balance	2 Supply	3 Irrigation	4 Payment	5 Other Users
1	4	1	3	5	2
2	3	1	4	5	2
3	3	2	5	4	1
4	1	2	3	5	4
5	3	4	5	2	1
6	4	3	2	1	5
7	3	1.5	4.5	4.5	1.5
8	3	2	5	4	1
9	2	5	1	3	4
Points	26	21	31	35	22
Rank	**3**	**1-2**	**4**	**5**	**1-2**

APPROVAL VOTING

PUBLIC AUTHORITIES No. of participants =7

Approval of plans

Participants	1 Balance	2 Supply	3 Irrigation	4 Payment	5 Other Users
1	+				+
2	+	+			+
3		+	+	+	+
4	+	+	+	+	+
5	+	+	+	+	+
6	+	+	+		+
7	+	+			
No. of approvals	6	6	4	3	6
Rank	**1-3**	**1-3**	**4**	**5**	**1-3**

USERS No. of participants = 9

Approval of plans

Participants	1 **Balance**	2 **Supply**	3 **Irrigation**	4 **Payment**	5 **Other Users**
1	+		+		
2	+	+			+
3	+	+			+
4	+	+			
5					+
6			+	+	+
7	+	+	+	+	+
8	+	+	+	+	+
9		+			+
No. of approvals	6	6	4	3	7
Rank	**2-3**	**2-3**	**4**	**5**	**1**

CIVIL SOCIETY No. of participants =5

Approval of plans

Participants	1 **Balance**	2 **Supply**	3 **Irrigation**	4 **Payment**	5 **Other Users**
1	+				+
2	+				
3	+				+
4	+				+
5	+				+
No. of approvals	5	0	0	0	4
Rank	**1**	**3-5**	**3-5**	**3-5**	**2**

Annex III. (Evaluation Sheets)

Evaluation sheets for AHP, Borda Count, and Approval Voting methods delivered to all participants in three interest sub-groups during two sessions

EVALUATION SHEET #1 (AHP – *Session I*)

Interest Group ____________________

Participant Name ____________________

Position ____________________

Criterions vs. Goal

	Political	Economic	Social	Environm	Technical
Political	1				
Economic		1			
Social			1		
Environmental				1	
Technical					1

Alternatives vs. Criterions

Political

1				
	1			
		1		
			1	
				1

Economic

1				
	1			
		1		
			1	
				1

Social

1				
	1			
		1		
			1	
				1

Environmental

1				
	1			
		1		
			1	
				1

Technical

1				
	1			
		1		
			1	
				1

Saaty's scale for pair wise comparisons in AHP

Judgment term	Numerical term
Absolute preference (element *i* over element *j*)	9
Very strong preference (*i* over *j*)	7
Strong preference (*i* over *j*)	5
Weak preference (*i* over *j*)	3
Indifference of *i* and *j*	1
Weak preference (*j* over *i*)	1/3
Strong preference (*j* over *i*)	1/5
Very strong preference (*j* over *i*)	1/7
Absolute preference (*j* over *i*)	1/9

An intermediate numerical values 2,4,6,8 and 1/2,1/4,1/6,1/8 can also be used.

EVALUATION SHEET #2 (Borda Count and Approval Voting – *Session 1*)

Interest Group ____
Participant Name ____
Position ____

Borda Count

Order alternative plans by importance for river basin
(1 – most important, 2 – second most important, …, 5 – least important)

Plan	Importance
1. Balance	
2. Supply	
3. Irrigation	
4. Payment	
5. Other users	

Approval Voting

Approve alternative plans as you wish (at least one must be approved!), regardless their importance. Just insert '+' in the right column of table below

Plan	Approved?
1. Balance	
2. Supply	
3. Irrigation	
4. Payment	
5. Other users	

EVALUATION SHEET #3 (Borda Count and Approval Voting – *Session 2*)

Participant Name ____

Borda Count

Order alternative plans by IMPORTANCE for river basin
(1 – most important, 2 – second most important, ..., 3 – least important)

Plan	Importance
1. Balance	
2. Supply	
5. Other users	

Approval Voting

Approve alternative plans as you wish (at least one must be approved!), regardless their importance. Just insert '+' in the right column of table below

Plan	Approved?
3. Balance	
4. Supply	
5. Other users	

References

Aguaron J, Escobar MT, Moreno-Jiménez JM (2003) Consistency stability intervals for a judgement in AHP decision support systems. Eur J Oper Res 145(2):382–393

d'Angelo A, Eskandari A, Szidarovszky F (1998) Social choice procedures in water resources management. Environ Manag 52(3):203–210

Blagojevic B, Srdjevic Z, Bezdan A, Srdjevic B (2016a) Group decision making in land evaluation for irrigation: a case study from Serbia. J Hydro Inf 18(3):579–598

Blagojevic B, Srdjevic B, Srdjevic Z, Zoranovic T (2016b) Heuristic aggregation of individual judgments in AHP group decision making using simulated annealing algorithm. Inf Sci 330:260–273

Bolloju N (2001) Aggregation of analytic hierarchy process models based on similarities in decision makers' preferences. Eur J Oper Res 128:499–508

Bryson N (1995) A goal programming method for generating priorities vectors. J Oper Res Soc 46:641–648

Cabrerizo FJ, Urena R, Pedrycz W, Herrera-Viedma E (2014) Building consensus in group decision making with an allocation of information granularity. Fuzzy Sets Syst 255:115–127

Chu A, Kalaba R, Springam K (1979) A comparison of two methods for determining the weights of belonging to fuzzy sets. J Optim Theory Appl 127:531–541

Crawford G, Williams C (1985) A note on the analysis of subjective judgment matrices. J Math Psychol 29:387–405

Dong YC, Xu JP (2016) Searching the consensus path with minimum adjustments. Book. In: Consensus building in group decision making. Springer, Singapore
Dong Y, Zhang H (2014) Multiperson decision making with different preference representation structures: a direct consensus framework and its properties. Knowl-Based Syst 958:45–57
Forman E, Peniwati K (1998) Aggregating individual judgments and priorities with the analytic hierarchy process. Eur J Oper Res 108:165–169
Golany B, Kress M (1993) A multicriteria evaluation of methods for obtaining weights from ratio-scale matrices. Eur J Oper Res 69:210–220
Ishizaka A, Labib A (2011) Review of the main developments in the analytic hierarchy process. Expert Syst Appl 38:14336–14345
Jonoski A, Seid AH (2016) Decision support in water resources planning and management: the Nile basin decision support system. In: Real-world decision support systems. Springer, Cham, pp 199–222
Kadziński M, Greco S, Słowiński R (2013) RUTA: a framework for assessing and selecting additive value functions on the basis of rank related requirements. Omega 41(4):735–751
Kou G, Lin C (2014) A cosine maximization method for the priority vector derivation in AHP. Eur J Oper Res 235:225–232
Lakicevic M, Srdjevic Z, Srdjevic B, Zlatic M (2014) Decision making in urban forestry by using approval voting and multicriteria approval method (case study: Zvezdarska forest, Belgrade, Serbia). Urban For Urban Green 13:114–120
Moreno-Jimenez JM, Aguaron J, Escobar MT (2008) The Core of consistency in AHP-group decision making. Group Decis Mak Negot 17:249–265
Mikhailov L (2000) A fuzzy programming method for deriving priorities in the analytic hierarchy process. J Oper Res Soc 51:341–349
Mikhailov L, Singh MG (1999) Comparison analysis of methods for deriving priorities in the analytic hierarchy process. Proc IEEE Int Conf Syst Man Cybern 1:1037–1042
Morais DC, de Almeida AT (2012) Group decision making on water resources based on analysis of individual rankings. Omega 40(1):42–52
Ramanathan R, Ganesh LS (1994) Group preference aggregating methods employed in AHP: an evaluation and intrinsic process for deriving members' weightgages. Eur J Oper Res 79:249–265
Saaty T (2003) Decision-making with the AHP: why is the principal eigenvector necessary. Eur J Oper Res 145:85–91
Saaty T (1980) The analytic hierarchy process. McGraw-Hill, New York
Srdjevic Z, Srdjevic B (2014) Modelling multicriteria decision making process for sharing benefits from the reservoir at Serbia-Romania border. Water Resour Manag 28(12):4001–4018 ISSN 0920-4741
Srdjevic B, Srdjevic Z (2011) Bi-criteria evolution strategy in estimating weights from the AHP ratio-scale matrices. Appl Math Comput 218:1254–1266
Srdjevic B (2007) Linking analytic hierarchy process and social choice methods to support group decision-making in water management. Decis Support Syst 42(4):2261–2273
Srdjevic B (2005) Combining different prioritization methods in analytic hierarchy process synthesis. Comput Oper Res 32:1897–1919
Zendehdel K, Rademaker M, De Baets B, Huylenbroeck V (2010) Environmental decision making with conflicting social groups: a case study of the Lar rangeland in Iran. J Arid Envir 74 (3):394–402

Agroforestry Systems and Innovation in Extra-Virgin Olive Oil Chain (EVOO) in Central Italy: A Multi-stakeholder Perspective

Andrea Pisanelli, Claudia Consalvo, Giuseppe Russo, Marco Ciolfi, Endro Martini, Marco Lauteri, Francesca Camilli, and Pierluigi Paris

Abstract Agroforestry is a land use practice in which woody perennials (trees or shrubs) are integrated with crops and/or livestock on the same land unit. Such practices have shaped key features of the rural landscape in Mediterranean countries, where trees have been traditionally and deliberately retained by farmers or included in cultivated or grazed lands. Global and European policies acknowledge the role that agroforestry can play to promote multifunctional agriculture providing products and delivering additional, highly important, ecosystem services. Nevertheless, it is also recognised that several constraints such as the lack of knowledge and expertise of farmers, land users and policy makers concerning agroforestry systems establishment and management hamper the adoption of agroforestry systems. In order to fill this gap, a European research project, funded within the EU's H2020 research and innovation programme, started in January 2017: Agroforestry Innovation Networks (AFINET). AFINET acts at EU level in order to take up research results into practice and to promote innovative ideas to face challenges and resolve problems of practitioners. To achieve this objective, AFINET proposes an innovative methodology based on the creation of a European interregional network, linking different Regional Agroforestry Innovation Networks (RAINs). RAINs are working groups created in nine strategic regions of Europe, interconnected and articulated through the figure of the Innovation Broker. RAIN in Italy is focused on multipurpose olive tree systems in the territory around Orvieto Municipality, Umbria Region, Central Italy. The network considers the extra-virgin olive oil value chain, from the olive

A. Pisanelli (✉) · C. Consalvo · G. Russo · M. Ciolfi · M. Lauteri · P. Paris
Institute of Research on Terrestrial Ecosystems, National Research Councill, Porano (TR), Italy
e-mail: andrea.pisanelli@cnr.it

E. Martini
Alta Scuola Associazione Culturale e Scientifica, Orvieto (TR), Italy

F. Camilli
Institute of Biometeorology, National Research Council, Florence, Italy

A. Theodoridis et al. (eds.), *Innovative Approaches and Applications for Sustainable Rural Development*, Springer Earth System Sciences,
https://doi.org/10.1007/978-3-030-02312-6_5

production on farm to the olive processing, producing oil and residues, at the oil mill. Through the RAIN, we expect to improve the management of olive orchards, promoting the adoption of agroforestry solutions and practices, and to improve the extra-virgin olive oil production process by identifying innovative uses of the bio-residues. This paper reports the results obtained in the first and second RAIN meetings, organized in September 2017 and January 2018, with the participation of 27 and 39 stakeholders from the local olive oil chain, respectively. The first meeting was mainly aimed to introduce the AFINET project and its tools and to implement a participative approach in order to identify bottlenecks and opportunities of the extra-virgin olive oil value chain, according to stakeholder knowledge and experience. The second meeting was focused on the identification of potential innovations that should be implemented within the olive oil value chain according to stakeholders' needs and perceptions.

Keywords Communication · Participative approach · Multi-stakeholders · Value chain

1 Introduction

In the last century, intensive farming systems have been promoted all over the world, causing several environmental constraints and agricultural ecosystem degradation linked to the loss of important ecological traits, such as the presence of the woody component. Therefore, farming systems should switch from intensive to extensive practices, that is to say, from the application of external inputs to the efficient use of the available resources (Gross 2016).

Agroforestry is the practice of deliberately integrating woody vegetation (trees or shrubs) with crop and/or animal systems to benefit from the resulting ecological and economic interactions (Nair 1993; McAdam et al. 2009). Such practices have shaped key features of the rural landscape in Mediterranean countries where trees have been traditionally and deliberately retained by farmers or included in cultivated or grazed lands providing secondary products such as fruits, fodder for livestock and wood for fuel, litter or timber as well as environmental benefits (Eichhorn et al. 2006). Agroforestry systems may increase productivity and profitability in comparison to monoculture (Graves et al. 2007) in a sustainable way and provide various environmental benefits (Palma et al. 2007), such as soil erosion control, decrease nitrogen leaching, increase carbon sequestration and improve biodiversity and rural landscape (European Commission DG Environment 2006).

European policies (CAP) acknowledge the role that agroforestry can play to promote multifunctional agriculture providing products and delivering additional, highly important, ecosystem services. Since the 2007–2013 programming period (EU REG No 1698/2005), agroforestry systems have been included in the forestry measures to be supported in the Rural Development Plans (RDPs) because of "their high ecological and social value" (Pisanelli et al. 2014). Furthermore, along with its growth strategy EU 2020, the EU has launched the agricultural European Innovation Partnership creating a specific focus group on "Agroforestry systems: Integrating woody crops into specialized crop and livestock systems" (EIP-AGRI 2017). This focus group aimed to investigate challenges, opportunities and existing practices and ways forward to further develop agroforestry systems in Europe. The focus group also suggested innovative actions and ideas involving farmers, advisers, industry and other stakeholders in smart practices to assess ecosystem services, sharing experiences and knowledge, optimal scales and landscape designs, creating agroforestry value chains and developing management guidelines of specific agroforestry systems, and to see the availability of lacking crop resources. It is also known that several constraints such as the lack of knowledge and expertise of farmers, land users and policy makers concerning agroforestry systems establishment and management hamper the adoption of agroforestry systems (Camilli et al. 2018; De Jalón et al. 2017).

In order to fill this gap, a European research project, funded within the EU's H2020 research and innovation programme, started in January 2017: Agroforestry Innovation Networks (AFINET, http://www.agroforestry.eu/afinet). AFINET acts at EU level in order to take up research results into practice and promote innovative ideas to face challenges and resolve problems of practitioners. In order to achieve this objective, AFINET proposes an innovative methodology based on the creation of a European interregional network, linking different Regional Agroforestry Innovation Networks (RAINs). RAINs are working groups created in nine strategic regions of Europe (Spain, the UK, Belgium, Portugal, Italy, Hungary, Poland, France and Finland), interconnected and managed through the figure of the "Innovation Broker". RAINs represent different climatic, geographical, social and cultural conditions and enclose a balanced representation of the key actors with complementary types of expertise (farmers, policy makers, advisory services, extension services, etc.). Within the AFINET scheme, each project partner is in charge to organize five RAIN meetings during the 3 years of the project.

The Italian RAIN is focused on multipurpose olive tree systems in the area around Orvieto Municipality, Umbria Region, Central Italy. The network considers the extra-virgin olive oil supply chain, from the olive production on farm to the olive processing, producing oil and residues, at the oil mill. In the study area, olive orchards represent one of the most common land use practices (Martini 2010). Through the RAIN, we expect to improve the management of olive orchards promoting the adoption of agroforestry solutions and practices and improving the extra-virgin olive oil production process by identifying innovative uses of the olive oil production residues.

This paper reports the results obtained in the first and second RAIN meetings, organized in September 2017 and January 2018 and attended by 27 and 39 stakeholders of the local olive oil supply chain, respectively. Both meetings adopted a participative approach involving stakeholders in discussing bottlenecks and opportunities to elicit the main agroforestry challenges and innovations that should be faced to improve the local olive oil value chain.

2 Material and Methods

A review of agroforestry systems applied to olive orchards in Italy was carried out consulting available scientific references. A description of the olive oil value chain in the studied area was performed consulting available databases and bibliographic sources. During the first RAIN meeting, AFINET project was introduced to stakeholders who, then, were invited to form three groups and join three discussion tables in order to identify bottlenecks, challenges and knowledge gaps affecting the olive oil supply chain. Each table focused on a specific issue related to the olive oil supply chain: climate and environment, socio-economy and policy. After 20 min of discussion, each group of participants moved from a table to another in order to take part to the discussion of all the issues. The facilitator at each table introduced the participative exercise and invited the stakeholders to discuss. Afterwards, the stakeholders were asked to write in a post-it note the main constraints and opportunities related to each issue. The first RAIN meeting was followed by an online questionnaire submitted few days later to the stakeholders who were asked which agroforestry challenges should be implemented in order to improve the local olive oil value chain. Such challenges were discussed with the stakeholders during the second RAIN meeting in order to exploit opportunities of the olive oil supply chain and identify possible innovations that should be taken into account.

3 Agroforestry Systems and Olive Orchards in Italy

The olive tree has been cultivated in Italy and in the Mediterranean basin since the dawn of civilizations. It has characterized the Mediterranean landscape, economy, food and diet for about three millennia, throughout innumerable historical events and socio-economic changes (Besnard et al. 2013). Since its origin, the olive tree has been mostly cultivated in association with crops (legumes and cereals) and grazing, with low plant densities in mixed agroforestry systems (Lelle and Gold 1994), a diversified farm production that was well adapted to the prevailing needs of rural populations and to the irregularities of the Mediterranean climate. This mixed cultivation system had been prevailing until the recent industrialization of modern agriculture, mainly oriented to crop specialization and the maximization of unit productivity (Sereni 1974), leading to a prevalent orientation of research towards high plant densities and reduced tree vigour. In the last decades, the keywords for the modern olive cultivation have been crop specialization and suppression of olive cultivation in agroforestry systems, considered as an index of poor technological and economic progress. Nonetheless, the traditional management of olive trees in agroforestry systems has resisted those changes and is still widespread in many areas of the country. Official national statistics (ISTAT) report that currently in Italy there are about one million hectares of olive trees. According to more recent estimates of the European agroforestry project

AGFORWARD (http://www.agforward.eu/index.php), about 200,000 hectares of olive trees are managed in agroforestry systems in Italy (den Herder et al. 2017).

Today, agriculture practices need to combine food security with environmental protection, safeguarding biodiversity and soil fertility and combating climate change. The modern management of the agroforestry systems of the olive orchards can address issues such as multifunctional sustainable agriculture throughout green mulching, grazing and intercropping. Olive trees produce the maximum amount of fruit yield when tree canopy intercepts 55% of the available light. If the olive trees have a too high planting density, the consequent self-shading causes a decrease in the fruit yield. Thus, 45% of residual solar radiation that is not used efficiently by olive trees – if water and soil nutrients are not limiting factors – can be exploited by the associated crops or understory vegetation and/or rearing animals.

The green mulching of olive grove and, more generally, of fruit woody crops is a technique which has received great attention in the previous decades, because of the important effects on the economy and the environment. In Italy, many fruit trees plantations, including olive tree groves, are often located in hilly areas which are at risk of soil erosion if the soil is mechanically cultivated or chemically treated to contain weeds. An alternative solution is the maintenance of a permanent and controlled herbaceous cover, with positive effects against soil erosion and on the physical, chemical and biological characteristics of the soil. In this way, green mulching influences soil fertility increasing the organic matter, with the accumulation of carbon (C) into the soil. Important additional benefits of green mulching are the elimination of the compacted ploughing soil layers, the increase in soil micro-porosity and the improvement of the soil structure promoting soil water storage and its availability to cultivated plants. Increasing the soil C reserve is a strategic action of agricultural and forestry sectors to reduce the C concentration in the atmosphere, thus reducing the greenhouse effects and the related climate change. The latter is already affecting the majority of Italian lands, with increasingly frequent droughts requiring a more efficient use of natural water resources in agriculture. There is a wide scientific and technical literature on olive tree grassing down, with research activities carried out in Italy and other Mediterranean countries. A Spanish study (Hernandez et al. 2005) showed fast increase of C and water reserves in the soil after applying green mulching in a rain-fed olive plantation in the central area of the Iberian Peninsula, with an average precipitation of 500 mm per year. For 5 years, green mulching with sub clover (*Trifolium subterraneum*) and vetch (*Vicia sativa*), chemical weeding without soil tillage and soil tillage were compared. Sub clover and vetch are two nitrogen fixing legumes, self-seeding, with a vegetative stasis in the middle of summer. At the end of the observation period, the olive production in the different soil managements did not show any differences on fruit yield, showing that green mulching is not competitive for soil water and nutrients with the olive tree. Furthermore, with the sub clover-vetch green mulching, there was a + 27% increase in available water into the soil profile and + 36% in soil C. These data unequivocally demonstrate the importance of green mulching with selected species in olive cultivation in adaptation to climate change.

Intensive livestock farming is a highly critical contribution of modern agriculture, as it is responsible for polluting emissions and serious risks of low food production and quality. Intensive livestock farming has a key role in the greenhouse effect, contributing to 50% of C emissions in the atmosphere of the entire agricultural sector.

Alternatively, rearing of domesticated husbandry in agroforestry systems can contribute to animal welfare, improving the quality of animal productions (meat, milk, eggs, etc.) and ensuring the supply of supplementary fodder resources to grazing animals from the arboreal component (acorns, fodder fronds, fallen fruits) in addition to the grass/pastureland. The animal dejections are disposed directly on site, and tree root systems can intercept the leached nitrogen, reducing the nitrate pollutions of soil water, water tables and connected water bodies. Furthermore, the C emitted from the animals can be stored in the woody biomass of the associated trees.

Animal grazing (especially sheep, cattle and goat grazing) under olive trees is still widespread in Italy. Unfortunately, there are no reliable statistics about the current use of grazing under olive trees with traditional low tree density and large tree size. Collection of such data is limited by the strong fragmentation of olive grove estates, as well as by the randomly applied AF practice. Much great attention has been provided by Italian research to free-range chicken breeding under the olive tree cover. In this regard, Dal Bosco et al. (2014) demonstrated the convenience and effectiveness of this innovative practice showing interesting perspective to be implemented, given the increasing consumers' attention to animal welfare and healthy poultry products. Furthermore, the strong fragmentation of olive tree estates allows the easy and profitable management of high-density poultry raring in a limited area, and this agroforestry practice is getting popular among farmers and consumers.

Intercropping or alley cropping in olive orchards can be very limited for light-demanding crops such as wheat, sunflower, alfalfa, clovers, etc. that can be associated with olive groves just with low tree densities, varying from a minimum interrow distance of 20 m up to 50 m (Paris, data not published). Such systems are more common in areas with gentle slope making the circulation of farm machinery for intercrop cultivation and harvesting easier. With regard to that, it is necessary to optimize the cultivation distances between olive tree canopy and intercrop (Razouk et al. 2016). Small vegetable gardens with olive trees associated with winter or summer vegetable crops are frequent, for self-consumption, in the peri-urban areas of rural towns. In this case, the degree of shading of olive trees on vegetable crops does not seem a limiting factor. In modern intensive and super-intensive olive plantations with high plant density (400 and 1667 trees ha^{-1} or 6×4 m and 4×1.5 m, respectively), it is important to select crops suitable for the high degree of shading. An example is the wild asparagus, very appreciated by the market in the Mediterranean countries, a niche vegetable crop of high commercial value. Wild asparagus, as a sufficiently shade-tolerant plant and highly resistant to drought, can be used in a sustainable and profitable association with olive trees either in modern orchards or in traditional groves in marginal areas (Rosati 2011, 2014). The variety of combinations between olive trees and crops is very wide; however, farmers' choices can be affected by multiple factors deserving a much in-depth research activities.

4 The Olive Oil Value Chain in Italy and Umbria Region

Olive oil is the base of Mediterranean diet, and 80% of the olive oil world production is concentrated in Spain, Italy and Greece in addition to Turkey, Tunisia and Morocco. In these countries, olive oil represents not only a resource for rural economies but also an important element of the cultural heritage and the rural landscape (Sabbatini 2014).

Italy is the second olive oil producer of the European Union and provides about 20% of the European production. The Italian olive oil value chain is based on about 900,000 small-scale producing farms (average farm size is about 1.2 ha), growing olive trees in about 1,000,000 ha.

According to Unaprol (2015), the Italian olive production has been decreasing during the last decades mainly because of the abandonment of most marginal fields. Moreover, the Italian olive oil industry has difficulty competing in the world markets, mainly because the strong competition arising from the Spanish olive oil industry that is increasingly involved also in processing and trade (Pomarici and Vecchio 2013).

In Italy olive tree cultivation is the most important agricultural activity in many rural areas, and Umbria can be considered one of the most interesting regions because of the high-quality production (both olive and extra-virgin oil) and the close connection with traditional knowledge and the local environment. In fact, olive trees in Umbria are still managed in an extensive way, often in marginal sites, with minimal mechanization and low external inputs such as chemical treatments.

Moreover, olive trees, as permanent crops, guarantee a partial tree cover reducing hydrogeological risk. Soil management usually keeps natural grassing undercover reducing soil carbon emission and increasing soil fertility (Bateni 2017). Intercropping with livestock can also be practised in olive orchards, increasing the complexity of the olive tree multifunctional system.

The regional olive oil chain involves about 30,000 farms growing olive trees covering about 27,000 ha and including 270 oil mills; the regional olive production is reported in Fig. 1. The variability of the olive yield is due to both the extreme climatic events that influence the Mediterranean region and the massive parasitic attacks of *Bactrocera oleae* (olive fly).

The regional extra-virgin olive oil production represents 5% of the national production, and 7% of high-quality oil is labelled with Protected Designation of Origin (PDO): Colli Assisi Spoleto, Colli Martani, Colli Amerini, Colli del Trasimeno and Colli Orvietani. In Umbria olive trees are cultivated in the whole region.

The main strengths of olive oil supply chain in Umbria are:

- Elevated landscape value of olive orchards
- High-quality level of the extra-virgin olive oil
- Great awareness and expertise of farmers and oil mill managers
- High cultural and traditional value

Fig. 1 Olive production (tons) in Umbria region

On the other side, the main weaknesses of the olive oil chain in Umbria are:

- High productive costs
- Low intensive management practices
- Small-scale farm dimension

In Italy there are 556 olive tree varieties adapted to the different ecological conditions of the country (http://www.frantoionline.it), and, only in the Umbria region, there are 51 cultivars of olive trees for oil production. Olive orchard management requires appropriate treatments such as soil management, fertilizations, pest treatment, pruning and harvesting. Cultivation can be conventional, integrated or organic. Pruning and harvesting are usually performed manually. After the harvest, olives must be brought to the oil mill within 48 h in order to avoid the fermentation process.

The olive oil production phase comprises the extraction of the oil from the olives and the production of additional by-products (water, pomace and husk) that require to be properly managed. The olive oil extraction can be performed according to different methods such as traditional pressing or centrifugation (with three-phase or two-phase) depending on the oil mill technologies. The traditional method by pressing (discontinuous process) is relatively obsolete and is generally replaced by centrifugation systems that contribute to lower costs, better olive oil quality and shorter storage time of olive before processing.

Continuous centrifugation with three-phase system requires a great amount of water and energy. The process generates a solid waste (olive husk or olive pomace), olive oil and wastewater. Continuous centrifugation with two-phase system allows separation of olive oil from olive pasta without addition of water, so the method generates only olive oil and a semi-solid waste called olive wet husk or wet pomace. Finally, continuous centrifugation with a two and half system brings together the advantages of the two different systems (it requires the addition of a small amount of water) and generates a solid fraction (olive wet husk or wet pomace) that includes part of the vegetation water and a small quantity of olive mill wastewater.

The by-product management is very important because the olive oil mill wastes have a great impact on soil and water environments because of the high phytotoxicity (phenol, lipid and organic acids), and their management is one of the main problems of the olive oil industry (Souilem et al. 2017). At the same time, such wastes also contain potential valuable resources such as a large portion of organic matter and a wide range of nutrients that could be recycled (Roig et al. 2006) to produce bioenergy (Caputo et al. 2003), compost (Tomati et al. 1995), pharmaceutical products (Vlyssides et al. 2004) and olive pâté (Grimelli 2012).

5 Bottlenecks, Challenges and Innovations in the Olive Oil Value Chain

The 27 stakeholders participating at the first RAIN meeting included farmers, multipliers such as members of trade associations and citizens, researchers and policy makers (Fig. 2). Stakeholders were asked to highlight the main bottlenecks, knowledge gaps and challenges in policies, climate and environment and socio-economic aspects related to the olive oil value chain. The main bottlenecks and opportunities are summarized in Table 1.

Concerning the policy issue, stakeholders remarked that most of the bottlenecks are related to the Common Agricultural Policy (CAP), mainly because grant

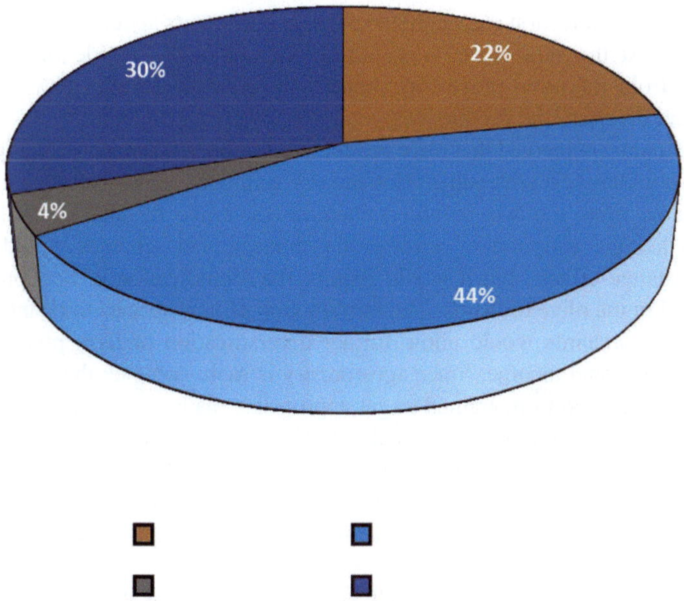

Fig. 2 Type of the 27stakeholders that participated at the first RAIN meeting in September 2017

Table 1 Bottlenecks and opportunities emerged at the participative workshop implemented during the first RAIN meeting

Issue	Bottlenecks	Opportunities
Policy	Common Agricultural Policy (CAP) complexity	Improve the support to agroforestry practices; promote the cooperation among stakeholders
Socio-economic	Low competitiveness of the local olive oil value chain	Educate citizens and consumers; identify new products
Climate and environment	Reduction of olive yield	Identify best management practices; promote farm diversification

mechanisms are perceived as too complicated and not efficient, especially to address the needs of small-scale farms. Stakeholders complained also about the bad link of policy to the real agricultural world, the poor communication of the EU regulations and the lack of control on the distribution of subsidies to the farmers. A distinguished support to agroforestry practices (i.e. a label for agroforestry products) could compensate the highest management costs of agroforestry practices, whereas the implementation of cooperation among stakeholders would improve the sharing of knowledge and facilitate the agreement on common strategies to be adopted and followed.

Stakeholders perceived that the limited competitiveness of the local olive oil value chain is the most important socio-economic bottleneck. In fact, stakeholders complained on the low price of high-quality extra-virgin olive oil mainly because consumers are not aware of the olive oil quality and most of them are not willing to spend more money for a higher price. Therefore, it would be relevant to promote adequate communication campaigns to educate consumers about olive oil beneficial properties. At the same time, it would be important also to identify innovative products from the olive processing residues in order to develop new local value chains and diversify the productions.

Stakeholders perceived that olive production has been decreasing mainly because of climate and/or pest adversities. Hot and dry summers reduce the olive yield, and, at the same time, too wet summers increase the olive fly attacks. Stakeholders thought that the adoption of agroforestry management schemes would mitigate climate change effects, but it would require the identification of best practices to be applied in the olive orchards. The introduction of new species to be intercropped in the olive orchards would allow further diversification of farm production and generate additional income. Since agroforestry is more complex than monoculture, farmers would need to be trained to get appropriate management skills.

A total of 16 stakeholders responded to the online questionnaire in which they were asked to list the most important challenges that should be implemented to contrast the bottlenecks and exploit the opportunities (Table 2). Challenges should mainly address best managements scheme, value chain improvement, communication and information.

The second RAIN meeting was attended by 39 stakeholders including farmers, farm advisors, multipliers, oil mill and technicians of the olive processing and

Table 2 Most relevant challenges that should be implemented in order to contrast bottlenecks

Challenges
Share information on optimal tree/crop/livestock combinations, in order to maximize productivity and ecosystem services
Identify fodder trees and nutritional value of nuts and fruits for specific animal species
Assessment of the profitability of agroforestry systems in the short term
Informing consumers and society about agroforestry and its environmental and economic benefits
Promote the establishment of demonstrative farms managing agroforestry systems
Increase the knowledge on the costs and benefits of specific agroforestry systems
Better understanding of the value chain (supply, demand and marketing opportunities) of agroforestry products
Improvement of policy support tools to promote agroforestry systems at farm level
Reducing legislative uncertainty regarding the tree planting on agricultural land

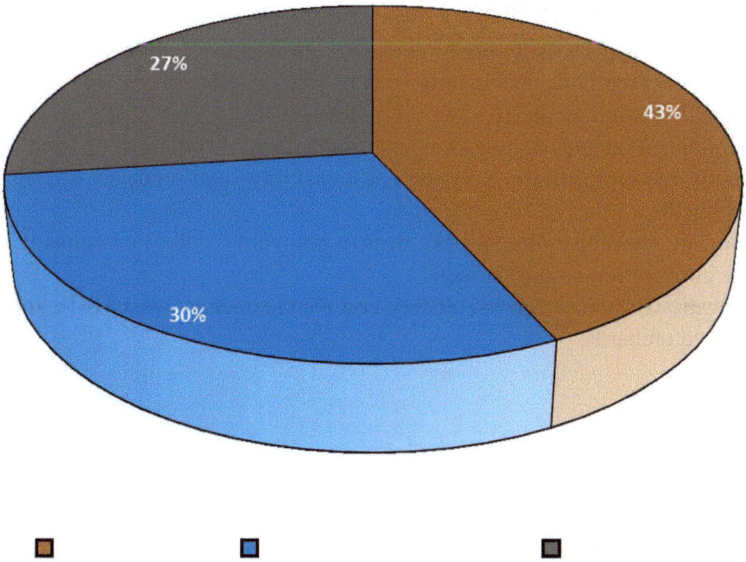

Fig. 3 Type of the 39 stakeholders that participated at the second RAIN meeting in January 2018

researchers (Fig. 3). During the meeting the stakeholders were invited to discuss about the above-mentioned challenges highlighted by the questionnaire.

Stakeholders perceived that the management of agroforestry systems is more complex than the conventional farming, but it certainly has lower environmental and landscape impacts. For this reason, farmers claimed the need to be compensated (i.e. with subsidies) for introducing and managing agroforestry systems in their farmlands. Stakeholders also claimed that it is important to increase the awareness about the massive use of chemical products in agriculture and the negative effects on the environment and human health, as well. This would be possible by increasing the adoption of organic management schemes at farm level. Climate change is particularly perceived as a threat to agriculture since it is now a fact and can no longer be considered as linked to exceptional events. Therefore, it is necessary to redefine, through applied research and research transfer at all steps of the supply chain, innovative cultivation models able to cope with the new climate patterns. In order to increase consumers' awareness about olive oil properties, stakeholders suggested interacting with citizens, in particular students, to highlight sustainable agricultural practices to manage olive orchards. Moreover, it would be appropriate to play out promotion and dissemination actions in shopping centers with distribution and tasting of food obtained in agroforestry systems. Research centres should deeply investigate how to optimize agroforestry systems establishment and management as well as the evaluation of the relative supply chain impacts as compared to conventional systems, on the basis, for example, of the Life Cycle Assessment (LCA) and the Carbon Footprint Analysis.

According to such challenges and suggestions, the possible innovations that should be taken into account are:

- Economical aspects and supply chain development and diversification: set up new process methods and technologies to provide olive processing by-products with added value while reducing oil mill waste.
- Communication and awareness raising: implement communication campaigns to demonstrate agroforestry systems sustainability and educate students and consumers.
- Policy and administration issues: identify innovative policy instruments to promote cooperation among farmers.
- Cultivation technical aspects: identify best management practices to be adopted in the olive orchards.

6 Conclusions

During the last two decades, the consensus and interest on agroforestry systems have been progressively increasing among public institutions at European level. Since the beginning of the new century, agroforestry and related issues have been included in the main European research programmes, and research projects, specifically focused on agroforestry, were funded both within the 6th Framework (i.e. "Silvoarable Agroforestry For Europe", SAFE project, 2001–2005) and the 7th Framework programme (i.e. "Agroforestry that Will Advance Rural Development", AGFORWARD project, 2014–2017). Those projects investigated and identified the most interesting agroforestry practices that could be adopted in different environments and socio-economic conditions in Europe.

At the same time, Common Agricultural Policy has included agroforestry as a supporting practice within the Rural Development Plans. Furthermore, since 2012 a European Agroforestry Federation has been constituted (EURAF, http://www.eurafagroforestry.eu) with the aim to promote agroforestry by any communication means, including lobbying for agroforestry adapted policies at the European scale. EURAF has about 280 members from 20 different European countries, and national agroforestry associations have been also created in most of the European countries.

Despite these efforts, agroforestry is still poorly applied at European farm level mainly because of a lack of knowledge among stakeholders. In the current EU H2020 research programme, AFINET project aims to bridge such gaps disseminating knowledge on agroforestry at wide level among European stakeholders. Stakeholders involved in the AFINET project acknowledge the high ecological value of agroforestry systems in promoting farm diversification, biodiversity conservation, landscape value and soil fertility. The most limiting factors to the adoption of such practices are related to the economic uncertainty to produce agroforestry goods and the management complexity of the systems. It is desirable that through AFINET communication and dissemination channels, such barriers will be partially removed favouring the increased adoption of agroforestry practices at European farm level.

Acknowledgement This paper is part of the AFINET project, with financial support from the European Union's Horizon 2020 research and innovation programme under grant agreement No 727872.

References

Bateni C (2017) Soil carbon stock in olive groves agroforestry systems under different management and soil characteristics. Master in Environmental Management of Mountain Areas Thesis Faculty of Science, University of Bolzano

Besnard G, Khadari B, Navascués M, Fernández-Mazuecos M, El Bakkali A, Arrigo N, Baali-Cherif D, Brunini-Bronzini de Caraffa V, Santoni S, Vargas P, Savolainen V (2013) The complex history of the olive tree: from Late Quaternary diversification of Mediterranean lineages to primary domestication in the northern Levant. Proc Biol Sci. https://doi.org/10.1098/rspb.2012.2833

Camilli F, Pisanelli A, Seddaiu G, Franca A, Bondesan V, Rosati A, Moreno MG, Pantera A, Hermansen JE, Burgess PJ (2018) How local stakeholders perceive agroforestry systems: an Italian perspective. Agrofor Syst. https://doi.org/10.1007/s10457-017-0127-0

Caputo AC, Scacchia F, Pelagagge P (2003) Disposal of by-products in olive oil industry: waste to energy solutions. Appl Therm Eng 23:197–214

Dal Bosco A, Mugnai C, Rosati A, Paoletti A, Caporali S, Castellini C (2014) Effect of range enrichment on performance, behavior and forage intake of free-range chickens. J Appl Poult Res 23:137–145. https://doi.org/10.3382/japr.2013-00814

De Jalón SG, Burgess JP, Graves A, Moreno G, McAdam J, Pottier E, Novak S, Bondesan V, Mosquera-Losada R, Crous-Dura'n J, Palma J, Paulo JA, Oliveira TS, Cirou E, Hannachi Y, Pantera A, Wartelle R, Kay S, Malignier N, Van Lerberghe P, Tsonkova P, Mirck J, Rois M, Kongsted AG, Thenail C, Luske B, Berg S, Gosme M, Vityi A (2017) How is agroforestry perceived in Europe? An assessment of positive and negative aspects by stakeholders. Agrofor Syst. https://doi.org/10.1007/s10457-017-0116-3

den Herder M, Moreno G, Mosquera-Losada R, Palma J, Sidiropoulou A, Santiago Freijanes J, Crous-Duran J, Paulo JA, Tomé M, Pantera A, Papanastasis V, Mantzanas K, Pachana P, Papadopoulos A, Plieninger T, Burgess PJ (2017) Current extent and stratification of agroforestry in the European Union. Agric Ecosyst Environ 241:121–132. https://doi.org/10.1016/j.agee.2017.03.005

Eichhorn MP, Paris P, Herzog F, Incoll LD, Liagre F, Mantzanas K, Mayus M, Moreno G, Papanastasis VP, Pilbeam DJ, Pisanelli A, Dupraz C (2006) Silvoarable systems in Europe - past, present and future prospects. Agrofor Syst 67:29–50

EIP-AGRI (2017) Agroforestry: introducing woody vegetation into specialised crop and livestock systems. Final report 2017. https://ec.europa.eu/eip/agriculture/sites/agri-eip/files/eip-agri_fg_agroforestry_final_report_2017_en.pdf

European Commission DG Environment (2006) Science for environment policy. News Alert Service, 41

Graves AR, Burgess PJ, Palma JHN, Herzog F, Moreno G, Bertomeu M, Dupraz C, Liagre F, Keesman K, van der Werf W, Koeffeman de Nooy A, van den Briel JP (2007) Development and application of bio-economic modelling to compare silvoarable, arable and forestry systems in three European countries. Ecol Eng 29:434–449

Grimelli A (2012) DMF2012: produrre patè d'olive direttamente dal decanter. Teatro naturale. http://www.teatronaturale.it/strettamente-tecnico/l-arca-olearia/12879-dmf2012:-produrre-pate-d-olive-direttamente-dal-decanter.htm

Gross K (2016) Ecology: biodiversity and productivity entwined. Nature 529:293–294

Hernandez AJ, Lacasta C, Pastor J (2005) Effects of different management practices on soil conservation and soil water in a rain fed olive orchard. Agric Water Manag 77:232–248

Lelle MA, Gold MA (1994) Agroforestry systems for temperate climates: lessons from Roman Italy. For Conserv Hist 38:118–126

Martini E (2010) Paysage et Environnement de l'Olivier dans la Region d'Umbria (Stratégie de Marketing et de Promotion). Fête de l'Olivier à Meknes – 2° édition: Meknes (Maroc), 17–19 Juin 2010. KNOLEUM, paysages de l'olivier: Etudes conduites par les 6 partenaires du projet Interreg III B Knoleum concernant la filière olévicole

McAdam J, Burgess P, Graves A, Riguero-Rodríquez A, Mosquera-Losada MR (2009) Classifications and functions of agroforestry systems in Europe. In: Rigueiro-Rodríguez A, McAdam J, Mosquera-Losada M (eds) Agroforestry in Europe: current status and future prospects. Springer Science + Business Media B.V, Dordrecht, pp 21–42

Nair PKR (1993) An introduction to agroforestry. Kluwer Academic Publishers, Dordrecht

Palma JHN, Graves AR, Bunce RGH, Burgess PJ, de Filippi F, Keesman KJ, van Keulen H, Liagre F, Mayus M, Moreno G, Reisner Y, Herzog F (2007) Modeling environmental benefits of silvoarable agroforestry in Europe. Agric Ecosyst Environ 119:320–334

Pisanelli A, Marandola D, Marongiu S, Paris P, Rosati A, Romano R (2014) The role of rural development policy in supporting agroforestry systems in EU. Book of abstracts of the 2nd European agroforestry conference "Integrating Science and Policy to Promote Agroforestry in Practice". Cottbus, Germany, 4–6 June 2014. isbn: 978-972-97874-4-7

Pomarici E, Vecchio R (2013) The Italian olive oil industry in the global competitive scenario. Agric Econ Czech 59(8):361–372

Razouk R, Daoui K, Ramdani A, Chergaoui A (2016) Optimal distance between olive trees and annual crops in rain fed intercropping system in northern Morocco. J Crop Sci Res 1:23–32

Roig A, Cayuela MA, Sánchez-Mondero MA (2006) An overview on olive mill wastes and their valorization methods. Waste Manag 26:960–969

Rosati A (2011) Asparagi nell'oliveto e la produzione raddoppia. Olivo e Olio 7/8:20–24

Rosati A (2014) Coltivare asparagi selvatici e allevare polli in un piccolo oliveto. Vita in Campagna 12: 44–48

Sabbatini V (2014) The Supply Function of Olive Oil: A Case Study of Italy. Procedia Economics and Finance 14:553–558

Sereni E (1974) Storia del paesaggio agrario italiano. Laterza, Bari

Souilem S, El-Abbassi A, Kiai H, Hafidi A, Sayadi S, Galanakis CM (2017) Olive oil production sector: environmental effects and sustainability challenges. In: Galanakis CM (ed) Olive mill waste 1st edition, recent advances for sustainable management isbn: 978-0-12-805314-0

Tomati U, Galli E, Pasetti L, Volterra E (1995) Bioremediation of olive mill waste-waters by composting. Waste Manag Res 13:509–518

Unaprol (2015) Filiera olivicola: analisi di scenario e studi di settore. http://www.unaprol.it/images/stories/allegati%20scenario/Pubblicazione%20finale%20III%20annualita.pdf

Vlyssides AG, Loizides M, Karlis PK (2004) Integrated strategic approach for reusing olive oil extraction by-products. J Clean Prod 12:603–611

Part II
Novel Technologies and Innovative Farming Systems to Support Rural Development

Epidemiological Investigation of Pseudorabies in Greece

Konstantinos Papageorgiou, Evanthia Petridou, Georgios Filioussis, Alexandros Theodoridis, Ioannis Grivas, Odysseas Moschidis, and Spyridon K. Kritas

Abstract Pseudorabies is an acute, frequently fatal disease that mainly affects pigs and incidentally other animals. Although pseudorabies virus (PRV) has been eradicated from many European countries, it is still endemic in East and Southeast parts of Europe. Greece belongs to the countries where the disease is enzootic. In this study, we investigated the presence of PRV in Greek farms. For that reason, 42 pig farms were selected from the entire Greek territory. Blood samples from different age groups had been collected from each farm and were tested by ELISA for the presence of antibodies against wild strains of PRV. The results of our study showed that 28.6% of the selected farms were positive for the presence of antibodies against wild-type strains of PRV and that factors such as the non-implementation of biosafety measures and the high density of pig farms in an area may affect the probability of a farm to become PRV positive. This study provided some useful information with regard to the presence of PRV in the domestic pigs in Greece. This information may assist in designing and implementing measures in order to control and eradicate the disease from the domestic pigs in Greece.

Keywords Pseudorabies · Epidemiology · Pigs · Eradication

K. Papageorgiou (✉) · E. Petridou · G. Filioussis · S. K. Kritas
Department of Microbiology and Infectious Diseases, School of Veterinary Medicine,
Faculty of Health Sciences, Aristotle University of Thessaloniki, Thessaloniki, Greece
e-mail: kvpapage@vet.auth.gr

A. Theodoridis
Laboratory of Animal Production Economics, School of Veterinary Medicine,
Faculty of Health Sciences, Aristotle University of Thessaloniki, Thessaloniki, Greece

I. Grivas
Laboratory of Anatomy, Histology and Embryology, School of Veterinary Medicine,
Faculty of Health Sciences, Aristotle University of Thessaloniki, Thessaloniki, Greece

O. Moschidis
Department of Business Administration, University of Macedonia, Thessaloniki, Greece

A. Theodoridis et al. (eds.), *Innovative Approaches and Applications for Sustainable Rural Development*, Springer Earth System Sciences,
https://doi.org/10.1007/978-3-030-02312-6_6

1 Introduction

Aujeszky's disease or pseudorabies is an acute, frequently fatal disease that mainly affects pigs and incidentally other domestic and wild animals. The term "pseudorabies" was used as a result of the disease's clinical resemblance to rabies. The disease was first described in 1813 in cattle, which were showing extreme pruritus. Aladar Aujeszky was the Hungarian veterinarian who first described and reproduced the disease in 1902, providing evidence that the etiologic agent was filterable (e.g. not a bacterium but a virus) (Mettenleiter et al. 2012). PRV has been classified in the genus *Varicellovirus,* subfamily *Alphaherpesvirinae* and family *Herpesviridae.* The virus particles have the typical architecture of a herpes virion (Mettenleiter 2000). The various strains of PRV differ in their infectivity and virulence in pigs as well as in their ability to be shed during infection. Those differences are associated with identified differences in their genome (Nauwynck et al. 2007). PRV pathogenesis depends on the age of the pig, the dose, the route of the inoculation and the virus strain. More specifically, the development and severity of the clinical signs of the disease diminish with increasing age. Neuronal signs are severe in pigs prior to weaning while fatteners are relatively resistant to nervous disease, showing mainly respiratory symptoms. In boars and sows, PRV infection is characterized by symptoms from respiratory and reproductive systems (Mettenleiter 2000; Papageorgiou et al. 2011a). Although swine is the only natural host of the virus, PRV can also infect a large number of species including cattle, sheep, goats, cats, dogs and wildlife (Pensaert and Kluge 1989; Banks et al. 1999; Bitsch and Munch 1971; Glass et al. 1994). There are no published data for PRV infection in humans (Tischer and Osterrieder 2010).

PRV is spread all over the world, particularly in regions with dense pig populations, including Europe, Asia and America. In Europe, PRV has been eradicated in Germany, Cyprus, Austria, Sweden, The Netherlands, Denmark, Czech Republic, Finland, France, Hungary, Luxemburg, Belgium, Switzerland, Slovakia and the UK as a result of the implementation of eradication programmes, but it is still endemic in East and Southeast of Europe (Hahn et al. 2010). PRV has also been eradicated from Canada, New Zealand and the USA (MacDiarmid 2000). More specifically, the USA has been classified as free of the disease since 2007. Although PRV has been eradicated from many countries throughout the world, the virus is still endemic in the populations of wild boar (Meng et al. 2009; Millicevic et al. 2016; Verpoest et al. 2014; Muller et al. 2011). Therefore, these populations should be considered as potential PRV source of infection for domestic pigs. In countries that are free of PRV, vaccination is prohibited.

Greece belongs to the countries where the disease is still enzootic. Up to 1973, only sporadic cases of Pseudorabies were diagnosed in bovine and sheep populations. According to an old serological study in 1969 (Papatsas et al. 1995), 20.8% of the collected blood samples of domestic pigs from several regions of Greece were positive to the presence of antibodies against PRV. But at that time, there was no serious and organized pig farming in Greece. In addition, although two recent Greek studies (Touloudi et al. 2015; Marinou et al. 2015) evidence the presence of PRV in 32% to 35% in wild boars, there is no recent data regarding the presence of PRV in the population of Greek domestic pigs. Here, we conducted an epidemiological study in order to investigate the presence of PRV in the Greek pig farms.

2 Materials and Methods

2.1 Characteristics of the Study Population

The study was carried out in Greece from October 2010 to October 2011. Greece produces up to 40% of the pork consumed within the country. The size of the country is approximately 132,000 km^2, 75% of which is covered by mountains. The pig population of the country is estimated to be around 1.2 million, with at least 88% of this population produced in 315 individually owned closed one-site system farrow-to-finish (FTF) farms of sizes larger than 100 sows. The majority of these farms are located in continental Greece and Crete, belong to individual persons, apply intensive indoor housing and have their own feed mill. As such farms reflect the commercial pig industry, the current study included FTF herds larger than 100 sows.

2.2 Selection of Herds

Forty-two (42) farrow-to-finish (FTF) pig herds were selected from the entire Greek territory at random, based on geographical criteria, in order to obtain representative data from the population herds. The sample represented more than 10% of the FTF farms. The herds were divided according to their size in two categories: up to 300 sows (small farms) and larger than 301 sows (large farms) (Table 1). As a first step, the selected farmers were contacted by phone in order to introduce the purpose of the study. Consequently, an appointment was arranged for the team in order to visit the farms, to collect data about the herds and the area where these holdings were located and to obtain the samples.

2.3 Collection of Herd Information

The data were obtained through face-to-face interviews of the owners of the selected farms and the following inspection of the pig units. The obtained information pertained to:

Table 1 Characteristics of commercial pig herds in Greece (>100 sows) and herd sampling for the study

Territory	Area (km^2)	Density (# of farms/ 1000 km^2)	Number of farms sampled /number of farms in territory *Herd-size category*		
			Small	**Large**	**Total**
East Macedonia and Thrace	*19,000*	*1.4*	*4/16*	*1/10*	*5/26 (19.2)*
Central and West Macedonia	*25,000*	*1.1*	*4/18*	*5/11*	*9/29 (31.0)*
Thessalia	*14,000*	*6.9*	*4/81*	*3/16*	*7/97 (7.2)*
Epirus and West Sterea Hellas	*15,000*	*6.1*	*2/66*	*6/24*	*8/90 (8.9)*
East Sterea Hellas	*20,000*	*1.9*	*7/21*	*4/15*	*12/36 (30.6)*
Peloponnesos and Crete	*30,000*	*0.9*	*1/25*	*1/12*	*2/37 (5.4)*
Total	*123,000*	*2.4*	*22/227 (9.7%)*	*20/88 (22.7%)*	*42/315 (13.3%)*

- Herd size, e.g. the number of sows on the premises. Farms with less than or equal to 300 sows were considered as small, while those with more than 300 sows as large.
- Pig herd area density, e.g. less dense (<20 farms per 1000 km^2) or more dense ($\geq$20 farms per 1000 km^2) areas.
- Direct distance from the closest pig farm, e.g. short (<6 km) or longer ($\geq$6 km) distance. All the selected farms did not include more than one neighbouring pig farm within 6 km radius. No other pigs, except breeding animals, and no semen were entering the farm.
- Purchase (or not) of breeding animals (gilts or/and boars) from genetic companies. No other pigs – except breeding animals – and no semen were entering the farm. In the herds that did not purchase breeders, gilts were produced within the farm by a grandparent stock of breeders.
- Practising (or not) of at least monthly quarantine in a distant building used exclusively for the newly purchased breeding animals.
- Practising (or not) of certain hygienic/biosecurity measures at farm. More specifically, (a) regular cleaning followed by disinfection and a stand-empty period of a minimum of 3 days between two production groups; (b) prevention of entry of lorries carrying pigs, feed or manure; (c) fence and barrier at entrance; and (d) prevention of entry of visitors, presence and use of a sanitary room to change clothes and use of unit protective clothing (but not always downtime and showering applicable).
- Practising (or not) of all-in/all-out (AIAO) flow in all production stages. A compartment was defined as a subdivision of a building with its own ventilation system. AIAO was considered to take place if the compartment (with its own

ventilation system) was filled up the same day, was emptied in one or two times and when a complete depopulation had taken place prior to restocking.

- Practising (or not) of vaccination for PRV.
- Presence (or not) of substantial economic problems in the farm that frequently interfere with routine management. Farmer's psychology and motivation for careful and extra work, as well as undisrupted supply of feed ingredient, medication and vaccines, were the immediate consequences of these economical problems.
- Presence (or not) of certain systemic clinical manifestations at the time of sampling, such as mortality, nervous, respiratory, gastrointestinal or/and reproductive health problems in one or more production stages in the farm.
- Production stage at which important clinical manifestations were present (or not) at the time of sampling (neonatal, nursery, grower and finishing stage).

2.4 *Sampling and Laboratory Testing*

A minimum of 8 blood samples from each out of 5 different age groups (i.e. 6-, 8-, 10-, 12- and 22-week old pigs) had been collected from each farm (in total 1723 blood samples). The blood samples for each age group were collected from pigs of different pens and, ideally, of different rooms. Sera were individually tested by anti-PRV-gB ELISA (IDEXX Laboratories, Westbrook, ME) for the presence of antibodies against the PRV and by anti-PRV-gE ELISA (IDEXX Laboratories, Westbrook, ME) for the differentiation of antibodies against the wild strains of PRV.

2.5 *Statistical Analysis*

Apparent prevalence of PRV-infected farms was estimated as the proportion of farms rearing at least one pig presenting antibodies against PRV, while its corresponding confidence interval (95% CI) was calculated around the estimated apparent prevalence using the exact method. Multiple correspondence analysis (MCA) has been used in order to investigate most profound interactions among several variables, particularly categorical, in order to localize dominant and more substantial trend in their structure. No prior precondition (e.g. distribution that should follow the data or models that we assume that apply to the population) is needed (Moschidis 2009, 2015; Moschidis and Chadjipadelis 2017).

The association of the herd and neighbourhood characteristics of the farms (predictors) with PRV status was investigated through the application of invariable logistic regression models with robust standard errors. Subsequently, a multivariable logistic regression model was built using as predictors the variables that presented a strong invariable association with PRV status (p-value <0.10). Multicollinearity between these latter predictors was estimated, selecting variables that exhibit tolerance greater than 0.4, following Lambert et al. (2012) methodological approach. The Hosmer-Lemeshow goodness of fit was estimated for the assessment of the final multivariable logistic regression model. The SPSS software (Version 22.0) was used.

3 Results

The sample used in this study represented 13.3% of the total commercial FTF pig herds larger than 100 sows in the country, including almost 10% of the small farms and 23% of the large farms. The characteristics of the total sampled herds are presented in Table 2. According to the data, 83.3% of the selected farms were located in areas with fewer than 20 farms (low herd density areas), 66.7% were located to close distance (<6 km) from other farms, 69.0% were purchasing gilts or/and boars from an outside source, 78.0% were not applying quarantine, 66.7% were suffering from substantial economic problems and 61.9% and 52.4% were suffering from respiratory and reproductive problems, respectively, while the main health problems in most of these herds were observed during nursery and grower stage (69.0% and 61.9% of total herds, respectively). Moreover, of the total number of selected herds (42), in 35 farms (83.3%), pigs were vaccinated against PRV.

Table 3 shows wild-type PRV-positive farms throughout the Greek territory. More specifically, the exposure of the farms to wild-type PRV was 28.6% with most of the positive holdings located in the region of East Macedonia and Thrace and Central and West Macedonia (Fig. 1).

The proportion of positive PRV farms for the level of each predictor is indicated in Table 2. More specifically, 53.8% of the PRV-positive herds were located in farm low-density areas, 91.7% had a direct distance of less than 6 km from other herds, 83.3% were purchasing gilts or/and boars from an outside source, 100% were not applying quarantine during the entrance of newly purchased pigs, 69.2% were suffering from substantial economic problems and 83.3% and 58.3% were suffering from respiratory and reproductive problems, respectively, while the main health problems in 83.3% and 91.7% of these herds were observed during nursery and grower stage, respectively.

The results of MCA are presented in Tables 4 and 5. Factorial axes F1 and F2 interpret 44.32% and 16.08% of inertia, respectively (e.g. both 60.4% of inertia) (Fig. 2). Points with CTR larger than 20 (1000/48) were preserved in axes.

In the first tendency (F1 axis) (Table 4), two roughly opposite groups of farms and characteristics were observed (denoted by opposite signs): (A) a group of farms that were practising basic hygienic measures and quarantine and did not suffer from major economic problems. These farms were not purchasing breeding animals from other farms. In these farms, respiratory, reproductive and general health problems were not issues, while nursery, growing and finishing phases were not affected. (B) A group of farms with PRV that were mostly suffering from economic problems and did not practise basic hygienic measures. In these farms, growing and finishing

Table 2 Characteristics of the sampled breeding farms in Greece (42 farms) and descriptive statistics for predictors tested for association with PRV-positive status

Predictors	Category	Farms (number)	*(%)*	Number of positive sites (%)	Characteristic within positive herds (%)
Size (no sows)	<300	22	*52.4*	6 (27.3)	50.0
	≥300	20	*47.6*	6 (30.0)	
Density (farms/1000 km^2)	<20	35	*83.3*	7 (20)	53.8
	≥20	7	*16.7*	5 (71.4)	
Distance (km)	<6	28	*66.7*	11 (39.3)	91.7
	≥6	14	*33.3*	1 (7.1)	
Gilt purchase	No	13	*31.0*	2 (15.4)	
	Yes	29	*69.0*	10 (34.5)	83.3
Quarantine	No	32	*78.0*	12 (37.5)	100.0
	Yes	9	*22.0*	0 (0.0)	
Biosecurity measures	No	19	*45.2*	10 (52.6)	83.3
	Yes	23	*54.8*	2 (8.7)	
AIAO	No	18	*43.9*	8 (44.4)	66.7
	Yes	23	*56.1*	4 (17.4)	
PRV vaccination	No	7	*16.7*	3 (42.9)	
	Yes	35	*83.3*	9 (25.7)	75.0
Economic problems	No	14	*33.3*	3 (21.4)	
	Yes	28	*66.7*	9 (32.1)	69.2
Mortality	No	32	*76.2*	6 (18.8)	
	Yes	10	*23.8*	6 (60.0)	50.0
Nervous signs	No	36	*85.7*	9 (25.0)	
	Yes	6	*14.3*	3 (50.0)	25.0
Respiratory signs	No	16	*38.1*	2 (12.5)	
	Yes	26	*61.9*	10 (38.5)	83.3
Gastrointestinal signs	No	27	*64.3*	7 (25.9)	
	Yes	15	*35.7*	5 (33.3)	41.7
Reproductive signs	No	20	*47.6*	5 (25.0)	
	Yes	22	*52.4*	7 (31.8)	58.3
Neonatal stage problems	No	39	*92.9*	9 (23.1)	
	Yes	3	*7.1*	2 (66.7)	16.7
Nursery stage problems	No	13	*31.0*	2 (15.4)	
	Yes	29	*69.0*	10 (34.5)	83.3
Grower stage problems	No	16	*38.1*	1 (6.3)	
	Yes	26	*61.9*	11 (42.3)	91.7
Finisher stage problems	No	29	*69.0*	5 (17.2)	
	Yes	13	*31.0*	7 (53.8)	58.3

Table 3 Exposure of Greek farms to wild-type PRV as detected by ELISA

	PRV gE-ELISA antibody-positive farms /number of farms sampled (%) *Herd-size category*		
Territory	**Small**	**Large**	**Total**
East Macedonia and Thrace	1/4	1/1	2/5 (40.0%)
Central and West Macedonia	2/4	2/5	4/9 (44.4%)
Thessalia	1/4	1/3	2/7 (28.6%)
Epirus and West Sterea Hellas	1/2	2/6	3/8 (37.5%)
East Sterea Hellas	1/7	0/4	1/11 (9.1%)
Peloponnesos and Crete	0/1	0/1	0/2 (0.0%)
Total	**6/22 (31.6%)**	**6/20 (27.3%)**	**12/42 (28.6%)**

Fig. 1 The stars on the map indicate the location of the selected farms on the Greek territory. The red and the green stars indicate the positive and negative farms, respectively, to the presence of antibodies against the wild-type strains of PRV

Table 4 Multiple correspondence analysis (MCA) results – F1 axis

Variable	ID	#F1	COR	CTR
Absence of general health problems	**C11**	638	802	73
Absence of problems in grower pigs	**D31**	465	796	66
Absence of respiratory symptoms	**C41**	438	771	58
Absence of major economic problems	**B51**	420	672	47
Animal quarantine during entrance	**B72**	487	564	38
Application of basic hygienic measures	**B82**	260	646	29
Absence of problems in nursery (weaned) pigs	**D21**	336	479	28
No purchase of breeding animal from outside	**B61**	332	493	27
Absence of reproductive symptom	**C61**	264	542	26
Absence of problems in finisher pigs	**D41**	192	688	20
Presence of major economic problems	**B52**	−211	672	23
Presence of general health problems	**C12**	−200	802	24
Presence of reproductive symptom	**C62**	−241	542	24
Presence of increased mortality	**C22**	−441	517	34
No application of basic hygienic measures	**B81**	−316	646	35
Presence of respiratory symptoms	**C42**	−270	771	36
Problems in grower pigs	**D32**	−287	796	41
PRV-infected	**A11**	−456	658	44
Problems in finisher pigs	**D42**	−430	688	46

CTR contributions, *COR* projections

Table 5 Multiple correspondence analysis (MCA) results – F2 axis

Variable	ID	#F2	COR	CTR
Problems in neonates	**D12**	679	398	68
No vaccination of sows for PRV	**B101**	426	454	63
No application of AIAO system	**B91**	219	364	47
East Sterea Hellas	**B15**	259	296	36
Small farms	**B21**	149	240	25
Presence of gastroenteric symptoms	**C52**	−186	212	27
Large farms	**B22**	−165	240	28
Application of AIAO flow system	**B92**	−182	364	39
Vaccination of pigs for PRV	**B112**	−424	391	62
High-density area	**B32**	−496	467	85
Epirus and West Sterea Hellas	**B14**	−578	801	132

CTR contributions, *COR* projections

pigs were mostly affected, and increased mortality, respiratory, reproductive and general health problems were mostly evident.

In the second tendency (F2 axis) (Table 5), two groups of farms and characteristics had been formed: (A) A group of small farms, mostly located in East Sterea Hellas, in which vaccination of sows against PRV and AIAO was not practised and

Total inertia 0.12536

Axis	Inertia	%Interpretation	Sum	Histogram of characteristic roots.
1	0,0555556	44,32	44,32	|**
2	0,0201522	16,08	60,39	|****************
3	0,0137212	10,95	71,34	|***********
4	0,0073600	5,87	77,21	|******
5	0,0068333	5,45	82,66	|*****
6	0,0055255	4,41	87,07	|****
7	0,0032446	2,59	89,66	|***
8	0,0024631	1,96	91,62	|**
9	0,0023905	1,91	93,53	|**
10	0,0019312	1,54	95,07	|**
11	0,0012693	1,01	96,08	|*
12	0,0010156	0,81	96,89	|*

Fig. 2 Table of inertia

neonates were mostly affected. (B) A group of large farms located in high-density areas (e.g. Epirus and West Sterea Hellas), practising vaccination of fatteners against PRV as well as AIAO system, in which gastroenteric symptoms were most often seen.

In logistic regression models, the variable that expresses the quarantine measures had to be excluded from the regression analysis due to absence of variation in the values for positive sites, i.e. the dependent variable (presence of virus) didn't vary within this variable. After excluding these variables, 17 predictors were used in the logistic regression in order to investigate their association with PRV status. The predictors that appear to have an invariable association with PRV status at significance level of 0.10 (p-value $\leq$0.10) are presented in Table 6. Results indicate that 9 out of the 15 predictors were strongly related to PRRSV status in the farms; however, two variables, namely, those that express respiratory health issues and health issues at grower stage, were excluded from the multivariable logistic model due to multicollinearity.

According to the results (Table 7), it appears that factors such as "pig herd area density" and "hygienic/biosecurity measures" play a key role in the probability of a farm to become PRV positive. More specifically, farms, which were located in low-density areas and were applying hygienic/biosecurity measures, had a predicted probability of being positive for PRV of 1.97%. However, their probability was increased to 26.7% when farms were located in low-density areas but were not applying hygienic/biosecurity measures. Farms that were located in high-density areas, applying hygienic/biosecurity measures, had a predicted probability of being positive for PRV of 30.8%. Finally, the probability of being positive for PRV was increased to 88.9% when farms were located in high-density areas and were not applying hygienic/biosecurity measures.

Table 6 Predictors associated (p-value ≤0.10) with PRV-positive status using invariable logistic regression assuming robust SE

Predictors	Odds ratio	95% CI	P-value
Farm size (≥ 300 sows per farm)	0.255	0.06–1.16	0.076
Density (≥farms/1000 km^2)	4.500	0.81–24.95	0.085
Distance from closest farm ≥6 km	0.119	0.01–1.07	0.057
Biosecurity measures	0.086	0.02–0.48	0.005
AIAO	0.263	0.06–1.11	0.069
Mortality	6.500	1.36–31.06	0.019
Grower stage problems	10.999	1.23–98.80	0.032
Finisher stage problems	5.600	1.28–24.42	0.022

Table 7 Predictors associated (p-value ≤0.05) with PRV-positive status using multivariable logistic regression

Predictors	b	SE(b)	Odds ratio	95% CI	Wald test	P-value
Intercept	−1.010					
Density	3.097	1.798	22.14	−0.42 – 6.62	1.72	0.085
Biosecurity measures	−2.894	1.174	0.06	−5.19 – −0.59	−2.47	0.014

Hosmer-Lemeshow goodness-of-fit test, p-value = 0.35

4 Discussion

Up to 1973, only sporadic cases of PRV in sheep, bovine and mink had been diagnosed in Greece. The first clinical report, followed by virus isolation of PRV in pigs, was on May 1974. Two more clinical cases with high mortality of suckling piglets had been reported on January 1976 and February 1977. In 1983, the import of breeding animals from other European countries caused outbreaks of the disease in the whole country (Papadopoulos 1989; Papadopoulos et al. 1996; Papatsas et al. 1995). According to the previous published study of 1969, antibodies were found in 20.8% of the tested serum samples. It is necessary to point out that that study refers to swine blood serum samples which were sporadically tested before the "industrialization" of pig farming and before the onset of vaccination programmes (the vaccinations for PRV in Greece started in the mid-1980s but never applied in a systematic way). In the present study, the majority of the positive farms (75%) was practising a vaccination scheme against PRV. The latter finding indicates that vaccination alone is not sufficient to eradicate the disease, unless it is accompanied by other measures such as the removal of the animals, which are found positive to the presence of antibodies against PRV. The aim of vaccination as a part of an eradication programme is not only to induce clinical protection but also to stop the transmission of the virus within and between herds by inducing herd immunity. Both attenuated and inactivated vaccines can be induced (Kritas 1994; Papageorgiou et al. 2011b). The development of marker vaccines and the use of diagnostic tests (differential ELISA) can play an important role in the eradication and control campaigns, as it was determined in the PRV eradication programme in the USA (Foley and Hill 2005; Papageorgiou et al. 2011b).

The proportion of positive PRV farms for the level of each predictor is indicated in Table 2. Of the PRV-positive herds, 91.7% had a direct distance of less than 6 km from other herds, 83.3% were purchasing breeding animals (gilts or/and boars) from sources outside the farm, 83.3% were not practising certain biosecurity measures and 100% of the PRV-infected farms did not apply quarantine for the newly purchased breeding animals! The analysis of the data also showed that 69.2% of the PRV-infected farms were facing substantial economic problems. The factor "economic problems" had been added to the questionnaire as a result of the crisis which affects the Greek economy in the last 8 years. It had been addressed to the owner but also to key personnel in a direct or indirect way. The reason was to include or detect purposefully or latently missing "gaps" in routine management, biosecurity or hygiene. For example, medication and vaccines may have been properly purchased in the farm, but their application was not correct due to underpaid personnel or not supervised closely due to bad psychology of the farmer. More specifically, among other things, the crisis led some farmers to abandon vaccination against PRV. It is important to keep in mind that although vaccination suppresses the manifestation of typical clinical signs of the disease, it doesn't eliminate the virus. One of the main characteristics of herpes viruses is latency. Latency is specified as a condition in which infectious virus is not produced, although viral DNA persists. For an unknown reason, most probably associated with stress factors, the virus is reactivated and subsequently may "come up" in the population. That was the reason for the re-emergence of the virus in many farms.

By using MCA, all variables can be simultaneously included and can reveal their most intense interactions, as well as dominant tendencies. Interpretation of the first axis shows that farms lacking basic hygienic measures suffer from major health problems (e.g. mortality, respiratory, reproductive and general health problems) particularly during fattening period. This is something expected as it is widely known that biosecurity measures prevent the entrance and circulation of pathogens within a farm. Apparently this is the reason why PRV was presenting such farms, where it may well contribute to all described symptomatology. As no detection of other pathogens had been attempted at present, it is not known whether the same applies for these as well, but it would be interesting to investigate it in the future by this analysis. Major economic problems were also associated with this group of farms, but it was not possible to distinguish whether they represented the result or the cause of the health problems.

The interpretation of second axis had added some more information to those derived by first axis. Thus, observations in small farms show that sow vaccination against PRV and application of AIAO flow system may be necessary to secure good health in neonates. In addition, observations in large farms show that fattener's vaccination against PRV and application of AIAO flow system may reduce PRV-associated symptoms (gastrointestinal symptoms are not considered typical for PRV) at least in high-density areas. It is known that specific maternal and active immunity may sufficiently protect neonates and fatteners, respectively, against PRV symptoms (Kritas et al. 1997; Papageorgiou et al. 2011b). The fact that increased mortality and neurological symptoms had not been observed in neonatal pigs and respiratory and reproductive clinical signs had not been observed in older pigs of these farms confirms the general good efficacy of vaccinations against PRV.

Furthermore, the results of the multivariable logistic regression analysis had end up with a quantitation of the correlation of two variables that are "pig herd area density" and "hygienic/biosecurity measures". Therefore, the probability of being PRV-infected can vary from 2% for a "clean farm" of a low herd density area to 89% for a "dirty farm" of a high herd density area. Although PRV is transmitted primarily between swine through nose-to-nose contact, under favourable conditions, the virus may spread by aerosols (Vannier 1988; Christensen et al. 1990). Thus, it is obvious that the higher the density of pig farms of an area, the more likely is that a farm will become positive for PRV. Factors such as movement of wild and domestic animals should also play a role in the spread of PRV in high-density areas.

This study provided some useful information with regard to the presence of PRV in the domestic pigs in Greece. In addition, the statistical analysis of the collected data shed light to the correlation of PRV-infected pig farms to factors such as pig area density and the application of certain biosecurity/hygienic measures. The ultimate objective of pseudorabies control is its eradication. Several PRV control and eradication programmes have been implemented in Europe and the USA (Andersson et al. 1997; Bech-Nielsen et al. 1995; Muller et al. 2003; Vannier 1988). Compared to other several European countries, which had already eradicated PRV, Greece has many important advantages (Papadopoulos et al. 1996):

- The low density of the pig population (7 pigs/km2 in Greece, when in Holland it is 400 pigs/km2, in Belgium 230 pigs/km2, in Germany 73 pigs/km2, in Italy, Portugal, Spain and France between 20 and 30 pigs/km2).
- The type of the units is principally farrow-to-finish having their own feed mill. Thus, entrance of virus in the farms can be better prevented when compared to the fattening type of units.
- As a country that imports most of its breeding stock, a PRV-free status of animals can be required from breeder countries.
- Vaccinations with live or inactivated gE-vaccines are regularly applied in the majority of the organized farms.

In the case that farmers wish to quit PRV vaccination, this should be done not based on clinical or post-mortem findings but on intense laboratory testing of the current and incoming stock. A qualified herd health management specialist on infectiology should direct such procedures plus all appropriate additional measures.

In conclusion, this study provides new information regarding the presence of PRV in Greek pig farms. More specifically, it is the first well-organized scheduled research for the epidemiological study of pseudorabies, a disease which causes serious economic problems in the Greek pig industry. The use of the obtained information may assist in the designation and implementation of measures in order to control and eradicate the disease from Greece.

References

Andersson H, Lexmon A, Robertsson JA, Lundeheim N, Wierup M (1997) Agricultural policy and social returns to eradication programs: the case of Aujeszky's disease in Sweden. Prev Vet Med 29(4):311–328

Banks M, Torraca LS, Greenwood AG, Taylor DC (1999) Aujeszky's disease in captive bears. Vet Rec 145:362–365

Bech-Nielsen S, Bowman GL, Miller GY, Orloski-Snider KA, Burkholder RH, Dodaro SJ (1995) Pseudorabies (Aujeszky's disease) eradication progress and program costs in Ohio, USA. Prev Vet Med 22:41–53

Bitsch V, Munch B (1971) Pseudorabies in carnivores in Denmark. I. the red fox (Vulpes vulpes). Acta Vet Scand 12:274–284

Christensen LS, Mousing j, Mortensen S, Soerensen KJ, Strandbygaart SB, Henriksen CA, Andersen JB (1990) Evidence of long distance airborne transmission of Aujeszky's disease (pseudorabies) virus. Vet Rec 127(19):471–474

Foley PL, Hill RE (2005) Regulatory considerations for marker vaccines and diagnostic tests in the U.S. Biologicals 33:253–256

Glass CM, McLean RG, Katz JB, Maehr DS, Cropp CB, Kirk LJ, McKeirnan AJ, Evermann JF (1994) Isolation of pseudorabies (Aujeszky's disease) virus from a Florida panther. J Wildl Dis 30:180–184

Hahn EC, Fadl-Alla B, Lichtensteiger CA (2010) Variation of Aujeszky's disease viruses in wild swine in USA. Vet Microbiol 143:45–51

Kritas SK (1994) Neuropathogenesis of wild-type and deleted Aujeszky's disease virus strains in pigs with and without specific maternal antibodies. PhD. Thesis. Faculty of Veterinary Medicine, University of Gent, Belgium

Kritas SK, Nauwynck HJ, Pensaert MB, Kyriakis SC (1997) Effect of the concentration of maternal antibodies on neural invasion of Aujeszky's disease virus in neonatal pigs. Vet Microbiol 55:29–36

Lambert M-È, Arsenault J, Poljak Z, D'Allaire S (2012) Epidimiological investigations in regard to porcine reproductive and respiratory syndrome (PRRS) in Quebec, Canada. Part 2: prevalence and risk factors in breeding sites. Prev Vet Med 104:84–93

MacDiarmid SC (2000) Aujeszky's disease eradication in New Zealand. Aust Vet J 78:470–471

Marinou KA, Papatsiros VG, Gkotsopoulos EK, Odatzoglou PK, Athanasiou LV (2015) Exposure of extensively farmed wild boars (Sus scrofa scrofa) to selected pig pathogens in Greece. Vet Q 35(2):97–101

Meng XJ, Lindsay DS, Sriranganathan N (2009) Wild boars as sources for infectious diseases in livestock and humans. Philos Trans R Soc Lond Ser B Biol Sci 364:2697–2707

Mettenleiter TC (2000) Aujeszky's disease (pseudorabies) virus: the virus and molecular pathogenesis-state of the art. Vet Res 31:99–115

Mettenleiter TC, Ehlers B, Muller T, Yoon K-J, Teifke JP (2012) Herpesviruses. In: Diseases of swine 10th edition. John Wiley & Sons, Inc., New York, pp 421–446

Millicevic V, Radojicic S, Valcic M, Ivovic V, Radosavljevic V (2016) Evidence of Aujeszky's disease in wild boar in Serbia. Vet Res 12:134. https://doi.org/10.1186/s12917-016-0758-9

Moschidis O (2009) A different approach of multiple correspondence analysis (MCA), than this of Specific MCA. J Math Soc Sci 47e année, n 186:77–88

Moschidis O (2015) Unified coding of qualitative and quantitative variables and their analysis with ascendant hierarchical classification. Int J Data Anal Tech Strateg 7(2):114–128

Moschidis O, Chadjipadelis T (2017) A method for transforming ordinal variables. In: Data science. Springer, Cham, pp 285–294

Müller T, Bätza HJ, Schlüter H, Conraths FJ, Mettenleiter TC (2003) Eradication of Aujeszky's disease in Germany. J Vet Med B Infect Dis Vet Public Health 50(5):207–213

Muller T, Hahn EC, Tottewitz F, Kramer M, Klupp BG, Mettenleiter TC, Freuling C (2011) Pseudorabies virus in wild swine: a global perspective. Arch Virol 156:1691–1705

Nauwynck H, Glorieux S, Favoreel H, Pensaert M (2007) Cell biological and molecular characteristics of pseudorabies virus infections in cell cultures and in pigs with emphasis on the respiratory tract. Vet Res 38:229–241

Papadopoulos O (1989) Control of Aujeszky's disease in Greece. The view from a country that imports a great part of its swine breeding stock. In: Vaccination and control of Aujeszky's disease. Kluwer Academic Publishers, Brussels, pp 227–229

Papadopoulos O, Kyriakis SC, Paschaleri-Papadopoulou E, Papatsas J, Kritas SK, Tsinas A, Tsiloyiannis V (1996) The present status of Aujeszky's disease in Europe and the plans for the Greek eradication programme. In: Proceedings of Bilateral Veterinary Medical School Meeting Belgrade-Thessaloniki, Kopaonik, Yugoslavia, pp 73–76

Papageorgiou K, Burriel AR, Filioussis G, Psychas V, Nauwynck HJ, Kritas SK (2011a) Aujeszky's disease (pseudorabies). An old threat in current pig industry? Part I. Pathogenetic information and implications. J Hell Vet Med Soc 62(1):29–37

Papageorgiou K, Burriel AR, Filioussis G, Christodoulopoulos G, Nauwynck HJ, Kritas SK (2011b) Aujeszky's disease (pseudorabies). An old threat in current pig industry? Part II. Epidemiology, immunity, prevention and the current situation in Greece. J Hell Vet Med Soc 62(2):125–131

Papatsas J, Paschaleri-Papadopoulou E, Koubati-Artopiou M, Kritas SK, Kyriakis SC (1995) Aujeszky's disease in pigs. update review and proposed measures for its control and eradication in Greece. J Hell Vet Med Soc 46:19–29

Pensaert MB, Kluge JP (1989) Pseudorabies virus (Aujeszky's disease). In: Virus infections of porcines. Elsevier, New York, pp 39–64

Tischer BK, Osterrieder N (2010) Herpesviruses – a zoonotic threat? Vet Microbiol 140:266–273

Touloudi A, Valiakos G, Athanasiou LV, Birtsas P, Giannakopoulos A, Papaspyropoulos K, Kalaitzis C, Sokos C, Tsokana CN, Spyrou V, Petrovska L, Billinis C (2015) A serosurvey for selected pathogens in Greek European wild boar. Vet Rec Open 2:e000077. https://doi.org/10.1136/vetreco-2014-000077

Vannier P (1988) The control programme of Aujeszky's disease in France: main results and difficulties. In: Van Oirschot JT (ed) Vaccination and control of Aujeszky's disease. Kluwer Academic Publishers, Boston, pp 215–226

Verpoest S, Cay AB, de Regge N (2014) Molecular characterization of Belgian pseudorabies virus isolates from domestic swine and wild boar. Vet Microbiol 172:72–77

Using the Multi-stakeholder Approach to Match Potential Innovations with Challenges Experienced by European Sheep and Goat Farms

Ian Rose, Daniel Martin-Collado, Stefano Orsini, Rafaelle Zanoli, David Yañez-Ruiz, Konstantinos Zaralis, and Georgios Arsenos

Abstract The decline in European sheep and goat farmers is a complicated agricultural challenge that requires solutions and innovation to reverse. The complexity of this challenge means that a multi-stakeholder approach is required to understand how to prioritise each component of the challenge and find appropriate solutions. The multi-stakeholder approach used in this research included semi-structured interviews with 47 sheep and goat farmers. Additionally, a survey and 3 workshop discussions were done with stakeholders from 28 research and industry organisations.

All stakeholders agreed that improvement of the market for sheep and goat products was the most important priority. Industry representatives and researchers, however, suggested breeding programmes and availability of labour as important, whilst farmers suggested cost of production, solutions for predators and animal health were more important. Additionally, innovations that industry representatives believe should be tested are for breeding, collection of farm data and reproduction despite marketing and processing being the main priority. This mismatch is probably

I. Rose (✉) · G. Arsenos
Laboratory of Animal Husbandry, School of Veterinary Medicine, Aristotle University of Thessaloniki, Thessaloniki, Greece
e-mail: gusrose@vet.auth.gr; arsenosg@vet.auth.gr

D. Martin-Collado
Agrifood Research and Technology Centre of Aragón, Zaragoza, Spain

S. Orsini
The Organic Research Centre, Elm Farm, Hamstead Marshall, Newbury, UK

R. Zanoli
Università Politecnica delle Marche Home, Department of Agricultural, Food and Environmental Sciences, Ancona, Italy

D. Yañez-Ruiz · K. Zaralis
The Spanish National Research Council, Granada, Spain

A. Theodoridis et al. (eds.), *Innovative Approaches and Applications for Sustainable Rural Development*, Springer Earth System Sciences,
https://doi.org/10.1007/978-3-030-02312-6_7

because these stakeholders do not have the capacity to test innovation or solutions related to marketing and processing of products.

The multi-stakeholder approach was effective at identifying key priorities and potential solutions that can be used to make sheep and goat farms more sustainable. More solutions related to product marketing and processing could have been identified if the research included stakeholders specialising in these areas.

Keywords Multi-stakeholder · Sheep · Goats · Interviews · Workshops · Innovation

1 Introduction

The European sheep and goat farmers experience many challenges with many farmers currently becoming discouraged and leaving the sector because of a combination of socio-economic reasons. Additionally, it is likely that sheep and goat farmers will be affected very differently by mounting pressure over land use and increasing pressure towards intensification compared to other livestock sectors, which have already overcome an intensification process. Overcoming these challenges is important because extensive and semi-extensive sheep and goat production systems are an exceptional way of utilising environmental resources which do not compete with other land uses such as production of crops for human consumption and maintain animal and plant diversity. Finding solutions to keep the sheep and goat industries sustainable, competitive and resilient first requires consensus across the sector stakeholders about the most important challenges and the strategies with highest probability of success.

The challenges facing European sheep and goat farms have various interacting technological, social-cultural, economic, institutional and political dimensions that make them complex (Schut et al. 2014). Additionally, there are interactions across levels that include diverse stakeholders adding more complexity (Funtowicz and Ravetz 1993). These diverse stakeholders include anyone related directly or indirectly to the challenge or are interested in finding a solution (McNie 2007). The participation in key stakeholders is critical to finding solutions to complex agricultural challenges (Giller et al. 2011).

Additionally, European sheep and goat farms are diverse in production aims, size, breeds and levels of intensification. This diversity also includes many environmental and socio-economic conditions with local, regional, national or international importance and can vary in sustainability and their needs for innovation. Therefore, the diversity of farms and their production aims makes selecting appropriate indicators to assess sustainability difficult.

Furthermore, identifying challenges and potential solutions cannot be based on the old top-down model but on partnership and co-development with all stakeholders (Dubeuf 2011). There are also many types of innovations available as solutions to challenges. Innovation is the development, introduction and application of a new or significantly improved product (good or service), a new marketing method or a new organisational method in business practice, workplace organisation or external relations where an economic or social benefit is assumed for individuals, groups or entire organisations (OECD/Eurostat 2005). These innovations can be technology or knowledge with different physical forms and processors (Sunding and Zilberman 2001). The relevance of potential innovations may not be the same across the diverse farm types in Europe (Dubeuf and Sayadi 2014). Therefore, inputs from stakeholders that represent the diversity of farms in Europe are required.

Getting consensus about what should be the focus now and in the future requires input from key industry stakeholders. Including key stakeholders into research and analysis using a multi-stakeholder approach aims to foster cooperation between researchers with next and end users of the research results. These next and end users include farmers and farmers' groups, advisers and government agencies that closely cooperate throughout the whole research project period. The multi-stakeholder approach aims to find the most relevant research, which outcomes are applicable and deal with the main threats that farmers are facing at farm level and market level. This paper describes how the multi-stakeholder approach was used to identify the challenges for sheep and goat farmer in Europe and possible solutions to these challenges.

2 Materials and Methods

This research is part of the Horizon 2020 project Innovation for Sustainable Sheep and Goats in Europe (www.isage.eu). The multi-stakeholder network included farmers, farmer groups, researchers and industry organisations that work directly with farmers or farmer groups. These stakeholders identified the key challenges that they believe are the most important for sheep and goat sectors in Europe. Additionally, potential solutions that could make sheep and goat industries resilient to these challenges in the future were identified. The challenges and potential solutions were recorded using semi-structured interviews, discussion groups and surveys. These surveys and interviews were done in the United Kingdom, France, Spain, Italy, Finland, Greece and Turkey.

2.1 Challenges and Priorities

The challenges and priorities for sheep and goat farmers were identified using reviews and interviews with sheep and goat farmers.

Review of challenges and priorities: Researchers and industry representatives from 12 organisations in the United Kingdom, France, Spain, Italy, Finland, Greece and Turkey identified the most important challenges and priorities in each country. The challenges and priorities in each country were identified using a review of literature, national documents and surveys of stakeholders.

Table 1 Number of sheep and goat farmers interviewed in each country

	Interviews	
Country	Sheep farmers	Goat farmers
Finland	5	0
France	4	2
Greece	4	3
Italy	5	2
Spain	10	6
United Kingdom	5	1
Total	33	14

Interviews with farmers: Sheep and goat farmers were interviewed using semi-structured interviews in Finland, France, Greece, Italy, Spain and the United Kingdom (Table 1). Farmers were asked about the importance of farm management, environmental conditions, animal productivity, health, nutrition, pasture management, social issues, succession, future plans and policy support measures. These questions were included in an interview guide used to help structure the interviews. The interviewers were explorative, and these questions were covered within four main topics:

(i) How changes in local area/environment have affected farms and their management
(ii) Constraints for the development of farms and how they will change farming in the future
(iii) Opportunities and threats for farmers
(iv) Priorities for sheep/goat farming that should be improved or studied

Interviewed farmers were selected from the following farm typologies:

(i) Intensive dairy farms (e.g. high input of purchased feedstuff)
(ii) Semi-extensive or extensive dairy farms (e.g. normally pasture-fed animals)
(iii) Intensive meat farms (e.g. high input of purchased feedstuff)
(iv) Semi-extensive or extensive meat farms (e.g. normally pasture-fed animals)
(v) Dual-purpose farms (farms where the farmer sees value in two or more different products, e.g. meat and wool, meat and dairy)

The results of the semi-structured interviews were organised into the following topics: farm structure, environment, farm/livestock management, technical issues, socio-economics, market and supply chain and policy. From these topics, strengths, weaknesses, opportunities and threats (SWOT analysis) were identified.

2.2 *Identification of Potential Solutions to Challenges*

Stakeholders including farmers, farmer organisations, industry groups and government and nongovernment research organisations (Table 2) identified potential innovative solutions to the identified challenges, following a participatory process

Table 2 Sheep and goat sector stakeholder organisations

Stakeholder type	Description	Number
Farmer/breeder group	Co-operative that assists and works with farmers or breeders directly, i.e. manages breeding programme, records data and advises on farm management	10
Farm	Commercial farms or breeders that work with research institutions	3
Industry	Large organisation that represents the commercial interest of farmers, including promotion and dissemination	4
University	Research group from a university that specialises in sheep and goats	5
Public research	Research group from a public organisation that specialises in sheep and goats	5
Farmer research	Non-profit, non-governmental research organisation funded by farmer levies	3

consisting of three workshops and one survey. Stakeholder organisations were from Greece, the United Kingdom, France, Italy, Spain, Finland and Turkey.

Firstly, an initial workshop was held in which 22 stakeholders from 18 of the organisations (Table 2) discussed and identified all potential innovation they thought could improve the sustainability of sheep and goat farms. The discussion included defining what innovation is and how it can be used across the sector. Secondly, after 2 months, stakeholders were surveyed to identify more potential innovations using an online survey using the Qualtrics survey platform (Qualtrics 2016). A total of 35 responses were received from 69 potential participants covering 95% of the participating stakeholders' organisations. Then, in a third stage of the participatory process, a revision workshop discussion including eight stakeholders from six organisations discussed and revised the list of innovations in a 1-day workshop. The organisations included a breeder group, farm, industry representative, public research, university and farmer research. These stakeholders grouped the list of innovations as either farm management, farm technology or product processing and marketing innovation. The stakeholders' description of the innovations from the initial workshop was used as the base to describe potential benefits and constraints of each innovation and included examples of how the innovation was already been used. The benefits of the innovation included a description of how they could be used to improve the environmental, economic and social sustainability of farms and sectors.

Finally, the list of grouped innovation including descriptions of benefits, constraints and potential benefits for sustainability was sent to the industry and research stakeholders previously to the final workshop event. The final stakeholder workshop was organised to discuss which innovations they believe are more urgent to investigate because they have the highest potential to improve sector sustainability. The stakeholders also discussed what they would measure to assess how these innovations improve the sustainability of sheep and goat farms and the sectors in Europe. The workshop discussion included 27 stakeholders from 15 of the organisations. The stakeholders were split into 3 groups of 6–11 participants. The groups were formed based on regions:

(i) Greece, Italy and Turkey
(ii) Spain
(iii) France and the United Kingdom

3 Results

3.1 Challenges and Priorities

Among the challenges and priorities, all countries identified better market access as a priority (Table 3). Suggested improvements included export opportunities, understanding public preferences and how to increase consumer awareness, development of niche markets, localised production and local supply chains and more balanced power relationships between market players, for example, through cooperation or market agreements. Additionally, all countries identified efficient breeding programmes focused on functional traits, genomic selection and autochthonous breeds as strategic priorities for the development of the sector. Meanwhile, innovative production practices and improved human capital in rural areas were reported as priorities in six countries. Product quality, hygiene and food safety were priorities in five countries. Environmental sustainability and improvement of the structural characteristics of the sector were mentioned only twice.

The interviewed farmers also identified that improvement of the market for sheep and goat products was the most common priority (Table 4). Although farmers identified many opportunities for alternative food chains and more consumption of sheep and goat products, they also identified that low margins in supply chains were a weakness and lower lamb consumption was a threat.

On-farm diversification, in particular processing and direct selling, represents a valuable source of income for most farmers, helping them to continue the agricultural production activity. Interestingly, many farmers considered increased training opportunities for post-production operations, including processing and marketing a

Table 3 Strategic priorities identified by each country

Priority	GR	IT	ES	FR	TR	UK	FI	Total
Market access and economic performance	X	X	X	X	X	X	X	7
Genetic improvement	X	X	X	X	X	X	X	7
Innovation in farming practices to improve productivity	X	X	X	X	X	X		6
Improvement of human and social capital	X	X	X	X	X	X		6
Product quality, hygiene and food safety	X	X	X	X	X			5
Environmental sustainability			X			X		2
System structure and exogenous factors	X	X						2
Total	6	6	6	5	5	5	2	35

GR Greece, *IT* Italy, *ES* Spain, *FR* France, *TR* Turkey, *UK* United Kingdom, *FI* Finland

Table 4 Results from the SWOT analysis

Strengths	Weaknesses
Good technical and veterinary services Good animal health and welfare (extensive and semi-extensive farms, local breeds) Low production costs (extensive farms) Farmers' livestock management skills Product quality, related with sustainable management practices such as grazing (especially extensive farms, organic farms)	High labour requirements Low margins from a supply chain perspective High production costs (intensive farms) Low farmers' marketing skills/knowledge Lack of available land for grazing and new entrants Predators (dogs, wolves, bears, badgers, buzzards, ravens, crows) Climate change for some farmers (especially water scarcity and flooding)
Opportunities	**Threats**
Alternative food chains and food labelling Farm diversification (e.g. tourism) Good market opportunities for the dairy sector Valorisation of wool production Being part of cooperatives and associations Contribution to local development and to environmental conservation Genetic improvement Training Increased farm self-sufficiency through on-farm production of inputs Migrants: increase in labour availability and meat consumption	Lack of generational turnover Environmental campaigns and misconceptions from the public about sheep/goat farming Urban encroachment/land being used for leisure activities Decrease in lamb meat consumption Decrease in farm subsidies

higher priority than for livestock and farm management. Also, many farmers suggested more investment in the wool sector as a response to the decrease in lamb meat consumption which is occurring in most of the countries involved.

Extensive and organic farmers identified for farm and livestock low production costs and good animal health as the main priorities. A number of intensive sheep farmers who used to manage farm extensively in the past switched to intensive production mainly due to a lack of available land for outdoor grazing. Overall there were no strong patterns between farm types (extensive/intensive systems) and the other themes discussed by the interviewed farmers.

The farmers identified environmental issues mainly related with predators and climate change as weak points. Predators seem to represent a problem for farmers in all the countries although with differences: farmers in the United Kingdom stated that their flocks have been suffering from an increase in dog attacks and they are not particularly worried about other predators, whereas wolves, foxes, bears, badgers, buzzards, ravens and crows represent a real problem for farmers located in Finland, France, Italy and Greece. This sometimes has implications on farm management, for example, some interviewed Finnish farmers have started keeping their flock inside during the summer and/or building fences.

The interviewed farmers discussed climate change in different ways. Farmers relate climate change with a number of different environmental management problems, especially water scarcity, flooding, increased animal diseases and manure management issues. A few interviewees believed that research could have a key role, but none of them mentioned any specific adaptive or mitigation measures that could be adopted on farm. Other farmers do not have a clear idea about climate change effects, as they either believe that climate goes in cycles or have not observed any significant change over time.

3.2 Solutions to Challenges

The potential innovations that can be implemented to make the industry resilient were grouped in three categories: farm management, farm technology and product processing and marketing (Table 5). The stakeholders identified potential solutions to improve marketing of sheep and goat products to increase consumption. Additionally, many innovations were related to farm management or technology that can be introduced to farms. These innovations were selected because they potentially make the industry more resilient by making it more profitable, reducing environmental impacts, improving the quality of life of farmers or improving the welfare of animals.

Table 5 Innovation that can be used to in response to industry challenges and priorities

Innovation category	Innovation
Farm management	
Pasture	Improve grazing practices, pasture and forage quality; by-products; matching animal requirements with supply
Health	Antibiotic alternatives; identification test (faecal/saliva); regionally integrated plans; welfare sensor in RFID ear tags; cortisol hair analysis
Reproduction	Improve frozen semen use; assisted reproduction techniques; reproduction plans
Breeding	Recording programmes; elite flocks; resilience traits; DNA collection and use
Farm technology	
Information and training	Integrated and easy to use data collection tools; access to abattoir feedback about carcass quality and health; tools to monitor body condition score and pasture availability; training on breeding programmes
Gadgets/apps	GPS; drones; temporary electric fencing; electronic identification systems; data collection; electronic microchip readers
Product processing and marketing	
Product processing	Low-fat/omega-3-enriched products; freeze drying for longer storage; vegetable rennet; new products (yoghurt, pudding); new, smarter packaging; new cuts for young consumers; innovation to increase halal slaughtering; animal welfare recommendations to improve product quality
Product marketing	Promote fresh sheep and goat products; improve use of environmental and social aspects of sheep farming in marketing; branding and provenance of products; attractive branding to differentiate products; certification; new recipe books; alternative markets; Reko-market

Table 6 Relation of priority innovations to sustainability

	Relation to sustainability			
Innovation type (*n*)	Economic	Environmental	Social	Welfare
Breeding and genetics (6)	6	1	2	2
Forage and feeds (4)	4	1	–	1
Handling technology (1)	1	–	–	–
Health and welfare (2)	–	–	–	2
Individual recording (6)	6	2	2	3
Training (3)	2	1	1	1
Pastures and grazing (2)	2	1	–	–
Products and marketing (2)	2	–	1	–
Reproduction (9)	9	–	–	4

Most of the stakeholders chose priority innovations based on how they can improve economic sustainability of the European sheep and goat farms (Table 6). The most important innovations were related to farm management with reproduction, breeding and genetics and individual recording the most popular selections. Only two innovations were selected related to marketing and new products. Although many of the innovations were chosen for economic benefits, many of the innovations may have environmental, social or animal welfare benefits.

4 Discussion

The multi-stakeholder approach used input from key stakeholders about how to make sheep and goat farms more sustainable. Farmers, industry representatives and researchers all reported that improvement of the market for sheep and goat products is the most important priority. Industry representatives and researchers, however, suggested breeding programmes and availability of labour as important, whilst farmers suggested the cost of production reduction, solutions for predators and animal health as more important priorities. These differences in priorities could be related to the role that each stakeholder has in the industry. Priorities for farmers tended to be related to the management of the farm whilst industry representatives and researchers tended to be related to the processing process. There was, however, a partial overlap between farmers and industry representatives and researchers about the priorities of the sheep and goat industries.

Stakeholders selected most of the innovation for their potential capability to improve farm economic performance. However, many innovations also have potential social, economic or animal welfare benefits, but these were not seen as the most important indicators. The focus on one dimension of the challenge, such as economy, is a common problem in multi-stakeholder research (Beintema et al. 2012). Despite including researchers specialised in social, environmental and animal welfare as stakeholders, the economic component was the end focus of most industry partners. This situation is similar to the "tragedy of the commons" dilemma (Hardin 1968) where people act according to their own interests, although at the common scale. In the sheep and goat sector, it would make sense to focus on sectorial challenges. Farmers will make the most logical decision based on their private interest, which in the time of crisis is to improve the profitability of their farm.

This mismatch highlights the difficulty of finding common targets across stakeholders with different interests. Although stakeholder participation is critical for collecting insights about different dimensions of the challenge (Faysse 2006), in practice this may be difficult because each stakeholder group may pursue their own interests rather than the interest of the whole sector (Leeuwis 2000). This also means that if innovations and solutions to sector challenges imply a decrease in farm productivity or additional costs, farmers will be less willing to implement the innovation or participate in research projects unless they are paid for their time and investments. Therefore, more discussion between these stakeholders may be required to come to a common consensus about what are the key sustainability issues that need to be measured when testing solutions.

Additionally, despite identifying many innovations that improve the marketability and processing of sheep and goat products, most of the priority innovations were related to the farm management because stakeholders either work with farmers to find solutions to improve farm management. These stakeholders, therefore, do not have the expertise or resources to test innovation related to marketing and processing of animal products. This requires more input from stakeholders related to this part of the sector.

The multi-stakeholder approach used in this study included many different stakeholders and adopted different methods. Semi-structured interviews were the most successful at getting in-depth information about priorities and challenges. The workshop discussions were successful at getting diverse ideas about all topics, but the input from participants was not always equal. Additionally, during the workshops, it was difficult to record in-depth information about the reason for the feedback. The online survey was the least successful at getting diverse opinions.

Successful multi-stakeholder research requires diverse methods including qualitative and qualitative data to be robust (Spielman 2005). Therefore the workshop surveys and semi-structured interviews could be complemented with more quantitative data about the problem faced by each stakeholder. Also, input from farmers was collected separately from those of researchers and industry representatives. Perhaps including all stakeholders together would have encouraged more interaction between the groups, potentially producing a deeper analysis of the challenge and solutions (Schut et al. 2015).

Farmers were interviewed separately from researchers and industry representatives to ensure that their responses were not influenced. The mixing of stakeholders needs to ensure that everyone is comfortable to contribute without feeling social pressure from others (International Institute for Sustainable Development 2014). Despite this, including farmers in discussions with industry and research representatives would improve the approach. Additionally including stakeholders related to the marketing and processing of sheep and goat products may have helped tease out more innovations that can be tested on farms to increase product consumption.

The workshop discussions were part of a larger meeting, and each discussion was between 2 and 2.5 h. Schut et al. (2015) suggest full-day meetings for complex agricultural issues to ensure that the appropriate themes are identified and discussed in full. Schut et al. (2015) also suggest that workshops, surveys and interviews should be used together across stakeholders to collect an exhaustive list. Therefore, a strongest involvement of farmers in the discussions with industry representatives and researchers could facilitate the identification of innovations.

5 Conclusions

Stakeholders of sheep and goat sectors in Europe reported that improvement of the market for sheep and goat products is the most important priority. Apart from this priority, there were different priorities for farmers compared to researchers and other industry stakeholders. Additionally, despite identifying improved markets as the biggest priority, the most important innovations that industry representatives believe should be tested are for breeding, collection of farm data and reproduction. This mismatch is probably because these stakeholders do not have the capacity to test innovation or solutions related to marketing and processing of products and therefore believe it is the biggest priority.

Acknowledgements This work is part of the iSAGE Project (http://www.isage.eu/), with financial support from the European Union's Horizon 2020 research and innovation programme under grant agreement No 679302. We also thank Professor Raffaele Zanoli for providing access to the Qualtrics survey platform (Qualtrics 2016), which was used for the development of the on-live survey used in this work.

References

Beintema N, Stads GJ, Fuglie K, Heisey P (2012) ASTI global assessment of agricultural R&D spending. International Food Policy Research Institute, Washington DC, Agricultural Science and Technology Indicators, Rome, Italy and the Global Forum on Agricultural Research, Rome, Italy

Dubeuf JP (2011) The social and environmental challenges faced by goat and small livestock local activities: Present contribution of research–development and stakes for the future. Small Rumin Res 98(1):3–8

Dubeuf JP, Sayadi S (2014) Multi-functionality issues for small ruminants: What changes are needed in territorial public policies and training?: Report of two round tables on territorial issues and training for the development of goat farming. Small Rumin Res 121(1):136–145

Faysse N (2006) Troubles on the way: an analysis of the challenges faced by multi-stakeholder platforms. Nat Res Forum 30:219–229

Funtowicz SO, Ravetz JR (1993) Science for the post-normal age. Futures 25:739–755

Giller KE, Tittonell P, Rufino MC, van Wijk MT, Zingore S, Mapfumo P, Adjei-Nsiah S, Herrero M, Chikowo R, Corbeels M, Rowe EC, Baijukya F, Mwijage A, Smith J, Yeboah E, van der Burg WJ, Sanogo OM, Misiko M, de Ridder N, Karanja S, Kaizzi C, K'ungu J,

Mwale M, Nwaga D, Pacini C, Vanlauwe B (2011) Communicating complexity: integrated assessment of trade-offs concerning soil fertility management within African farming systems to support innovation and development. Agric Syst 104:191–203
Hardin G (1968) The Tragedy of the commons. Science 162:1243–1248
International Institute for Sustainable Development (2014) Rapid Rural Appraisal (RRA)
Leeuwis C (2000) Reconceptualizing participation for sustainable rural development: towards a negotiation approach. Dev Chang 31:931–959
McNie EC (2007) Reconciling the supply of scientific information with user demands: an analysis of the problem and review of the literature. Environ Sci Policy 10:17–38
Qualtrics (2016) Version 3. Copyright © 2017 Qualtrics. Qualtrics and all other qualtrics product or service names are registered trademarks or trademarks of Qualtrics, Provo, UT, USA. http://www.qualtrics.com
Schut M, van Paassen A, Leeuwis C, Klerkx L (2014) Towards dynamic research configurations: a framework for reflection on the contribution of research to policy and innovation processes. Sci Public Policy 41:207–218
Schut M, Klerkx L, Rodenburg J, Kayeke J, Hinnou LC, Raboanarielina CM, Adegbola PY, van Ast A, Bastiaans L (2015) RAAIS: Rapid Appraisal of Agricultural Innovation Systems (Part I). A diagnostic tool for integrated analysis of complex problems and innovation capacity. Agric Syst 132:1–11
Spielman DJ (2005) Innovation systems perspectives on developing-country agriculture: a critical review, ISNAR Discussion Paper 2. IFPRI, Washington, p 58
Sunding D, Zilberman D (2001) The agricultural innovation process: research and technology adoption in a changing agricultural sector. Handb Agric Econ 1:207–261

Effects of High Stocking Grazing Density of Diverse Swards on Forage Production, Animal Performance and Soil Organic Matter: A Case Study

Konstantinos Zaralis and Susanne Padel

Abstract Mob grazing is regarded as a grazing management practice to increase soil organic matter, pasture productivity and nutrient cycling. There are different perspectives in the literature regarding the definition of *mob grazing*, but it is generally accepted that mob grazing is characterised by high stocking densities of livestock which are moved frequently from paddock to paddock (e.g. with the aid of electric fences), trampling forage into the soil as they graze. It has also been recognised that biodiverse pastures have the potential to build up carbon levels in the soil much more effectively than conventional (usually monocultures) or less diverse pastures; in turn all can enhance animal productivity and maintain good herd health. This paper reviews the concept of mob grazing and the benefits of diverse swards and provides evidence whether high stocking density as a grazing strategy can increase soil organic matter and enhance overall animal performance. The grazing rotation applied in the farm during the study year was rather short to fulfil the expectations of a mob-grazing system, but stocking density was high (115 t LW ha^{-1}). The results show that high stocking grazing density of biodiverse pastures has a remarkable effect on the build-up of the soil organic matter and that biodiverse pastures serve as a viable alternative to conventional pastures as they can maintain animal productivity at high levels.

Keywords Mob grazing · Intensive grazing · Dairy cows · Organic milk · Diverse swards · Soil organic matter

K. Zaralis (✉)
School of Agricultural Technology, Food Technology and Nutrition, Highest Technological Educational Institute of Western Macedonia, Florina, Greece
e-mail: kzaralis@teiwm.gr

S. Padel
The Organic Research Centre, Newbury, Berkshire, UK

A. Theodoridis et al. (eds.), *Innovative Approaches and Applications for Sustainable Rural Development*, Springer Earth System Sciences,
https://doi.org/10.1007/978-3-030-02312-6_8

1 Introduction

Over the last 60 years, the drive for high levels of production had led many farmers to overlook the importance of maintaining soil organic matter and the potential benefits associated with the use of herbal leys and mixed swards in their pastures (Foster 1988). Soil organic matter serves as a reservoir of nutrients for plants, provides soil aggregation, increases nutrient exchange, retains moisture, reduces compaction, reduces surface crusting and increases water infiltration into soil, which in turn all can improve animal productivity and maintain good herd health. The build-up of soil organic matter can be influenced by the way in which the sward is managed in terms of grazing (Wright et al. 2004; Wang et al. 2014) but also by the diversity of the plant species of the swards (Döring et al. 2013). The "Holistic Resource Management" (or the Savory Grazing Method; SGM) that was introduced in the 1980s by Allan Savory (1983) has renewed farmers' interest in increasing soil organic matter on farm land through grazing techniques. According to Savory, SGM is a management technique that has been developed to overcome the desertification problem for America and should not be seen or should not be used as a name for any particular grazing/stocking method (Allen et al. 2011). However, since its introduction, SGM has been altered and intensified by producers until it has given rise to a method of grazing known as ultrahigh stocking density grazing or mob grazing (Redden 2014). Mob grazing in the UK is usually applied in diverse swards as these leys are known to promote microbial activity resulting in increased soil carbon levels and building of humus in the soil. The objectives of this study were to (a) provide an insight to the concept of mob grazing and the beneficial effects of diverse swards on animal performance and soil organic matter and (b) quantify the effects of a mob-grazing technique on forage productivity, species diversity, forage utilisation as well as animal performance and soil quality on a case study farm.

1.1 Mob Grazing and Claimed Benefits

Literature appears to be scarce when it comes to scientific publications related to mob grazing as grazing management technique and its associated claimed benefits. The Forage and Grazing Terminology Committee (FGTC 1991) defines mob grazing as "In the management of a grazing unit, grazing by a relatively large number of animals at a high stocking density for a short time period", while Allen et al. (2011) in an extensive categorisation of grazing methods use the term "mob stocking" instead of "mob grazing" and refer to it as "a method of stocking at a high grazing pressure for a short time to remove forage rapidly as a management strategy". While both definitions aim to describe this grazing technique, they lack the specifics needed to adequately define mob grazing in terms of "stocking density" (i.e. heads/ha) or "short time period" (Anderson 2015). Among farmers, other terms are also used such as "short duration grazing" or "tall grass grazing" (R. Richmond, personal communication, 2014). Redden (2014) reports that stocking densities in mob grazing (i.e. ultrahigh stocking density grazing) often exceed that of 200.000 kg of LW/ha up to 1.000.000 kg of LW/ha, but there is not an international standard on this nor on the frequency by which animals should move to a new paddock within a day. Typically, in a mob-grazing system as applied in the USA or Americas, each paddock is grazed only once per growing season (Redden 2014). However, in the European continent, the available grazing rangeland of typical dairy/beef farm is much smaller in size and therefore is not possible for a farmer to apply one grazing rotation over the growing season. The inconsistency or inadequacy in the terminology being used to describe the concept could reflect the fact that mob grazing has emerged from Allan Savory's Holistic Resource Management under similar, but not identical, concepts from different practitioners around the world. This inconsistency could have also led to different techniques being researched from those being used by farmers as proposed by McCosker (2000) for the cell grazing technique.

It is generally accepted that the overall goal of mob grazing is to improve soil organic matter and diversification of the vegetation composition in the long-term (Clatworthy 1984; Savory and Butterfield 1999; Redden 2014; Bartimus et al. 2016). Restricting grazing animals to very small paddocks in order to achieve high stock densities results in even distribution of grazing pressure, hoof action and excreta across a pasture forcing animals to graze the entire area of the pasture they are allotted (Peterson 2014; Peterson and Gerrish 1995). It is regarded that even distribution of excreta can potentially reduce the need for artificial fertilisers, reducing production costs, while nutrients become readily available for use by the soil and plants (Peterson and Gerrish 1995). In addition, there is a growing notion that trampling of significant quantities of forage onto the soil surface by intensified hood action incorporates plant litter, live plant material and excreta into the soil increasing soil organic matter inputs and nutrient cycle efficiency (Savory 1983). This reportedly results in increased utilisation and increased harvest efficiency which can increase pastures carrying capacity by 25–100% (Savory and Parsons 1980; Stuth et al. 1981; Gompert 2010). Moreover, the advocates of mob grazing claim that as animals are moved from paddock to paddock multiple times each day or at least once per day often results in that much of the vegetation remains ungrazed (or not fully grazed) allowing the whole host of plant species to grow taller and form large, complex and deep root systems which in turn contribute to the increased organic matter in the soil (Savory and Butterfield 1999). Practitioners in the USA and UK suggest a wide variety of benefits from mob grazing including increased forage production, improved distribution of livestock grazing and increased soil function and plant diversity (Gompert 2010; Peterson 2010; Richmond 2011; Chapman 2012). Uneaten plant stems are trodden onto the soil surface, and these stalks act both as mulch and as a feed source for the soil microorganisms, building new soil in the process. It is regarded that the long recovery time leads to high volumes of above-ground forage, a mixture of leaf, seed and stem. It is also claimed that the high stocking density results in more than 50% of the plant being trampled into the ground by the animals (Chapman 2012; Richmond 2011). Practitioners who use mob-grazing methods report increases in vegetation production of 100–300% as well as improved plant diversity which they believe to be a result of improved soil function and fertility (Gompert 2010).

Published work on such grazing systems to date has mainly been carried out in arid areas. According to Savory and Butterfield (1999), this management has been shown to provide environmental improvements on previously overgrazed areas in Africa, Australia and America, through the return of organic matter to the soil. Clatworthy (1984) was the earliest publication which reported that planned grazing under mob stocking principles in Rhodesia doubled the number of animals which an arid area could carry, compared with a government grazing system (not defined), with no deterioration of the plant community. In another arid region, South Idaho, Weber and Gokhale (2011) demonstrated a statistically significant increase in soil moisture retention under "holistic planned grazing" (i.e. 3-day grazing at high stocking density) compared with both total rest of land and with a 30-day grazing with a lower stocking density.

The claimed benefits of mob grazing on soil organic matter and animal performance have not been studied in scientifically robust experiments, and this gap in scientific knowledge is reflected in the literature. While research has shown that grazing increases productivity of grasslands compared to non-grazed grasslands (Wright et al. 2004; Patton et al. 2007; Wang et al. 2014), none of the published work so far has used stocking densities in trials as high as those used in the mob-grazing technique. In addition, research conducted on the effects of grazing at lower densities among varied grazing systems has produced varied results, not all of which support the purported benefits of mob grazing (Redden 2014). In the UK, there is a small but growing interest in the mob-grazing technique especially because it is claimed to increase soil organic matter, but no quantitative has ever been carried out to date; there is therefore some uncertainty about the levels of production that may be achieved.

1.2 Benefits of Diverse Swards

There is ample of evidence in the literature showing the potential of diverse pastures to meeting the dual goal of reducing the environmental impact and the financial performance of sheep, dairy and beef farms (Andueza et al. 2013; Roca-Fernández et al. 2016; Romera et al. 2017; Vogeler et al. 2017). In a comprehensive review on the potential use of legume-based grasslands for livestock production in Europe, Lüscher et al. (2014) point out that such grasslands undoubtedly constitute a major element for more sustainable and competitive ruminant production systems. Seither et al. (2012) showed that diverse swards generally produce biomass of higher nutritive value than dominated grass swards.

In the UK, information about the performance of diverse swards and the benefits of mixing species comes from the relatively recent research project (Döring et al. 2013). In replicated field experiments in multiple locations across the UK, the performance of 12 legume and 4 grass species that were sown in monocultures or as a mixture was compared to farmer-chosen ley mixtures. The project provided evidence that species-rich legume-based leys can maximise pasture productivity and other ecosystem services, while functional diverse plant species can be optimised and fine-tuned to farm-specific needs (Döring et al. 2013). Specifically, diverse swards had increased above-ground biomass and greater stability of biomass production compared to monocultures, while productivity increased over time. In addition, they had greater resilience to adverse weather, climate and management conditions. According to Döring et al. (2013), mixes with high agronomic productivity contain both lucerne (*Medicago sativa*) and white clover (*Trifolium repens*), while the overall performance of a mix ley improves by including a third or fourth legume species. The three best multifunctional mixtures all contained black medic (*Medicago lupulina*), red clover (*Trifolium pratense*) and lucerne (Döring et al. 2013). Red clover is generally more productive than white clover, but less persistent and less tolerant to high grazing pressure than white clover (Gaudin et al. 2013). Due to its long roots, red clover is regarded as drought tolerant (Knight et al. 2008). Lucerne also produces high-quality feed, when dried or ensiled, although it appears to be relatively sensitive during establishment; thus rerepeated grazing or grazing at early stages of development should be avoided (Rochon et al. 2004; Blanco et al. 2010). In terms of grass species, ryegrass (*Lolium perenne*) and Italian ryegrass (*Lolium multiflorum*) are recommended for high yield, while festulolium, which is a cross between a fescue (*Festuca*) and ryegrass, provides a combination of high-quality forage with good winter hardiness, persistence and stress tolerance (Döring et al. 2013).

Milk production in dairy cows fed forage from legumes is often equal to, if not greater than, in cows fed diets that do not contain legumes (Rouille et al. 2016). Higher milk production (+0.8 kg/day), milk solids production (+0.04 kg/day) and pasture DM intake (+1.5 kg DM/day) were observed in trails with dairy cows grazing multispecies swards than mixed swards particularly during summer where herb proportion was the highest (Roca-Fernández et al. 2016); the authors attribute these outcomes to the cumulative effect of improved pasture nutritive value and increased pasture DM intake. Lindström et al. (2013) have shown that legumes and herbs (i.e. red clover, white clover and chicory) can provide considerable higher amounts of minerals per kg DM of grazed forage compared to grasses, which is particularly important for pasture-fed animals especially in organic or low-input conventional farms. Other studies have shown that feeding dairy cows on multispecies swards containing forbs presents an opportunity to reduce N losses without compromising milk yield (Bryant et al. 2017). The use of diverse swards appears therefore to be a cost-effective way to reduce nitrogen leaching, which is relevant for a dairy sector facing regulatory constraints (Romera et al. 2017). In in vivo digestibility trials with sheep, Andueza et al. (2013) have found that swards rich in forbs are of higher digestibility in early season compared to swards rich in grasses, indicating nutritional advantages of mixed swards. Most of the mixed leys used on farms also include herbs which are a good source for minerals (Pirhofer-Walzl et al. 2012), while other plants have medicinal properties. For example, chicory and sainfoin are known to have anthelmintic properties and can help to reduce drug use for internal parasitic control at a farm level, while species with high tannin content can enhance the rumen bypass protein intake in ruminants (Hoste et al. 2015). In this study we assessed the establishment and the productivity of diverse swards compared with ryegrass-white clover mixtures on a commercial dairy farm and to determine how the evolution of the species mix varies with grazing conditions and soil type.

2 Materials and Methods

2.1 Selection of the Case Study Farm

Because mob grazing involves high stocking density for a short period of time with long recovery times between consequent grazing, there is some uncertainty as to how applicable is this grazing management approach under UK conditions. To evaluate the effect of mob grazing on soil organic matter and animal performance, this study collected soil, forage and animal production data from a case farm in which mob grazing on diverse swards is being used since 2007 as a method to increase the organic matter of soil. The particular farm was selected for the case study because the farmer has studied in the USA (scholar of Nuffield Farming Scholarships Trust) practical aspects of management relevant to the issue of increasing soil carbon by using mob grazing as an approach to address this. The farmer believed that longer intervals between rotational grazing are likely to be best suited to the diverse swards that are established in his farm than to those based on ryegrass which are currently typical of UK dairy farms. According to the farmer, mob grazing holds the key to improving and maintaining soil fertility and forage productivity in his organic system. The farmer notes: "The best way to rebuild soil carbon levels is by the rotational grazing of bio-diverse pastures. The stable environment under the ley allows the biology to establish in the soil, whilst promoting the plant's ability to exude large amounts of sugars through its roots (up to 70% of what it produces). This provides a ready food source for the microbes, resulting in the ability to increase soil carbon levels by one per cent every three years (20t carbon/ha/yr). From my studies I have determined that the fertility of a soil is its ability to hold and recycle nutrients and water in a plant available form. To do this, a soil needs to be biologically active and fed a range of foodstuffs – a combination of rapidly digestible green plant material/animal slurries and slower digestible crop residues and farm yard manure. The biology in the soil is responsible for breaking down this material, releasing the nutrients from it, and building humus".

The farm is located in the Cotswolds, near Gloucestershire, UK, and it is a 220 ha mixed dairy/arable farm at approximately 260 m above sea level. The farm has a long history of arable use in many fields and was converted to organic production in 2005. The herd consists of 189 Friesian-Shorthorn cross dairy cows that are spring calving, with a lactation period of 300–310 days. Full-time housing of the cows is limited to 2 months (i.e. December and January). Kale (*Brassica oleracea*) and

Table 1 Seed mixture composition (% w/w) used in the study farm

Plant species	% weight
Grasses	
Cocksfoot *(Dactylis glomerata)*	7.8
Creeping red fescue *(Festuca rubra)*	5.4
Crested dogstail *(Cynosurus cristatus)*	1.6
Italian ryegrass *(Lolium multiflorum)*	7.7
Meadow fescue *(Phleum pratense)*	7.0
Perennial ryegrass *(Lolium perenne)*	7.8
Meadow grass *(Poa pratensis)*	1.6
Tall fescue *(Festuca pratensis)*	3.9
Timothy *(Festuca arundinacea)*	5.3
Yellow oatgrass *(Trisetum flavescens)*	1.6
Legumes	
Alsike clover *(Trifolium hybridum)*	1.6
Sainfoin *(Onobrychis viciifolia)*	20.2
Sweet clover *(Melilotus officinalis)*	5.4
Birdsfoot trefoil *(Lotus corniculatus)*	2.3
Red clover *(Trifolium pratense)*	2.3
White clover *(Trifolium repens)*	2.3
Herbs	
Burnet *(Sanguisorba minor)*	6.2
Chicory *(Cichorium intybus)*	5.4
Ribgrass *(Plantago lanceolata)*	0.8
Sheep's parsley *(Petroselinum crispum)*	2.3
Yarrow *(Achillea millefolium)*	1.5

fodder beet (*Beta vulgaris*) are grown for additional winter grazing. The farmer introduced a mob-grazing approach on diverse swards with the aim of increasing soil organic matter. Leys were reseeded as part of the rotation every 5 years with a diverse sward mixture that includes ten different grass species, six species of legumes and five herb species (Table 1). The best method for establishing the long-term diverse ley in the farm was found to be sowing under a spring cereal crop.

2.2 Data Collection and Sampling

Data were collected from April to September during the study year, while preliminary assessments were also carried out the year before.

Determination of Pasture Productivity, Forage Composition and Feed Intake Estimation

Herbage yield and composition of the diverse swards were assessed on a monthly basis in the same field, which was representative of the type and the phenological stage of the swards across the farm. The square-metre quadrat method was used to determine the productivity of the grazed pasture and thus allowed estimates of dry matter (DM) intake of the grazing cows. Briefly, a one-square-metre quadrat was placed randomly three times across the ungrazed plot (i.e. this was the next plot to be grazed by the animals within the next 12 h). All the vegetation within the quadrat area was cut to approximately 5 cm height, and the cut of above-ground biomass from each quadrat was collected in separate bags; fresh weight was recorded. The same procedure was followed also in the residual forage in the grazed plot (i.e. this was the plot the animals had just grazed).

Dry matter content of the herbage production both in the ungrazed and grazed plots was determined in house using a forced air oven for drying samples for 16 h (or overnight) at 100 °C. Differences in DM productivity between ungrazed and grazed plots were used to estimate forage DM intake of the grazing cows. Subsamples of the harvested forage were separated into four categories as follows: (a) grass, (b) clover, (c) other legumes and broadleaves and (d) senescent material for determination of sward composition. Additional herbage subsamples were analysed by wet chemistry for metabolisable energy (ME) and crude protein (CP) content by Sciantec Analytical Services, Stockbridge Technology Centre, Bishopdyke Rd, Selby YO8 3SD, UK.

Monitoring of Farm Records and Additional Calculations

At the end of the monitoring period, the farmer provided data on milk production and milk composition, grazing records (i.e. area and livestock numbers grazed daily over the monitoring period) as well as information on supplementary feeding records such as forage and concentrate supplementation, amounts and periods fed. These data in addition to chemical analysis data were used to estimate the ME intake of the cows over the monitoring season from the given field. Calculated data from the sampled field were extrapolated to provide an estimate of the total forage production in the whole farm for the study period.

Soil Samples

Historic data on the organic matter content of soil from three different fields are available from 2007 to 2012. Soil samples were taken again in autumn of 2013 and 2014 and in spring of 2015 from the same fields to assess the differences in soil organic matter. As changes in soil organic matter are likely to be slow, maximising the time between consecutive samplings may increase the likelihood of detecting a change.

3 Results and Discussion

3.1 Pasture Productivity and Herbage Composition

The productivity of the swards during the monitoring period averaged 10.3 tonnes per ha. Herbage composition and monthly productivity of the diverse swards grazed by the cows from April to September are presented in Fig. 1. Herbage production increased from April to August, while the composition of the herbage in summer months remained relatively constant as shown in Fig. 1. Clover production accounted for about 24%, while grass production accounted for 71% of the total herbage production. The productivity of other legumes and "broad leaves" represented 6% of the total production, and senescent material remained below 3% of the total production. Preliminary data collected in 2013 indicated a similar productivity of the grazed sward, but herbage composition fluctuated between months (data not shown).

Forage samples, representative of the whole filed, which were collected at farm visits during April–September 2014 were analysed for chemical composition. The grazing diverse sward had an average of 19.1% DM, 10.8 MJ of ME, 21.3% of CP and 376 g of NDF indicating a good-quality forage. The average ME content was marginal as normal values for this type of forage are 11–13 MJ of ME per kg DM, but CP content was high and NDF within the expected levels.

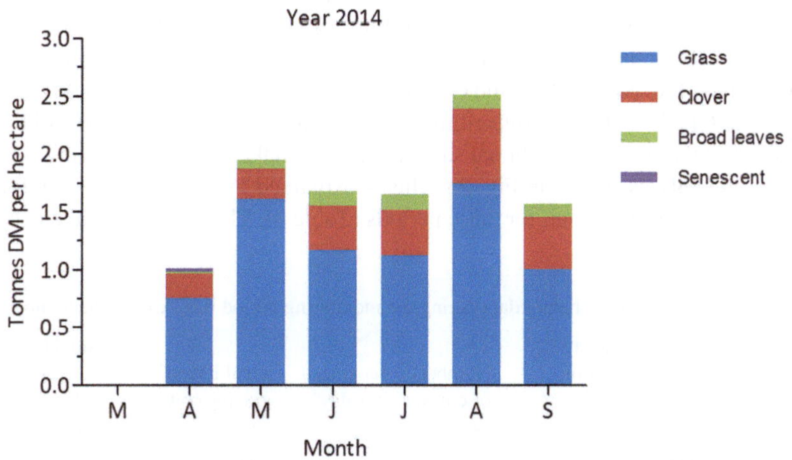

Fig. 1 Monthly above-ground biomass production (DM t*ha^{-1}) and vegetation composition in the study area

3.2 Grazing Data, Feed Intake and Cow Productivity

Grazing data from the 3rd of April to the 25th of September 2014 are presented in Table 2. On average 181 milking cows grazed a diverse sward field of total area of 12.5 ha for a total of 43 days in monthly rotation intervals. The duration of the grazing varied from 6 to 10 days based on herbage availability. The cows were moved on twice a day after each milking, grazing two adjacent plots of an average size of 0.9 ha delimited by electric fences. The average stocking density over the grazing period was 115 tonnes of livestock per hectare. The resting period of the diverse sward between consecutive grazing averaged about 21 days with 16 and 25 days the shortest and the longest, respectively. These resting periods do not coincide with the principles of "mob grazing" where resting periods are of long duration (i.e. more than 50 days) but the stocking density was relatively high (Table 2). In year 2013 the farmer was applying a 40–50-day rotation management allowing the pastures to recover for longer, but the total forage productivity was similar to 2014.

Daily ME requirements were calculated for an average LW of 550 kg and include ME requirements for maintenance (i.e. 65 MJ ME), reproduction (i.e. 26 MJ ME) and monthly milk yield based on monthly average milk fat and milk protein content. The estimated grazed intake per cow per day in each month and the calculated ME intake are presented in Table 3. Over the period the average daily grazed intake per cow was 17 ± 1.9 kg DM, but it fluctuated from as little as 10.9 kg DM in July up to 23.8 kg DM in August. The average daily concentrate supplementation per cow was 2.9 ± 0.29 kg DM ranging from 4.3 kg DM in April to 2.2 kg DM in September. The estimated ME intake from the forage in addition to the ME intake from the supplementary feed (i.e. Natural Organic Green HDF 18 Nuts (BOCM PAUL LTD), 862 g DM, 18% CP, 13 MJ ME) covered the daily ME requirements of the cows in most months, but there was a nutritional shortfall in ME intake during the grazing periods in June and July as shown in Fig. 2a. This is explained by the relatively low forage DM intake that is estimated for these periods (Table 4). The low DM intake is likely

Table 2 Summary of the grazing data during the monitoring period (i.e. April to September)

Grazing period		Grazing duration	Number	ha grazed	Total LW of	Stocking density
From	To	(days)	of cows	per day[a]	grazing cows (t)[b]	(t LW/ha)
03 Apr	09 Apr	6	150	2.08	82.5	79.2
05 May	12 May	6	180	2.08	99.0	95.0
02 Jun	11 Jun	8	189	1.56	104.0	133.1
05 Jul	23 Jun	10	189	1.25	104.0	166.3
09 Aug	19 Aug	6	189	2.08	104.0	99.8
12 Sep	25 Sep	7	189	1.79	104.0	116.4
On average		7	181	1.81	99.6	115.0

[a]The total area was not grazed at once but it was divided into two plots
[b]Assuming a cow LW of 550 kg

Table 3 Average monthly milk production and composition during April to September 2014 and calculated ME requirements per cow

Month	Milk fat %	Milk protein %	Milk yield kg	ME req for milk
April	3.47	3.15	24	113
May	3.40	3.22	26	122
June	3.53	3.21	25	119
July	3.49	3.19	22	104
August	3.69	3.27	20	97
September	3.98	3.43	17	87

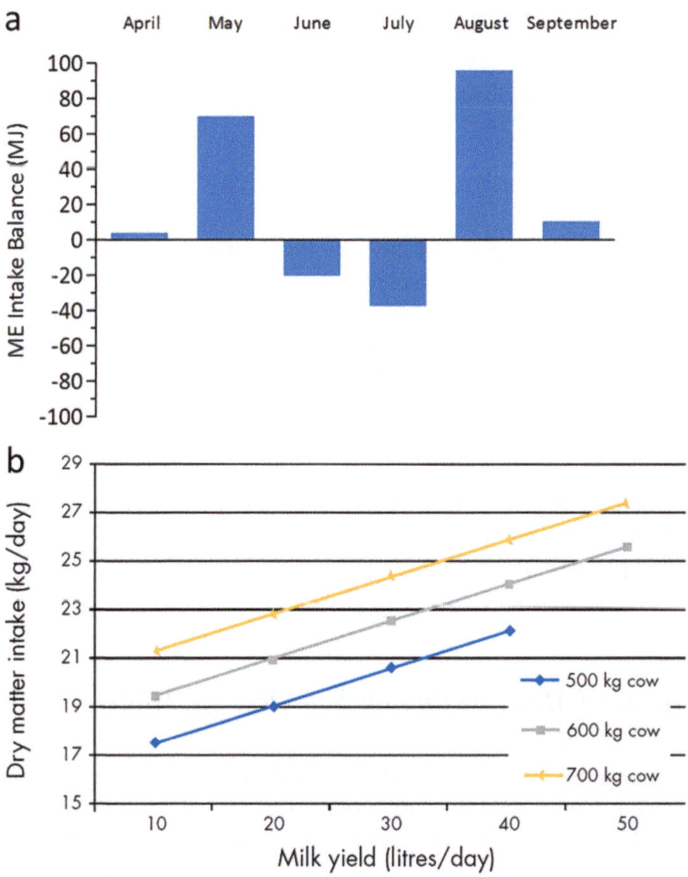

Fig. 2 Monthly estimated ME intake balance of the cows from April to September 2014 (Panel **a**); relationship between dry matter intake and daily milk yield (Panel **b; Source:** AHDB, UK feed into Milk 2005)

Table 4 Estimated feed (kg DM) and energy intake (MJ) per cow per day during the grazing period from April to September 2014

Month	Estimated grazed intake		Supplementary feed intake[a]				
	DM intake (kg)	ME intake (MJ)	Kg DM	ME intake (MJ)	Total ME intake	Total ME requirements	Energy balance
03–09 April	14.0	152	4.3	56	208	204	4
05–12 May	22.6	244	3.0	39	283	213	70
02–11 June	13.9	150	3.0	39	189	210	−21
05–23 July	10.9	118	3.0	39	157	195	−37
09–19 August	23.8	257	2.2	28	285	188	96
12–25 September	14.8	160	2.2	28	188	178	11

[a]Natural Organic Green HDF 18 Nuts (BOCM PAUL LTD), 862 g DM, 18% CP, 13 MJ ME
[b]Assuming a cow LW of 550 kg

attributed to the low forage availability (see Fig. 1) which in turn is attributed to the fact that the farmer applied a relatively short grazing rotation scheme and the pasture was not allowed to recover fully. Nevertheless, milk yield did not seem to have been compromised (Table 4) by the relatively low intakes estimated for these days which suggests that subsequent grazing in the next field in rotation allowed for good DM intakes. It is well established that milk production has a linear positive relationship with DM intake as cows produce more milk at higher intakes (see Fig. 2b). Over the monitoring period, the daily DM intake per cow averaged 19.6 kg DM, while the daily milk yield averaged 22.3 kg. These intake and productivity data are consistent with each other and are in accordance with the predictions postulated by the literature and illustrated in Fig. 2b.

3.3 *Effects of Mob Grazing on Soil Organic Matter*

It is well recognised that grassland soils with low organic matter content are characterised by poor fertility, are prone to compaction and flooding and are droughty and lacking soil microorganisms. As the organic matter rises and the soil becomes more fertile, the land grows more forage, and the capacity of the land to carry higher stocking rates increases. It has been suggested that rotational grazing of biodiverse pastures has the potential to build up carbon levels in the soil. The underlying hypothesis is that the stable environment under a diverse ley promotes the plant's ability to exude large amounts of sugars through its roots which in turn provides a food source for microbial activity resulting in increased soil carbon levels and building humus. The advocates of mob grazing suggest that this grazing system, by allowing plants to grow taller, results in the formation of large, complex and deep root systems and when they die off, they leave high amounts of organic matter in the soil. It is also advocated that trampling of significant quantities of forage onto the soil surface provides a better environment for the microorganisms and other soil life and increasing the soil organic matter.

In the case study farm, the mob-grazing approach on diverse swards was introduced with the aim of increasing soil organic matter. Despite the fact that monitoring

Table 5 Soil analysis results in three different fields (i.e. Big Aero, Lanes Estate, Pinchins)

Analysis factor[a] Field	Big Aero		Lanes Estate		Pinchins	
Year	2007	2015	2012	2015	2012	2015
Standard soil analysis						
Soil pH	7.2	6.9	7.9	7.9	7.7	7.8
Phosphate (mg/l)	13	7	16	15	6	10
Potash (mg/l)	123	131	247	180	107	154
Magnesium (mg/l)	107	138	106	101	123	124
Physical soil structure (%)						
Sand (%)	7	15	21	17	15	20
Silt (%)	55	39	46	41	43	36
Clay (%)	38	46	33	42	42	44
Macronutrients						
Organic matter %	**4.4**	**9.8**	**5.3**	**7.8**	**5.7**	**8**
Microbial activity	13	25	33	22	27	23
Sulphate (mg/l)	35	56	37	29	49	58
Total phosphorus	906	901	1025	1244	841	1069
Chemical						
CEC (meq/100)	41.2	36.0	28	30.5	32.9	30.7
Calcium (%)	84.7	78.9	88.1	88.2	87.7	86.1
Magnesium (%)	1.6	3.8	3.6	4	3.1	4.9
Ca/Mg ratio	53	21	24	22	28	18
Potassium (%)	0.8	1.2	1.8	1.9	1.1	1.6
Sodium (%)	0.2	0.4	0.5	0.3	0.4	0.7
Hydrogen (%)	0.0	0.0				
Others (%)	11.5	15.7	8.1	5.6	8.7	6.7
Trace elements (mg/l)						
Iron	33	64	40	57	49	65
Molybdenum	0.4	0.40	0.10	0.3	0.3	0.4
Copper	1.2	1.90	1.7	2.3	1.6	1.6
Selenium	0.68	0.68	0.39	0.38	0.53	0.46
Zinc	1.9	2.0	2.3	2.7	1.4	1.7
Manganese	11.5	22.8	11.8	14.1	16.6	24.6
Cobalt	0.3	0.3	0.2	0.2	0.3	0.4
Boron	1.10	1.60	1.5	1.4	1.3	1.3
Conductivity	2029	1913	2099	1977	2046	2056

[a]Samples were analysed by Kingshay Analysis Services, Kingshay, Bridge Farm, West Bradley, Glastonbury, Somerset BA6 8 LU

of the performance of the diverse swards was conducted only in one field (i.e. Big Aero), soil samples were collected in 2015 from three different fields (i.e. Big Aero, Lanes Estate, Pinchins) and compared with earlier results from 2007 to 2012. The results of the soil analyses in these fields are shown in Table 5.

Soil organic matter increased by 122.7%, 47.2% and 40.4% in Big Aero, Lanes Estate and Pinchins fields, respectively. The relative higher increase in soil organic matter in the Big Aero is attributed to the fact that samples collected in 2015 are compared with those collected in 2007 (i.e. 8 years earlier) and not in 2012, which is the case in the other fields (i.e. 3 years earlier). Yet, this is a marked improvement with more than double the levels of organic matter reserve. The build-up of the soil organic matter is also remarkable in the other fields as well. The overall soil analysis data suggest that soil improvement management through rotational high stocking grazing of biodiverse pastures appears to have a beneficial impact on soil organic matter. Microbial activity in the soil does not seem to have been improved considerably over the years, but it can be accelerated by biotreatment of slurry or farmyard manure in short term. In all fields tested, soil trace element status is generally low which may indicate a potential need for zinc and copper supplementation in the ration of the herd.

4 Conclusions

The results of this case study show that biodiverse pastures are sufficiently productive to serve as a viable alternative to conventional pastures (i.e. grass/clover pastures) as they can maintain animal productivity at high levels. Although the farmer claims that the grazing system he applies in his farm falls within the principles of "mob grazing", the average 21-day rotation he applied during the study period is regarded as rather short to allow plants to grow to a desired height that fulfils the expectations of mob grazing. However, it should be acknowledged that grazing rotations were longer in the previous years, while stocking density always remains high. This study indicated that the build-up of the soil organic matter is remarkable and suggests that soil improvement can be achieved through high stocking rotational grazing of biodiverse pastures.

Acknowledgements We thank Rob Richmond for this valuable input to this work. This study was undertaken in the European research project "SOLID; Sustainable Organic and Low-input Dairying" (Agreement no. 266367; http://www.solidairy.eu/), with financial support from the 7th Framework Programme.

References

Allen VG, Batello C, Berretta EJ, Hodgson J, Kothmann M, Li X, McIvor J, Milne J, Morris C, Peeters A, Sanderson M (2011) An international terminology for grazing lands and grazing animals. Grass Forage Sci 66:2–28

Anderson B (2015) When mob grazing is not mob grazing. Progressive Forage, Progressive Publishing. Available on-line: https://www.progressiveforage.com/forage-types/grasses-and-grazing/when-mob-grazing-is-not-mob-grazing

Andueza D, Picard F, Jestin M, Aufrère J (2013) The effect of feeding animals ad libitum vs. at maintenance level on the in vivo digestibility of mown herbage from two permanent grasslands of different botanical composition. Grass Forage Sci 68:418–426

Bartimus HL, Montgomery TG, Philipp D, Cater J, Coffey KP, Shanks BC (2016) 118 Mob grazing effects on cattle performance in Southeast Arkansas. J Anim Sci 94:55–55

Blanco M, Casasús I, Ripoll G, Panea B, Albertí P, Joy M (2010) Lucerne grazing compared with concentrate-feeding slightly modifies carcase and meat quality of young bulls. Meat Sci 84:545–552

Bryant RH, Miller ME, Greenwood SL, Edwards GR (2017) Milk yield and nitrogen excretion of dairy cows grazing binary and multispecies pastures. Grass Forage Sci 72:806–817

Chapman T (2012) Are mob grazed cattle the perfect arable break? A Nuffield Farming Scholarships Trust report. John Oldacre Foundation, UK

Clatworthy JN (1984) The charter estate grazing trial: results of the botanical analysis. Zimb Agric J 81(2):41–48

Döring TF, Baddeley JA, Brown R, Collins R, Crowley O, Cuttle SP, Howlett SA, Jones HE, McCalman H, Measures M, Pearce BD, Pearce H, Roderick S, Stobart R, Storkey J, Tilston E, Topp K, Watson CA, Winkler L, Wolfe MS (2013) Using legume-based mixtures to enhance the nitrogen use efficiency and economic viability of cropping systems – Final report (LK09106/HGCA3447)

FGTC (The Forage and Grazing Terminology Committee) (1991) Terminology for grazing lands and grazing animals. Pocahontas Press, Inc, Blacksburg or J Prod Agric 5:191–201

Foster L (1988) Herbs in pastures. Development research in Britain, 1850–1984. Biol Agric Hortic 5:97–133

Gaudin ACM, Westra S, Loucks CES, Janovicek K, Martin RC et al (2013) Improving resilience of northern field crop systems using inter-seeded red clover: a review. Agronomy 3:148–180

Gompert T (2010) The power of stock density. In: Proc.: Nebraska Grazing Conference. Kearney, NE

Hoste H, Torres-Acosta JFJ, Sandoval-Castro CA, Mueller-Harvey I, Sotiraki S, Louvandini H, Thamsborg SM, Terrill TH (2015) Tannin containing legumes as a model for nutraceuticals against digestive parasites in livestock. Vet Parasitol 212:5–17

Knight P, Rayns F, Rosenfeld A, Heading E, Will T et al (2008) An investigation into the adoption of green manures in both organic and conventional rotations to aid nitrogen management and maintain soil structure 2008

Lindström BEM, Frankow-Lindberg BE, Dahlin AS, Wivstad M, Watson CA (2013) Micronutrient concentrations in common and novel forage species and varieties grown on two contrasting soils. Grass Forage Sci 68:427–436

Lüscher A, Mueller-Harvey I, Soussana JF, Rees RM, Peyraud JL (2014) Potential of legume-based grassland-livestock systems in Europe: a review. Grass Forage Sci 69:206–228

McCosker T (2000) Cell grazing – the first 10 years in Australia. Trop Grassl 34:207–218

Patton BD, Dong X, Nyren PE, Nyren A (2007) Effects of grazing intensity, precipitation, and temperature on forage production. Rangel Ecol Manag 60:656–665

Peterson D (2010) How mob grazing impacts your soil. Retrieved from: https://www.drovers.com/article/how-mob-grazing-impacts-your-soil

Peterson D (2014) Soil health and production benefits of mob grazing. Retrieved from: http://www.forestrywebinars.net/webinars/soil-health-and-productionbenefits-of-mob-grazing

Peterson PR, Gerrish JR (1995) Grazing management affects manure distribution by beef cattle. In: Proc. Am. Forage Grassland Council, Lexington, pp 170–174

Pirhofer-Walzl K, Rasmussen J, Høgh-Jensen H, Eriksen J, Søegaard K, Rasmussen J (2012) Nitrogen transfer from forage legumes to nine neighbouring plants in a multi-species grassland. Plant Soil 350:71–84

Redden MD (2014) Grazing method effects on forage production, utilization, and animal performance on Nebraska sand hills meadow

Richmond R (2011) The benefits to agriculture and the environment of rebuilding soil carbon. A nuffield farming scholarships trust report. John Oldacre Foundation, UK

Roca-Fernández AI, Peyraud JL, Delaby L, Delagarde R (2016) Pasture intake and milk production of dairy cows rotationally grazing on multi-species swards. Animal 10:1448–1456

Rochon JJ, Doyle CJ, Greef JM, Hopkins A, Molle G, Sitzia M, Scholefield D, Smith CJ (2004) Grazing legumes in Europe: a review of their status, management, benefits, research needs and future prospects. Grass Forage Sci 59:197–214

Romera AJ, Doole GJ, Beukes PC, Mason N, Mudge PL (2017) The role and value of diverse sward mixtures in dairy farm systems of New Zealand: an exploratory assessment. Agric Syst 152:18–26

Rouille B, Delaby L, Delagarde R, Caillaud D, Desarmenien D, Daveau B, Guibert S (2016) Benefits of using legumes as forage or fodder to feed French dairy cattle. Fourrages 227:181–187

Savory, A., and S.D. Parsons (1980) The Savory Grazing Method. Rangelands. 2:234–237

Savory A (1983) The savory grazing method or holistic resource management. Rangelands 5:155–159

Savory A, Butterfield J (1999) Holistic management – a new framework for decision making. Island Press, Washington, DC

Seither M, Wrage N, Isselstein J (2012) Sward composition and grazer species effects on nutritive value and herbage accumulation. Agron J 104:497

Stuth JW, Kirby DR, Chmielewski RE (1981) Effects of herbage allowance on efficiency of defoliation by the grazing animal. Grass Forage Sci 36:9–15

Vogeler I, Vibart R, Cichota R (2017) Potential benefits of diverse pasture swards for sheep and beef farming. Agric Syst 154:78–89

Wang Y, Duan W-X, TU C, Washburn S, Cheng L, Hu S (2014) Soil carbon, nitrogen and microbial dynamics of pasturelands: impacts of grazing intensity and planting systems. Pedosphere 24:408–416

Weber KT, Gokhale BS (2011) Effect of grazing on soil-water content in semiarid rangelands of Southeast Idaho. J Arid Environ 75:464–470

Wright AL, Hons FM, Rouquette FM (2004) Long-term management impacts on soil carbon and nitrogen dynamics of grazed bermudagrass pastures. Soil Biol Biochem 36:1809–1816

Cloud Computing for IoT Applications in Climate-Smart Agriculture: A Review on the Trends and Challenges Toward Sustainability

Eleni G. Symeonaki, Konstantinos G. Arvanitis, and Dimitrios D. Piromalis

Abstract Climate-smart agriculture is an approach for guiding actions required to make agricultural activities more productive and sustainable with regard to resource sharing, cost saving, and efficient agro-system construction. Agricultural informatization, through the adoption of new technological trends such as cloud computing and Internet of Things (IoT), is an extensive prospect to realize sophisticated applications aiming in the development of services in rural areas for the benefit of climate-smart agriculture. Especially the integration of cloud computing and IoT technologies for applications in agriculture, through the automation and digital management of the entire agricultural production and food chain, is an innovative research field of significant concern. In this paper an attempt is made to survey the technological background of these technologies, as well as to review the main trends and identify the challenges regarding the adoption of cloud-based IoT applications in the agricultural sector for the benefit of sustainability in climate-smart agriculture.

Keywords Cloud computing · Internet of Things · Agricultural informatization · Smart agriculture · Sustainable development

---

E. G. Symeonaki (✉)
Department of Natural Resources Management & Agricultural Engineering, Agricultural University of Athens, Athens, Greece

Department of Industrial Design & Production Engineering, University of West Attica, Egaleo, Greece
e-mail: esimeon@uniwa.gr

K. G. Arvanitis
Department of Natural Resources Management & Agricultural Engineering, Agricultural University of Athens, Athens, Greece

D. D. Piromalis
Department of Industrial Design & Production Engineering, University of West Attica, Egaleo, Greece

A. Theodoridis et al. (eds.), *Innovative Approaches and Applications for Sustainable Rural Development*, Springer Earth System Sciences,
https://doi.org/10.1007/978-3-030-02312-6_9

1 Introduction

Agriculture is, beyond any doubt, one of the most important sectors of primary industry, yet it is characterized as sensitive, unstable, complex, dynamic, and highly competitive. In the twenty-first century, the sector of agriculture is going to face great challenges as, according to the Food and Agriculture Organization (FAO) of the United Nations, agricultural productivity should be increased by 60% in order to ensure a safe food supply which would adequately satisfy the nutritional needs of the constantly growing world population (Fig. 1). This goal has to be achieved despite the fact that the required resources are already stretched, as the amount of available agricultural land is declining due to increasing urbanization, soil erosion, and high salinity levels, while 70% of the world's freshwater supplies is consumed for agricultural purposes. In addition, it is required for agriculture to address the issues which arise from the global climate change, concerning the reduction of its greenhouse gas emissions (the agricultural sector generates about one-quarter of global GHGs), as well as the adjustment to extreme weather conditions which impact the quantity and quality of the crops (FAO 2013). To successfully encounter the increased global nutritional needs as well as the climate change issues, the sector of agriculture has to become "climate-smart."

Climate-smart agriculture (CSA) was defined by Lipper et al. as "an approach for transforming and reorienting agricultural development under the new realities of climate change" (Lipper et al. 2014). Yet, the definition which is most commonly used was presented by FAO at the Hague Conference on Agriculture, Food Security and Climate Change in 2010 and describes CSA as "agriculture that sustainably

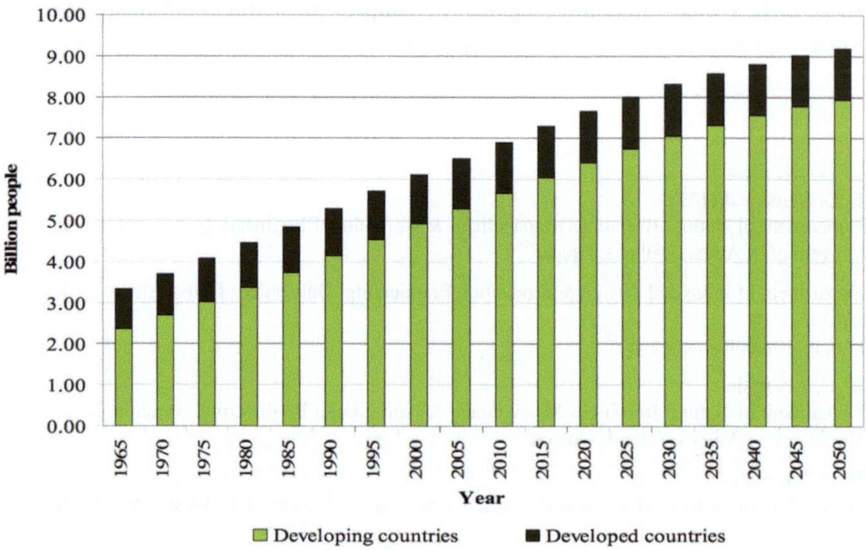

Fig. 1 Global population growth according to U⍰ Secretariat (2007)

increases productivity, enhances resilience (adaptation), reduces/ removes GHGs (mitigation) where possible, and enhances achievement of national food security and development goals." According this definition, the main goal of CSA is food development and security, while productivity, adaptation, and mitigation may be characterized as the interlinked pillars which are required in order to achieve this goal (FAO 2013).

CSA provides the means to relate actions both on-farm and beyond the farm, by incorporating elements concerning policies, institutions, investments, and technologies as follows (FAO 2013):

- Farm, cultivation, and livestock management for handling resources efficiently as well as increasing production and resilience
- Ecosystem and landscape management in conserving ecosystem services that are essential in order to increase resource efficiency and resilience at the same time
- Services for farmers and land managers so as to enable them to implement changes which are necessary for the efficient management of climate risks/ impacts and mitigation actions
- Changes which enhance the benefits of CSA in the wider food system, including value chain interventions as well as demand-side measures

One way of addressing these issues and increasing the quantity together with the quality of agricultural production is by using cutting-edge technologies in order to establish more "intelligent" and interconnected farms through agricultural informatization.

The implementation of information communication technology (ICT) is a major asset for the sustainable growth in agriculture. Until the recent past decades, the research focused on agricultural infrastructure development and information services. In order for this situation to be changed and promote a rapid development of agricultural informatization, it is essential to apply technologies which provide reliable, cheaper, and user-friendly ICT tools in agriculture.

Cloud computing, as a trend of future information technology applied in various fields, may play a significant role in agricultural informatization by bringing some new prospects to information management and services. Evolving cloud computing technology in agriculture is an extensive opportunity to carry out industry agricultural applications aiming in the sustainable development in rural areas.

Meanwhile, the innovative technology of the Internet of Things (IoT) is highly related to cloud computing as IoT acquires compelling computing tools through cloud computing and cloud computing encounters the optimum channel of practice based on IoT. Thus, the integration of these technologies, using radio-frequency identification (RFID) and wireless sensor networks (WSN), for data acquisition and monitoring corresponding to cultivations, as well as cloud computing applications for transferring, storing, and processing these data using the Internet, is predicted to bring revolutionary changes to agriculture, through the automation of agricultural production.

The theoretical and applied study of cloud computing for IoT agricultural applications in terms of sustainability appears to be of high significance. In this paper an attempt is made to review the technological background of cloud computing and IoT with emphasis in the most important approaches, regarding their application in the agricultural sector for the benefit of climate-smart agriculture toward the sustainable development in rural areas.

2 Conceptual Framework

2.1 Cloud Computing Technology Overview

Cloud computing, often called simply "the cloud," has been characterized as "the fifth utility service," for it provides readily available on-demand services, similar to any other utility which is available in modern society (Buyya et al. 2009).

Although the popularity of the term "cloud computing" was launched in 2006 by Amazon.com, it has emerged as a concept since the 1960s (Kleinrock 2005; Armbrust et al. 2009; Wheeler and Waggener 2009). As it appears from literature, there are various perspectives concerning the definition of "cloud computing" (Buyya et al. 2009; Armbrust et al. 2009; Vouk 2008; Vaquero et al. 2011; Mell and Grance 2010), among which the most recognized and generally accepted is the one stated by Mell and Grance (2010) on behalf of the US National Institute of Standards and Technology (NIST). In particular, according to NIST, cloud computing is defined as "a model for enabling ubiquitous, convenient, on-demand network access to a shared pool of configurable computing resources (e.g., networks, servers,

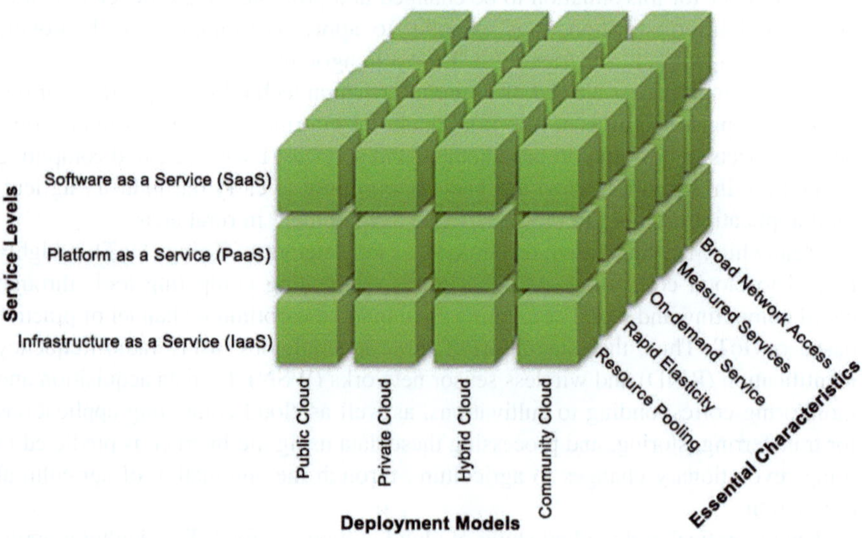

Fig. 2 Anatomy of cloud computing according to NIST

storage, applications, and services) that can be rapidly provisioned and released with minimal management effort or service provider interaction."

Furthermore, NIST suggests that the anatomy of cloud computing consists of five essential characteristics, three service layers, and four deployment models as shown in Fig. 2 (Mell and Grance 2010; Craig Wood 2010). Preferably, a cloud should involve all of the five essential characteristics, while as far it concerns the four deployment models, public cloud is the one that "cloud computing" has been initially referred to and is most commonly used. The other deployment models constitute variations of the public cloud, sharing similar characteristics and service layers (Rountree and Castrillo 2014). The three service layers refer to the services offered by the cloud providers and are described, depending on user requirements, as follows:

(a) *Infrastructure as a Service (IaaS)*: Due to recent developments in network management and virtualization, cloud infrastructure provides processing, storage, traffic monitoring, and redirecting, as well as other forms of lower-level hardware and network resources in a virtual way via the Internet network. These resources are offered upon the demand of the end user and are charged per use, differentiating from traditional hosting services in which physical servers or parts of them are offered and charged on a periodical basis (Rountree and Castrillo 2014; Leavitt 2009). Moreover, end users have control over operating systems, deployed applications, storage, and in some cases limited control over selected network components (e.g., host firewalls), without the need of managing or controlling the underlying cloud infrastructure (Mell and Grance 2010). Some researchers suggest to further divide IaaS into hardware as a service (HaaS) and data as a service (DaaS) (Wang et al. 2008), but it is more common for IaaS to be regarded as a whole concept.
(b) *Platform as a Service (PaaS)*: This service layer is more advanced than the IaaS one, for it acts as an integrated design, development, testing, and deployment platform and provides the end user with programming and execution environments such as operating systems, programming languages, databases, and web servers. End users are given the ability to develop their own applications directly onto the cloud infrastructure, after creating them by using programming languages and APIs supported by the provider. In addition, end users have total control over their deployed applications without being responsible for the management or control of the underlying cloud infrastructure such as network, servers, operating systems, or storage (Mell and Grance 2010). This approach reduces most of the system administrative responsibilities that are traditionally assigned to the developers, enabling them to concentrate on more productive issues (Lawton 2008). Finally PaaS carries some other appealing features including embedded instruments for measuring the deployed application usage as well as established online communities for cooperation, interaction, and problem-solving purposes (Rountree and Castrillo 2014).
(c) *Software as a Service (SaaS)*: Through this service, end users are provided, upon their demand, with complete turnkey solutions of software applications or even

more of sophisticated systems, such as CRM or ERP, directly via the Internet network (Leavitt 2009). These kinds of solutions are hosted as services in the cloud and are delivered via browsers, correspondingly to user subscriptions. SaaS is also characterized by a multi-tenant architecture, according to which all users share a single codebase maintained by the provider. Authorization and authentication security policies are employed in order to guarantee the partition of user data. Due to this sharing mechanism, the cost of the services provided is far more appealing compared to the traditional off-the-shelf and bespoke ones (Wang et al. 2008). Such an approach eliminates the workload of installing, running, as well as maintaining applications on local computers and reduces the cost of software purchases due to on-demand pricing policies.

Summing up, cloud computing is designed to integrate a perfect system, which is able of distributing high computing power to the end users by combining computer, storage, and network technologies with regard to load balance and reutilization.

2.2 Internet of Things Technology Overview

The Internet is considered to be one of the most significant technological achievements, having a strong impact on all aspects of humanity (Ezechina et al. 2015). The Internet of Things (IoT), sometimes called "Internet of Objects," is a new concept which is related to the interoperability of objects and their capacity to access and exchange information about themselves via the Internet.

The term "Internet of Things" was initially used about 15 years ago, by the Auto-ID Labs of the Massachusetts Institute of Technology (MIT) during their research regarding on-networked radio-frequency identification (RFID) infrastructures (Atzori et al. 2010; Mattern and Floerkemeier 2010). Although, since then, the concept of IoT has extended beyond the purposes of RFID technologies and nowadays the term "Internet of Things" is widely used, the research in literature cannot result in a unique definition but in several alternative ones, some of which focus on the things which become connected, while others emphasize on Internet-related aspects or semantic challenges in the IoT (Atzori et al. 2010). Nevertheless, what all of these definitions have in common is the perception that while the original Internet version was concerning data created by people, its consequent version is related to data created by things. According to this aspect, Internet of Things could be best defined as "a world-wide network of interconnected objects uniquely addressable, based on standard communication protocols" (DG Infso, EPoSS 2008). In Fig. 3 the concept of IoT is presented in terms of interconnection, suggesting that it enables anything to be connected anytime, at anyplace, with anything and anyone ideally using any networks and any services (Saranya and Nitha 2015).

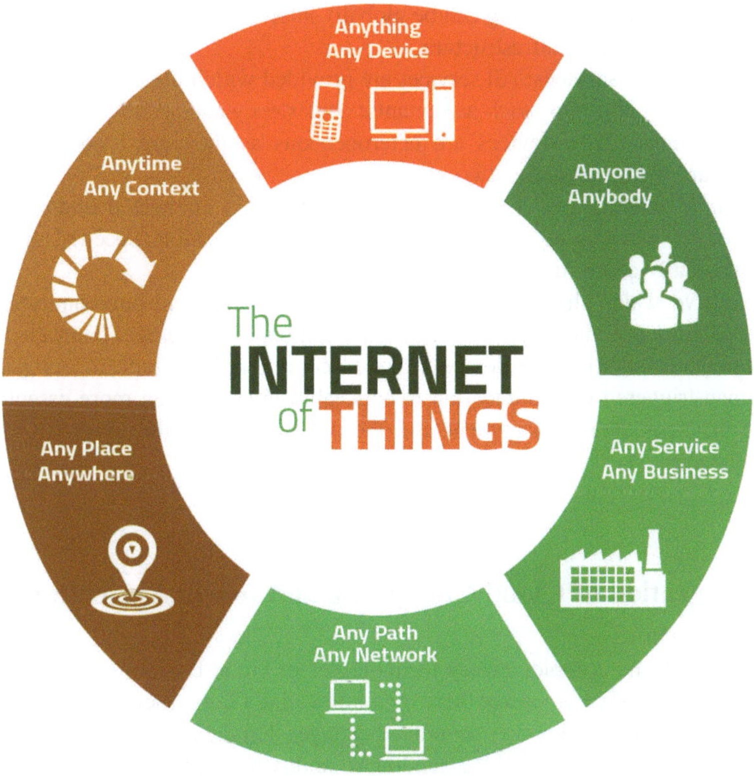

Fig. 3 The concept of the Internet of Things (IoT)

The innovative technology of IoT has already found numerous and diverse applications, which are continuously extending, to virtualize all aspects of everyday life. Up to date about 5 billion objects are "smartly" interconnected, while predictions indicate that this number will increase at approximately 50–100 billion by the year 2020 (Gubbia et al. 2013).

The future success of IoT is depending on standardization, which will provide reliability, compatibility, and interoperability for the efficient deployment of smart environments in a global scale (Grandinetti et al. 2013). The concept of network convergence using IP-based technologies for the IoT is essential and relates to the usage of a common multiservice IP network which supports a variety of applications and services. Using IP for the intercommunication and control of sensors and small devices enables the convergence of extensive, IT-oriented networks with specialized real-time networked applications. Several companies and organizations of leading technologies in energy and communications are working so as to specify the standards for the new IP-based technologies for the IoT. In this context the fundamental characteristics of the IoT as defined by the European Research Cluster on the Internet of Things (IERC) are as follows (Vermesan and Friess 2014):

- Interconnectivity: Anything can be globally interconnected with any information and communication infrastructure.
- Things-related services: All services are provided within the constraints of physical and virtual things, such as semantic consistency and security.
- Heterogeneity: The devices are heterogeneous as they are based on various hardware platforms and networks.
- Dynamic interactions: The number and state (e.g., connected and/or disconnected) of devices as well as their context (e.g., location, speed) can change dynamically at any time and any place.
- Extensive scale: The number of devices to be handled and interconnected is at least an order of magnitude larger than the number of the devices connected to the Internet. In addition the device-triggered communication ratio as compared to human-triggered communication is noticeably greater. Even more essential are the efficient data management and their semantics for the purposes of each application.

3 Integration of Cloud Computing and IoT in Agriculture

Since present, the complementary technologies of cloud computing and IoT have developed in separate; however there is an innovative IT argument in which they are combined, offering in this way several advantages. On the one hand, cloud computing may benefit from the IoT by extending its ability to delivering new services for a large number of real-world scenarios in a dynamically distributed way, while on the other hand, IoT can benefit from cloud computing with regard to its unlimited virtual resources and capabilities in order to compensate any technological constraints such as storage, processing, or energy. The complementary characteristics of cloud computing and IoT, as arising from literature, are reported in Table 1 (Dash et al. 2010; Fox et al. 2012; Parwekar 2011; Rao et al. 2012).

Essentially, the integration of cloud computing and IoT is a cutting-edge technology which may have great positive impact in agriculture as it can provide efficient management of resources and higher production by facilitating the storage, management, access, and dissemination of information. In particular cloud computing acts

Table 1 Characteristics of cloud computing and IoT

Cloud computing	Internet of Things (IoT)
Ubiquitous (resources usable from everywhere)	Diffusive (things placed everywhere)
Virtual resources	Real-world things
Virtually unlimited computational capabilities	Limited computational capabilities
Virtually unlimited storage	Limited storage
Internet as a service	Internet as a point of convergence
Big data management	Big data source

as an intermediate layer between the things and the applications, eliminating any functionalities and complexities which could arise from the implementation of the latter. Furthermore, a primary reason for adopting this framework in agriculture is to support farmers in making decisions and drawing strategies related to new challenges about cultivations in terms of sustainability.

A brief overview of the features provided in agriculture through the integration of cloud computing for IoT is presented as follows:

- Data acquisition and remote storage: Several of the available data acquisition tools, such as radio-frequency identification (RFID) sensors and wireless sensor networks (WSN), can be effectively integrated with cloud computing applications for temperature humidity, luminosity, soil moisture monitoring, etc. (Hori et al. 2010). Moreover cloud computing provides high storage capacity suitable for the backup of such large scale of data and information relevant to cultivations, offering in addition a suitable infrastructure for decision support as well as mutual information and experience sharing among the worldwide agricultural community (Feng 2010).
- Low-cost access to ICT resources: Provisioning of access to extensive ICT resources which are offered upon the demand of the user and are charged per use. In this way farmers can access the required resources on-demand from the cloud, instead of investing in owning expensive ICT hardware infrastructure (Prasad et al. 2013).
- Online agriculture expert consultation: In the situations when farmers are not able to solve any occasional problems they may face at the different stages of agricultural production, the integration of cloud computing for IoT offers a sufficient alternative, as online expert advice could be found in the repository of the cloud databases. Therefore, farmers could face any problem instantly, as they receive immediate and accurate response (Wenshun 2011).
- Land record automation: Due to the accessibility of large-scale storage infrastructure, land records are being digitized worldwidely. Cloud computing storage facility offers a feature of entering and storing a land record along with any descriptions which are relative to that particular region, such as production history, soil analysis result, etc. Various public or private operators are responsible of the data accuracy after the proper verification of facts and figures (Hori et al. 2010).
- Weather forecasting: Weather forecast and analytics for specific time periods may be provided to farmers so that they can take decisions related to cultivations (TongKe 2013).

A typical framework of cloud computing for IoT in smart agriculture is shown in Fig. 4. According to this framework, the management platform is the “core” of cloud computing and data storage. It involves the management of access to the application interface by the users, the computing and processing that are involved in the customization services, as well as the organization and coordination of the service nodes in the data center. The essential networking services are provided so as the acquired object information data to be efficiently and securely transferred. The

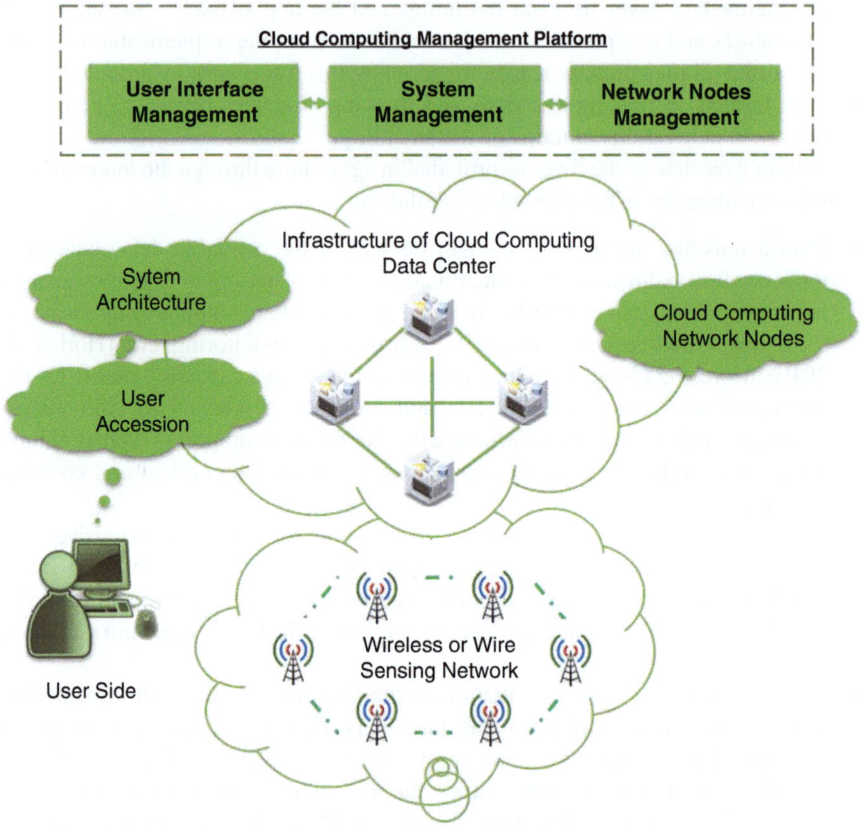

Fig. 4 Typical framework of cloud computing for IoT in smart agriculture

communication is established between RFID or wireless sensor networks (WSNs) and gateways or cloud proxy machines, interconnected to the network by applying protocols and standards based on Zigbee, LoRaWAN, LPWA, Bluetooth, Wi-Fi, as well as GPRS/3G/4G technologies. Any remaining communication required may be established in a service-oriented approach via electronic XML/EDI messages. The application includes also optical cable and other wire communication protocols and technologies (Ken 2012).

4 Paradigms of IoT Agricultural Applications Based on Cloud Computing

The integration of cloud computing for IoT applications in agriculture, through the automation and digital management of the entire agricultural production and food chain, is an innovative research field of significant concern for the scientific community. This cutting-edge technology can not only transform the methods of the agricultural production management in terms of sustainability but also strengthen the safety of the agricultural products quality.

Although several techniques regarding cloud-based IoT applications, especially designed to meet the needs of agricultural sector, have been introduced during the last years, these are mostly used fragmentary by a small number of innovative farmers, while their wide adaptation by the total of the agricultural community is still rather limited. Some of the most noteworthy applications, found in literature, are briefly overviewed as follows:

(a) *Agriculture cloud based on MAD-Cloud architecture*: It offers specialized services to farmers regarding crop cultivation, fertilizer usage, pest control, etc. In addition, researchers in the field of agriculture can access the cultivation history of multiple regions and add their suggestions regarding innovative agricultural techniques. Last but not least, existing cloud infrastructures like networks, servers, etc. can be used, and various services are supported regarding the interaction with the cloud by using IoT features such as sensors, mobile devices, GPS, etc. MAD-cloud framework as shown in Fig. 5 is based on a layered architecture consisting of three layers (Chandraul and Singh 2013):

   - MAD-Data Acquisition Layer (MDAL): This layer is deployed as SaaS, providing a variety of interface services to several types of users with diverse devices. It adopts Internet and IoT technologies which provide a number of services required by farmers, agriculture experts, or state officials. The addition or acquisition of data is performed via the applications or the device service interfaces.
   - MAD-Data Processing Layer (MDPL): This layer is deployed as PaaS including library platforms which store all data acquired from various devices into a uniform format, performs computations on large data sets, and provides reports to the end users. The layer is subdivided into four modules which perform services of data security, data processing, expert decision supporting, and data reporting.
   - MAD-Data Storage Service Layer (MDSSL): This layer is deployed as IaaS, providing database infrastructure facilities for the storage of the large amounts of data which is essential in agriculture sector. Additionally this layer supports the sharing and usage of the stored data depending on the needs of the users.

(b) *Cloud Agro System*: It is a cloud computing-based system designed to monitor the overall information related to agricultural activities. Up to date IT tools provide online language translation mechanisms in order to overcome language and tradition limitations in worldwide agricultural community. As a result, any type of information stored on the cloud is presented to the user's language of choice, supporting any decisions related to crop production. Furthermore, through sophisticated features, such as online questionnaires, researchers can exploit the experience of farmers worldwide in the development of new agricultural tools and techniques (Patel and Patel 2013).

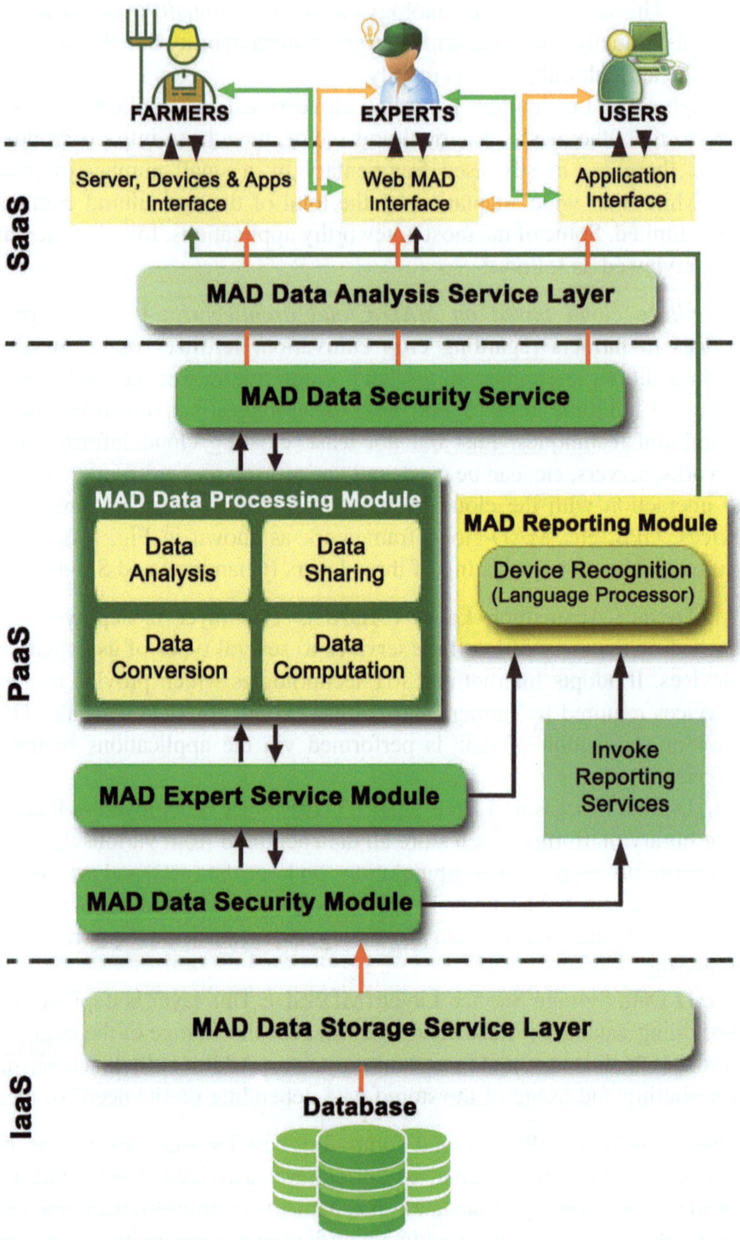

Fig. 5 MAD framework architecture

(c) *Agricultural mobile cloud-based platform*: It is a concept model of Mobile Cloud Computing (MCC) technology aiming to provide the users with extended and seamless functionality despite any probable limitations of mobile device resources, by using simple handheld devices such as laptops, tablets, and smartphones which support Wi-Fi, GPRS, 3G, or 4G technologies. In particular MCC allows the end user devices to maintain a very thin layer architecture for a minimum number of applications, while it forwards all computational processing workloads to the virtual machines on the cloud, overcoming in this way any mobile processing power and data storage limitations, extending the battery life, as well as increasing the security levels.

In this specific platform, a mobile server is established including application service providers (ASPs) which offer to farmers on-demand software services via a network architecture. The developer is connected to the ASP, while the end users are connected to the mobile infrastructure which provides application features and services designed to be user-friendly. An important asset in this platform is the image processing feature which supports the process and analysis of the field status just by capturing the leaf images (Prasad 2013).

The mobile cloud framework proposed in Fig. 6 is focusing on the farmers' needs and assists them to achieve relatively better cultivation and marketing through expert decision support.

(d) *WSAN linked to agricultural cloud computing system*: This approach integrates wireless sensor actor networks (WSANs) with cloud computing services to assist farmers optimize the usage of available resources in agricultural production.

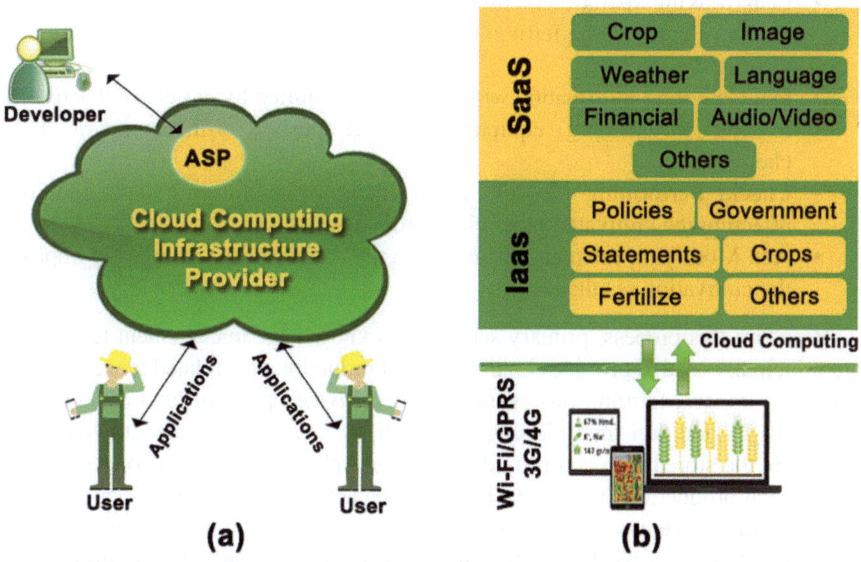

Fig. 6 Mobile cloud computing (**a**) system model and (**b**) proposed services

Environmental data from the fields are acquired by the sensors and process with the help of a decision support unit for actuating the process. Sensor nodes which acquire environmental data and a group of actor nodes which operate according to the decision taken by the decision support system are interconnected with wireless medium. The layered architecture shown in Fig. 7 consists of three groups (Mahesh et al. 2014):

- Sensing Group: It consists of various sensors placed in the fields for acquiring real-time data about temperature, humidity, weather, soil conditions, etc. The sensed data are forwarded at first to the sensor data storage unit and then to the cloud for further processing.
- Cloud Service Group: This group consists of layers for data acquisition, data processing, and data storage. These layers support database infrastructure facilities, library modules, and interface services for decision support through expert knowledge in order to be used for building high-level agricultural applications
- Actuator Group: It consists of actor node sets which operate in the environment according to the action commands received by the decision unit. In particular the action commands which are generated by the actor node controllers according to the digital data deriving from the cloud result in analog signals that are sent to the hardware devices in order to perform the intended actions.

(e) *PDCA cycle-based agricultural cloud services*: According to this perspective, the process of agriculture production is regarded as a PDCA (plan-do-check-act) cycle as shown in Fig. 8.

- Plan: Making operation plans about agricultural production activities such as planting, cultivation, fertilizing, irrigation, human and material resources, yield, etc.
- Do: Gathering information which has been acquired by agricultural activities relating to farming equipment, insect pests, fertilizers, agricultural chemicals, etc.
- Check: Performing progress management by monitoring the conditions affecting the cultivated plots via sensors or cameras
- Act: Modifying the plans according to the data acquired in order to improve the cultivation results

Based on this process, primary sensing and knowledge management techniques are principally used to provide cloud services. Production data related to weather and soil, crop images, farming observations, and cultivated plots of land are routinely collected. In addition analysis engines, such as data miners, are employed for analyzing the stored data and provide advice as well as suggestions concerning the agricultural production. Moreover, maintenance actions are performed via software updates and service amendments on a single system in the cloud center resulting in the elimination of disparities in the software versions used by different users (Hori et al. 2010).

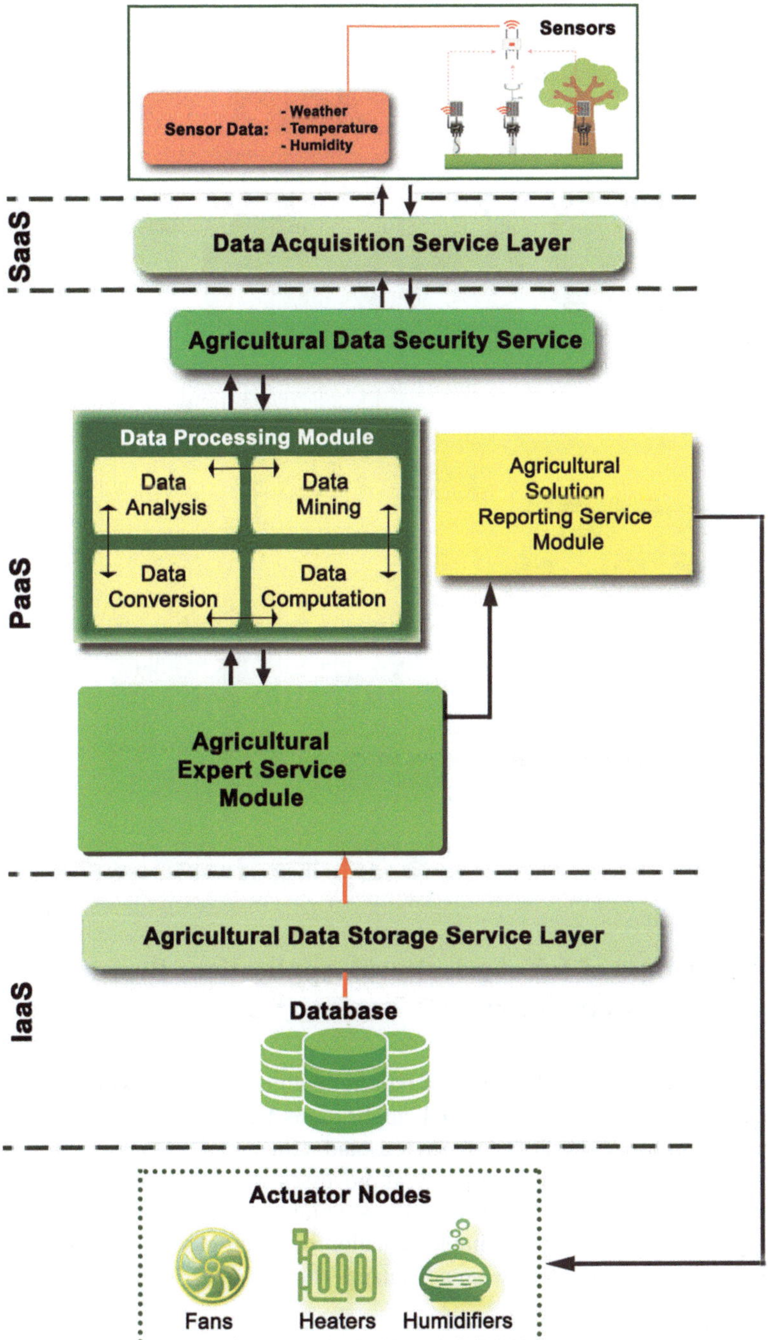

Fig. 7 Agricultural WSAN cloud-based IoT application

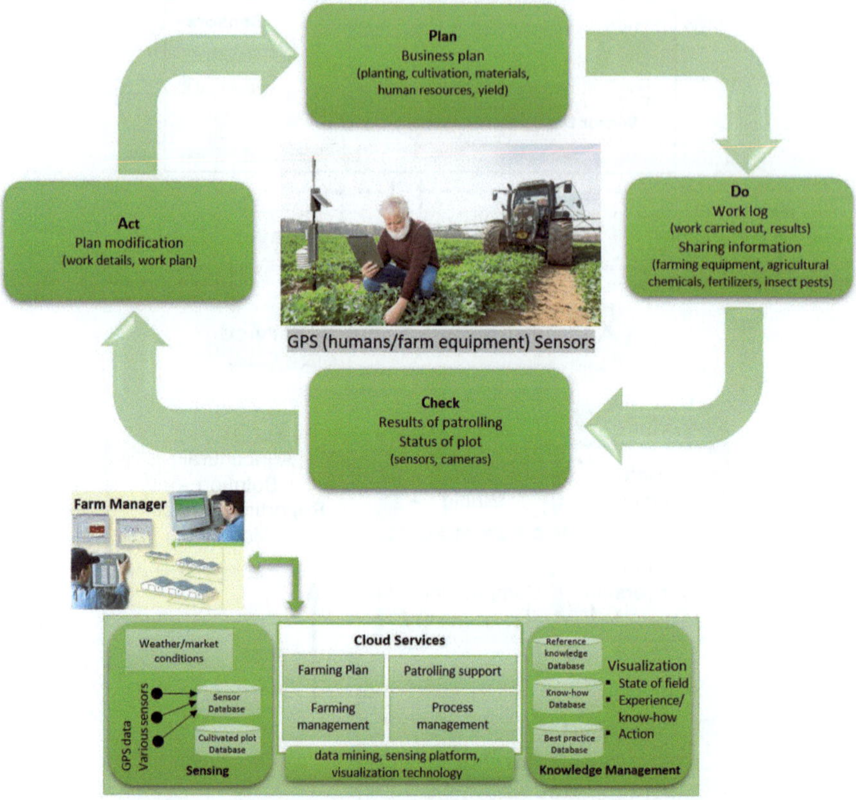

Fig. 8 PDCA cycle and cloud services in agriculture

5 Challenges of IoT Agricultural Applications Based in Cloud Computing

Agriculture, and especially climate-smart agriculture, is a field which may highly benefit from the technologies of IoT applications based in cloud computing with regard to resource sharing, cost saving, and efficient agro-system construction. Moreover, the integration of agricultural processes with such cutting-edge technologies can give a significant impetus to production, marketing, and sales of agricultural goods. In particular, the implementation of cloud-based IoT technologies in agriculture presents the following advantages (Prasad et al. 2013; Gao et al. 2011):

- Data management is performed by the service provider guarantying better and efficiently organized information resources.
- Stakeholders can access information at any time or location from the e-data bank databases.

- Communication and interaction among users worldwide are expeditive, effortless, and of sufficient security.
- Maintenance infrastructure requirements are drastically reduced as service providers are responsible for all technical issues.
- Farmers and researchers are motivated to get more involved into the field of climate-smart agriculture as all communication attempts are result oriented.
- The problem of rural-urban migration can be reduced as cloud computing services are provided remotely at any time. This aspect might also aid in unemployment control.
- Due to the mass involvement of different stakeholders, cloud-based IoT applications in agriculture can boost sustainable growth and economic development.

Despite the fact that the adoption of cloud-based IoT technologies in the agricultural sector presents several advantages, their actual implementation seems to be premature and lacks consistent integration, as more sophisticated solutions are still in experimental stage, while on the other hand, operational usage is mostly focusing on basic applications. Taking into consideration the present conditions, certain concerns and challenges are to be encountered regarding the efficient usage of cloud-based IoT applications in agriculture.

At first security and privacy vulnerabilities consist a critical barrier to the large-scale deployment of cloud-based IoT applications in agriculture. Since the integration of communication and computing with the rapidly increasing interconnected things provides the ability to remotely control their operations, several security issues arise regarding the interactions between the physical and cyber domains. A presumptive extension of the already applied IT security frameworks to the IoT is going to lack in efficiency as contrary to present networks, where assets are regularly protected inside firewalls with access control devices and several objects employed in the agricultural cloud-based IoT applications operate in highly vulnerable and unprotected environments where sensitive data might be compromised to eventual hacking attacks (e.g., sensors, tractors, drones, and other devices used in the open fields). The necessity of protecting these objects and ensuring their long lifespan increases the necessity of providing cloud-based security solutions with resource-efficient, thing-to-cloud interactions which will be suitable for the complex dynamic networks in agriculture (Medaglia and Serbanati 2010; Ashktorab and Taghizadeh 2012).

Secondly, the majority of agricultural cloud-based IoT applications, in order to be suitable for the open fields, especially in distant rural areas, require mobility support and widespread geographical distribution through an extended number of nodes, in addition to location tracking with the minimum margin of error as possible. Furthermore the strong need of real-time data processing imposes constant and high-speed network connectivity for the support of unobstructed streaming. In this context, micro-cloud or fog computing, as an evolved version of cloud computing technology, may enable new IoT applications and services which meet the aforementioned constraints of the agricultural environment, by applying new sophisticated data management as well as analytics services and taking advantage of the network features to their full extent (Yi et al. 2015).

Finally, generalizing the scope of the Internet-connected things beyond the sensing services, it is not hard to imagine the transformation of the entire agricultural sector into a smart web of context-sensitive remotely identified, sensed, and controlled virtual objects that will be shared and reused in different contexts, projecting the vision of climate-smart agriculture for the global sustainable development. In order for this vision to be realized, some more challenges to be faced include:

- Enhancing the implementation of IoT technologies in the agricultural sector by simplifying the usage of existing applications and making them more cost-effective in order to be turned into user-friendly and affordable for the majority of farmers
- Overcoming the lack of homogeneity of object data and IoT devices by establishing among them a certain degree of interoperability through open technologies for standards, applications, and platforms, specially addressing to the agricultural sector
- Providing broad, reliable, constant, high-speed, and low-cost network coverage for wireless communication in rural areas in order to ensure that the majority of farmers can benefit of cloud-based IoT technologies
- Establishing interaction with third-party databases containing information ranging from soil, water, and air analytics, insects and pest control, meteorological history, and forecast satellite imagery to logistics and retail
- Establishing training centers for offering guidance concerning computer skills and cloud computing services usage in order to minimize the high extent of computer illiteracy in rural areas as most farmers are not able to understand the functions of software and the Internet or complete tasks and solve problems on a computer without assistance

6 Conclusions

Since the first appearance of the concept in the 1960s, cloud computing has been established as one of the major trends of the future Internet technologies, enabling the provisioning of new services through the reduction of ownership cost and resources management. With the virtualization of objects being the next trend in the context of Internet of Things (IoT), the convergence of these two technologies is providing a flexible environment for services and innovations in climate-smart agriculture for the benefit of sustainability.

In this paper, an attempt was made to survey the most important trends regarding cloud-based IoT technologies and their applications in the agricultural sector with emphasis on the current situation as well as on the issues which require further attention.

The research in literature came up with findings about techniques and paradigms which indicate that cloud-based IoT technology provides sophisticated agricultural equipment, weather observation and forecasting applications, as well as cultivation control methods. Furthermore, it enables a "smart" interconnection with and among objects, leading in this way to the vision of "anytime, at anyplace, with anything and anyone" communications and enhancing the exchange of knowledge and experience through communication and information sharing. Modernization of agriculture by communicating rapidly the knowledge about new techniques improves the utilization of natural resources, reduces climate dependency, assists in environment and ecosystem protection, and promotes sustainable development.

Nonetheless, although cloud-based IoT technologies for agriculture receive an increasingly high level of attention by scientists and researchers, their operational implementation in the open fields seems to be mostly experimental and fragmentary. As, given the interest shown, more feasible and affordable solutions which address the challenges of cloud-based IoT technologies implementation in agriculture are expected to be widely available, transforming the entire sector into a smart web of context-sensitive interconnected objects that will be remotely identified, sensed, and controlled, eventually, farms are about to be transformed into self-adaptive autonomous systems involving smart interconnected objects that will be able to operate, make decisions, or even more "learn" without the need of on spot or distant interventions by farmers.

Consequently, it is strongly believed that the future growth of agriculture is depended on the adaptation of new technologies with a focus on farmer needs. The wide implementation of cloud-based IoT technologies in the agricultural sector is essential, so as to drastically enhance climate-smart agriculture activities for the benefit of sustainability.

Acknowledgment Authors would like to thank the Agricultural University of Athens as well as the University of West Attica, for providing the required resources and funds in order to undertake this research project.

References

Armbrust M, Fox A, Griffith R, Joseph AD, Katz R, Konwinski A et al (2009) Above the clouds: a Berkeley view of cloud computing, University of California at Berkeley, Tech. Report No.UCB/EECS-2009-28 Online at: www.eecs.berkeley.edu/Pubs/TechRpts/2009/EECS-2009-28.pdf. 1 Apr Accessed 2018

Ashktorab V, Taghizadeh SR (2012) Security threats and countermeasures in cloud computing. Int J Appl Innov Eng Manag (IJAIEM) 1(2):234–245

Atzori L, Iera A, Morabito G (2010) The internet of things: a survey. Comput Netw J, Elsevier, Amsterdam 54(15):2787–2805

Buyya R, Yeo CS, Venugopal S, Broberg J, Brandic I (2009) Cloud computing and emerging IT platforms: vision, hype, and reality for delivering computing as the 5th utility. Fut Gener Comput Syst (25)6. Elsevier, Amsterdam, pp 599–616

Chandraul K, Singh A (2013) An agriculture application research on cloud computing. Int J Curr Eng Technol, INPRESSCO 3(5):2084–2087

Craig Wood K (2010) Definition of cloud computing, incorporating NIST and G-Cloud views, online at http://www.katescomment.com/definition-of-cloud-computing-nistg-cloud. Accessed 1 Apr 2018

Dash SK, Mohapatra S, Pattnaik PK (2010) A survey on application of wireless sensor network using cloud computing. Int J Comput Sci Eng Technol 1(4):50–55

DG Infso, EPoSS (2008) Internet of things in 2020: a roadmap for the future, European Commission DG Infso & European Technology platform on smart systems integration, p 32

Ezechina MA, Okwara KK, Ugboaja CAU (2015) The internet of things (Iot): a scalable approach to connecting everything. Int J Eng Sci, Elsevier, Amsterdam 4(1):09–12

FAO (2013) Climate-smart agriculture sourcebook, online at: http://www.fao.org/docrep/018/i3325e/i3325e.pdf. Accessed 1 Apr 2018

Feng D (2010) Innovative research on rural information service mode based on cloud computing. Sci Technol Inf 18:214–217

Fox GC, Kamburugamuve S, Hartman RD (2012) Architecture and measured characteristics of a cloud based internet of things. In: International Conference in Collaboration Technologies and Systems (CTS), IEEE, pp 6–12

Gao J, Bai X, Tsai W (2011) Cloud testing-issues, challenges, needs and practice. Softw Eng Int J 1 (1):9–23

Grandinetti L, Pisacane O, Sheikhalishahi M (2013) Pervasive cloud computing technologies: future outlooks and interdisciplinary perspectives: future outlooks and interdisciplinary perspectives. IGI Global, Hershey

Gubbia J, Buyyab R, Marusic S, Palaniswami M (2013) Internet of things (IoT): a vision, architectural elements, and future directions. Futur Gener Comput Syst 29:1645–1660

Hori M, Kawashima E, Yamazaki T (2010) Application of cloud computing to agriculture and prospects in other fields. Fujitsu Sci Tech J, Fujitsu Global, Japan 46(4):446–454

Ken C (2012) Internet of things technology applied in field information monitoring. Adv Inf Sci Serv Sci AISS 4(12):405–414

Kleinrock L (2005) A vision of the internet. ST J Res 1(2):4–5

Lawton G (2008) Developing software online with platform-as-a-service technology. IEEE Comput Soc 41(6):13–15

Leavitt N (2009) Is cloud computing really ready for prime time? Comput, IEEE Comput Soc 42 (1):15–20

Lipper L et al (2014) Climate-smart agriculture for food security. Nat Clim Chang J 4:1068–1072

Mahesh DS, Savitha S, Anvekar DK (2014) A cloud computing architecture with wireless sensor networks for agricultural applications. Int J Comput Netw Commun Secur 2(1):34–38

Mattern F, Floerkemeier C (2010) From the internet of computers to the internet of things. Informatik-Spektrum, Springer Berlin Heidelberg 33(2):107–121

Medaglia CM, Serbanati A (2010) An overview of privacy and security issues in the internet of things, the internet of things. Springer, New York, pp 389–395

Mell P, Grance T (2010) The NIST definition of cloud computing, communications of the ACM, Special Publication 800–145, NIST, US

Parwekar P (2011) From internet of things towards cloud of things. In: 2nd International Conference in Computer and Communication Technology (ICCCT), IEEE, pp 329–333

Patel R, Patel M (2013) Application of cloud computing in agricultural development of rural India. Int J Comput Sci Inf Technol 4(6):922–926

Prasad S (2013) AgroMobile: a cloud-based framework for agriculturists on mobile platform. Int J Adv Sci Technol 59:41–52

Prasad MSVKV, Kumar GJ, Naidu VVS, Nagaraju GJ (2013) Use of cloud computing in agriculture sector, a myth or reality. Int J Eng Res Technol (IJERT) 2(10):38–42

Rao BP, Saluia P, Sharma N, Mittal A, Sharma SV (2012) Cloud computing for internet of things & sensing based applications. In: 6th International Conf. in Sensing Technology (ICST), IEEE, pp 74–380

Rountree D, Castrillo I (2014) The basics of cloud computing. Elsevier, Amsterdam, pp 123–149

Saranya CM, Nitha KP (2015) Analysis of security methods in internet of things. Int J Recent Innov Trends Comput Commun 3(4):1970–1974

TongKe F (2013) Smart agriculture based on cloud computing and IOT. J Converg Inf Technol 8 (2):210–216

Vaquero L, Rodero-Merino L, Morán D (2011) Locking the sky: a survey on IaaS cloud security, computing. Arch Inf Numer Comput 91(1):93

Vermesan O, Friess P (2014) IERC cluster SRIA 2014 – internet of things – from research and innovation to market deployment. River Publishers Series in Communication, Denmark/ Netherlands

Vouk MA (2008) Cloud computing-issues, research and implementations. CIT J Comput Inf Technol, University of Zagreb, Croatia 4:235–246

Wang L, Tao J, Kunze M, Castellanos AC, Kramer D, Karl W (2008) Scientific cloud computing: early definition and experience. In: High Performance Computing and Communications, HPCC'08, 10th IEEE International Conf., Dalian, China

Wenshun C (2011) Application and developing prospect of cloud computation in the agricultural informationization. Agric Eng 2(1):40–43

Wheeler B, Waggener S (2009) Above campus services: shaping the promise of cloud computing for higher education. Educ Rev 44(6):52–53

Yi S, Hao Z, Qin Z, Li Q (2015) Fog computing: platform and applications. In: IEEE Workshop on Hot Topics in Web Systems and Technologies (HotWeb), Washington, DC, USA, pp 73–78

User-Technological Index of Precision Agriculture: Data Collection and Visualization

Jan Masner, Jan Jarolímek, Michal Stočes, Pavel Šimek, Jiří Vaněk, and Vladimír Očenášek

Abstract User-Technological Index of Precision Agriculture (UTIPA) is a comprehensive system based on mutual sharing of opinions and experience within community of people related to precision agriculture – farmers, technology suppliers, and researchers. The main goal of UTIPA is to present the calculated index level for a particular technology (method) for precision agriculture and compare it to the indexes of other technologies. The index is based on evaluation of technological advancement and applicability for agricultural practice. The paper discusses methods for collecting data from questionnaires in general and elaborates on the technical solution developed specifically for data collection for UTIPA. The system also allows all participants who filled in the entry questionnaire to access the results. There are various visualizations and data views available.

Keywords Precision farming · Survey · Questionnaire · Farmers · Web · Technologies

1 Introduction

Besides precision agriculture, several terms are used for the same concept among the professional society nowadays. It includes *precision farming*, *smart agriculture*, *smart farming*, and *agriculture 4.0*. The concept of precision agriculture has been in the interest of the professional public since the 1990s. It generalizes the effort to identify solutions, tools, and processes that can improve productivity and profitability while protecting the environment (Cambouris et al. 2014). Precision agriculture plays a vital role in increasing production and is seen as part of the agricultural

J. Masner (✉) · J. Jarolímek · M. Stočes · P. Šimek · J. Vaněk · V. Očenášek
Department of Information Technologies, Czech University of Life Sciences Prague, Prague, Czech Republic
e-mail: masner@pef.czu.cz

A. Theodoridis et al. (eds.), *Innovative Approaches and Applications for Sustainable Rural Development*, Springer Earth System Sciences,
https://doi.org/10.1007/978-3-030-02312-6_10

process efficiency and environment-friendliness. The adoption of GPS guidance systems for tractors has been part of precision agriculture since its conception. Currently, precision agriculture encompasses many advanced modern technologies and approaches such as high-precision positioning systems (like GPS), automated steering systems, geomapping, sensors and remote sensing, integrated electronic communications, and variable rate technology (CEMA 2018). ICT technologies in general are one of the most important enablers (Kai and Perumal 2010).

The precision agriculture approaches are spreading through all areas of agricultural production (Morota et al. 2018), and crop production is no longer the only domain. The current efforts are trying to manage the resources of the whole farms from crops to the final product. On the other side, the utilization of precision agriculture and its operations and technologies in terms of widespread implementation is still under development (Paustian and Theuvsen 2017).

In summary, the concept of precision agriculture is based on observations and measurements followed by the appropriate responses – for example, through the introduction of new technology or by changing manufacturing processes. Precision agriculture technologies allow farmers to identify problems and opportunities and apply solutions with far greater accuracy (Lindblom et al. 2016).

A key factor in deciding whether a particular technology should be incorporated to practice is the understanding of agricultural production processes as well as the technology itself. Workers in agriculture management must choose among various options for applied research and technology, and in this decision-making process, there is a necessity to merge previous experience of the staff and the introduction of new technologies and procedures (Kumhala et al. 2003).

The question of implementation of the particular technology is partly an economic decision. The most important factors are profitability and investment rate of return. Therefore, it is important to inform farmers about economic benefits of precision agriculture technologies (Katalin et al. 2014).

Another factor that affects the decision process is usability. Technologies evolve very fast nowadays. More and more nontechnical users use them. Therefore, precision agriculture technology development needs to also focus on the field of human-computer interaction (Lindblom et al. 2016). Usability in association with precision agriculture is an area that is neglected within scientific literature. The usability is partly subjective attribute and needs evaluation from the user's perspective (Benda et al. 2015).

It is vital to establish effective decision models and support resources for the process of decision-making between particular technologies. The basic premise for appropriate decision-making is quick availability of quality data (Schimmelpfennig and Ebel 2016). To sum up, there are many factors that affect the decision-making process, i.e., economic profitability, investment rate of return, technological advancement, reliability, functionality, and usability. Farmers need information, whether particular technologies or methods are worth exploring.

This paper presents a User-Technological Index of Precision Agriculture. It is a complex system for evaluation of technologies and methods. Its objective is to provide users (farmers, suppliers, and researchers) with knowledge on the use of modern technologies in agriculture. Primarily, it is based on a point assessment of selected technologies (methods) of precision agriculture from the viewpoints of technological sophistication and usability for agricultural practice. It evaluates the principles of a given technology and not specific products and manufacturers.

The data for evaluation is collected from farmers, technology suppliers, researchers, and general public. The optimal way to get the data is to run a survey. Therefore, there is a need to spread questionnaires and get them back. Questionnaires can be disseminated in two major forms – printed on paper or electronically via the Internet. As Gordon and McNew (2008) suggest, it is important to make informed decisions as to which technology is the best one to implement. Many authors and studies suggest the Internet as the best carrying medium (Andrews et al. 2003; Lumsden 2005; Van Selm and Jankowski 2006). On the other side, web-based surveys generally exhibit a lower response rate (Fan and Yan 2010; Hamilton 2009).

2 Materials and Methods

The questionnaire for UTIPA is compiled from three general questions (email, country, and background) and ratings for several technologies. The rating is based on individual knowledge and experience. Each technology is evaluated in two individual criteria: *technological advancement* and *applicability in practice*. Both criteria consist of five-point scales where higher mark means the technology is better for a given criterion. The general principles of technologies are evaluated. Specific products, brands, or manufacturers are not reflected (Jarolímek et al. 2017).

The *technological advancement* criterion represents mainly a level of development and sophistication. It is evaluated whether the technology is in an experimental phase or in active use by farmers, whether it is mass produced by several manufactures or it is a technology of one individual producer. The criterion of *applicability in practice* includes evaluation of economic efficiency, quality, quantity of production, and ease of use. The technology needs to have well-proven economic benefits. The usability is evaluated based on experiences of the respondents.

The example for the evaluation can be the "Use of UAVs (drones) for crop imaging." The technology is well developed and proven to work (Pavlíček et al. 2018). So, it is very likely to score high in the *technological advancement* criterion. On the other side, the batteries do not last for a long period of time and need frequent recharging. Besides, drones are not cheap and need expertise to be operated. Therefore, the score in the *applicability in practice* is expected to be lower.

An important characteristic of the evaluated technology is also its unfamiliarity among the respondents. The survey provides also an option called "cannot judge."

The survey is disseminated in both major ways. The printed version is spread during conferences and seminars and periodically sent to Czech farmers. The web-based version of the questionnaire is shown in Fig. 1.

The Department of Information Technologies runs in cooperation with the Ministry of Agriculture of the Czech Republic a survey among Czech farmers about their

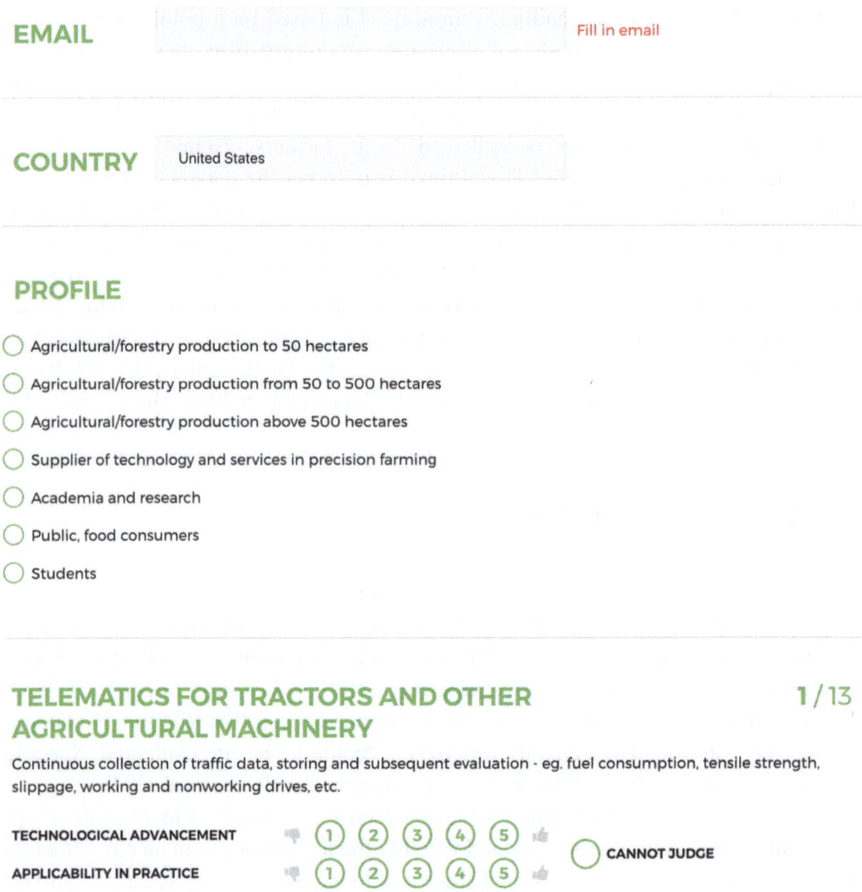

Fig. 1 The online version of the questionnaire – general questions and example of a technology evaluation

ICT equipment every 1 or 2 years since 2008. Questionnaires are primarily spread over the Internet via emails. Aside from that, some of the questionnaires are sent in printed form via classic post. The UTIPA questionnaire was included within last year's dissemination. The return rate of the printed questionnaires is still on a significantly higher level. In addition, the level of ICT equipment among Czech farmers is lower than a national average. The level of knowledge is significantly low in long term as well (Šimek et al. 2014). As mentioned above, the return rate of the printed form of questionnaire is higher than the electronic form. Therefore, the questionnaire for UTIPA is spread in printed form besides the electronic as well. The data gathered this way is inserted into the system manually later.

The process of survey dissemination is monitored, and various data are gathered. The online version of questionnaire as well as the whole application uses Google Analytics. It allows to analyze users' behavior on a website. All links that are used in promoting the questionnaire have a special URL attribute. This attribute helps to precisely identify the source of a user's visit. We can monitor various sources and campaigns such as emailing, QR codes, articles, etc. The monitoring of users continues during filling in the questionnaire and whole usage of the application.

One of the objectives of the UTIPA system is to form a community of people related to precision agriculture. Data are not collected via the entry questionnaire exclusively. There are too many technologies to be included in it. Respondents need to be motivated to sign in and evaluate more technologies. Another goal of the UTIPA is to track a development of the technologies over time. Therefore, the users should be motivated to come back to the application in the future and rate them again.

3 Results

The User-Technological Index of Precision Agriculture is a complex system that includes a methodology for collection, processing, and presentation of data and software which is available via a web interface. The software is also optimized for mobile devices. The user interface is designed using responsive web technologies, which allows the use of the website on different devices (mobile, tablet, desktop) via a web browser. It is also available as a native application for Android operating system. The data collection system for the questionnaire was maximally simplified.

3.1 Data Collection and the Entry Questionnaire

The system of data collection for UTIPA works on a simple principle. Each participant fills the email address, the basic information, and the ratings for selected technologies (approximately 10, based on the current occasion). To validate the contribution, an email with a confirmation link is sent to the given address. Once the provided link is clicked, the data and user are verified; and the participant can create a password. The user is automatically signed in and can immediately access the collected data in a given format. The whole process from the users' point of view, including the collection from the printed version of the questionnaire, is shown in Fig. 2.

Many surveys usually only spread a questionnaire and collect data. Results are delivered to the participants only occasionally. On the contrary, UTIPA works on the principle of "what data I provide is the type of data I gain access to." Therefore, each participant who filled the questionnaire has access to the results and can benefit from them.

We have observed that many questionnaires are left unfinished. Therefore, while the user is filling the form, all data are continuously saved on background via AJAX. This approach allows us to monitor the user's behavior and investigate the reasons

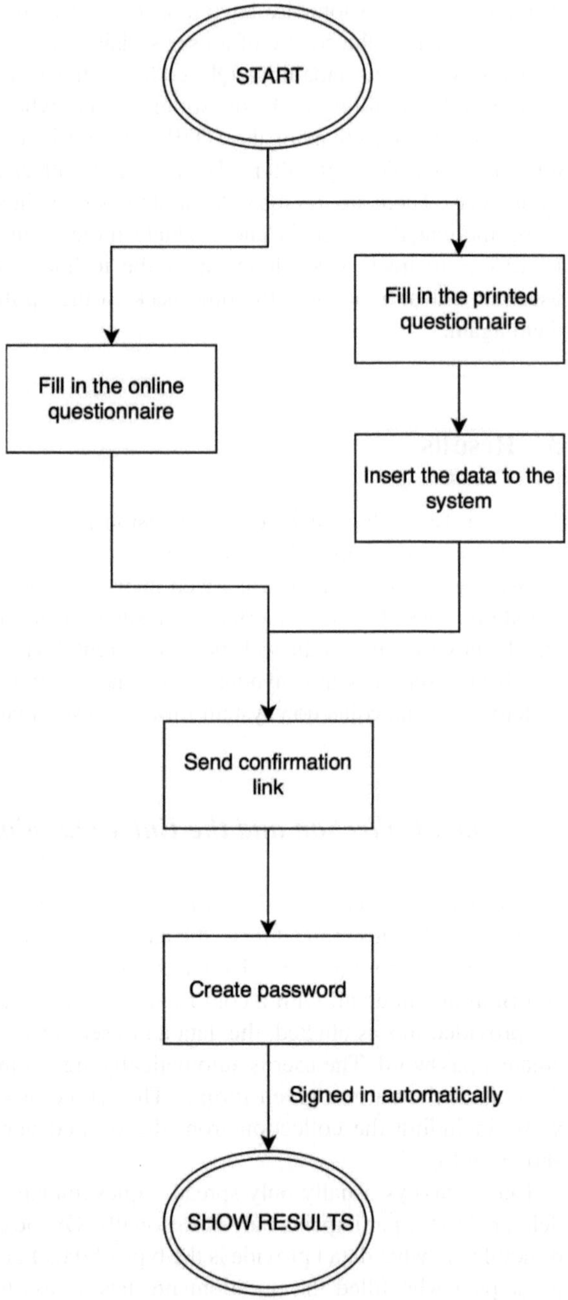

Fig. 2 The process of filling questionnaire and access to the data

behind incompletion. Furthermore, we have the data even though the participants did not finish the questionnaire for whatever reason. Moreover, users can continue filling it in later. We save a special hash using browser cookies. The same hash is saved in the application's database on a server side. So, when a user comes back using the same computer (and a same browser), they can continue to fill in the same questionnaire. The data they have already filled are automatically loaded. On top of that, the user can come back to the questionnaire using a different computer or browser. After the email is filled, the system recognizes that it is already known. The user is then offered with a possibility to receive a special activation link to continue filling the previously saved questionnaire. The whole process is shown in Fig. 3.

3.2 The UTIPA Index

The main contribution of the whole system is a calculation of the UTIPA index. Primarily, it is calculated for each technology and serves for various comparisons. The index consists of two parts, the numeric value and an additional character. The numeric part of the index has values between 0 and 5 and reflects the degree of usefulness and sophistication of the technology. The numeric value can be supplemented with the character which can be either "u" or "t" and expresses better ranking in favor of *applicability in practice* or *technological advancement* – the location of the point in the chart compared to the diagonal line (see Fig. 5). The methodology for calculation of the index was published by (Jarolímek et al. 2017). There is also a supplementary G-UTIPA index. It is calculated across a certain group of respondents including all technologies.

3.3 Data Visualizations

After successful email confirmation, all participants have access to the results and various data visualizations. There are currently six possible views on the data: basic statistics, technology comparison, country comparison, comparison according to respondent characteristics, unfamiliarity of technologies, and particular technology overview. Each view can visualize the data in a different way and follows the principle of "what data I provide is the type of data I gain access to." Therefore, users can only see results for the technologies that they have already rated.

Basic Statistics

The statistics view offers basic results in a form of table. The data can be filtered and ordered. Each technology is linked to its separate page with the overview. The view is shown in Fig. 4.

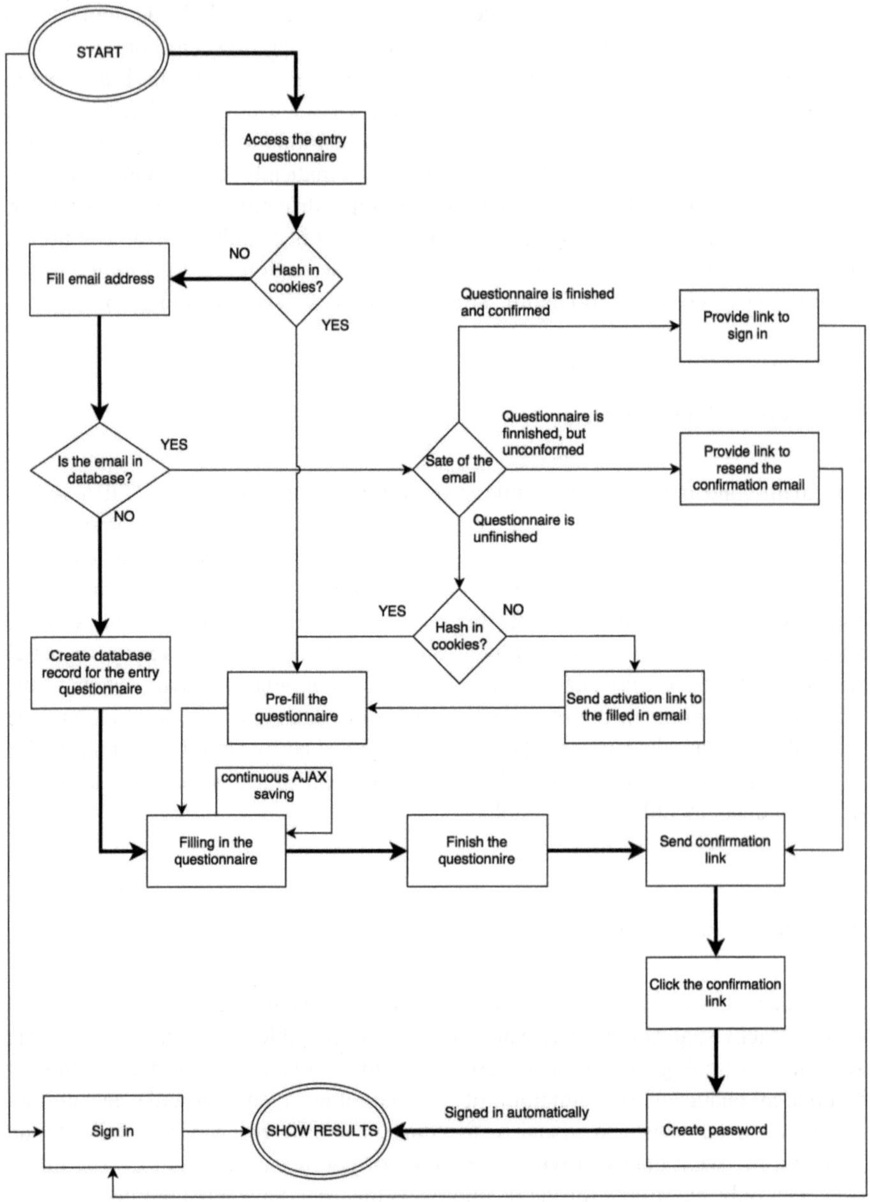

Fig. 3 The process of filling in the questionnaire with the return possibilities

Technology Comparison

The technology comparison view shows comparison of technologies between each other. As shown in the figure, the X-axis indicates *applicability in agricultural*

Order by: Title

List of technologies	Applicability in practice (My answer)	Technological advancement (My answer)	UTIPA index	Count
Automatic control of precision sowing according to sowing map	3.9 (0)	3.5 (0)	0.74U	27.27 %
Automatic regulation of fertilization dosage based on fertilization map	4.2 (0)	3.6 (0)	0.78U	14.56 %
Automatic regulation of row seeding based on seeding map	3.7 (3)	3.3 (4)	0.7U	20 %
Automatic setting of fertilization dosage based on current state of vegetation	3.9 (5)	3.4 (4)	0.73U	15.05 %
Autonomous guiding of agricultural machinery	4.1 (0)	3.4 (0)	0.75U	15.12 %
Guided drive of tractors and agricultural machinery on parcel	3.8 (0)	3.6 (0)	0.74U	15.71 %
Guided drive of tractors and agricultural machinery on parcel with 2 cm accuracy	4.1 (3)	3.8 (1)	0.79U	15.38 %
Mapping harvest yields of root crops	3 (3)	3.1 (4)	0.61T	22.22 %
Mapping the quality of harvested crops during harvesting of cereals	2.9 (3)	3.3 (2)	0.62T	18.18 %
Telematics for tractors and other agricultural machinery	3.9 (4)	3.4 (1)	0.73U	18.69 %
Use of unmanned vehicles for the management of crop production	3.2 (3)	2.6 (3)	0.58U	18.18 %

Fig. 4 Basic statistics view

practice, and the Y-axis indicates *technological advancement*. Each point in the chart represents a certain technology. When the number is hovered over, additional information is displayed. It contains name of the technology, exact values for both evaluation criteria, and the computed UTIPA index. In addition, users can compare own evaluation with the calculated values. It is indicated by a red circle as shown in Fig. 5. Users can additionally adjust the index calculation by setting specific filters. The view can be limited to certain countries (origin of respondents), characteristics of respondents, and particular technologies. The technology on the chart can be included in one of the fundamental segments: potential, practice, vision, and need. This is indicated by the background of the chart with an appropriate coloring.

Country Comparison

The country comparison view is based on the technology comparison. The main goal of the view is to compare the technologies according to the respondents' countries of origin. Within the view, each numbered circle represents the technology, as in the original view, and in addition, it is multiplied in the chart for each selected country.

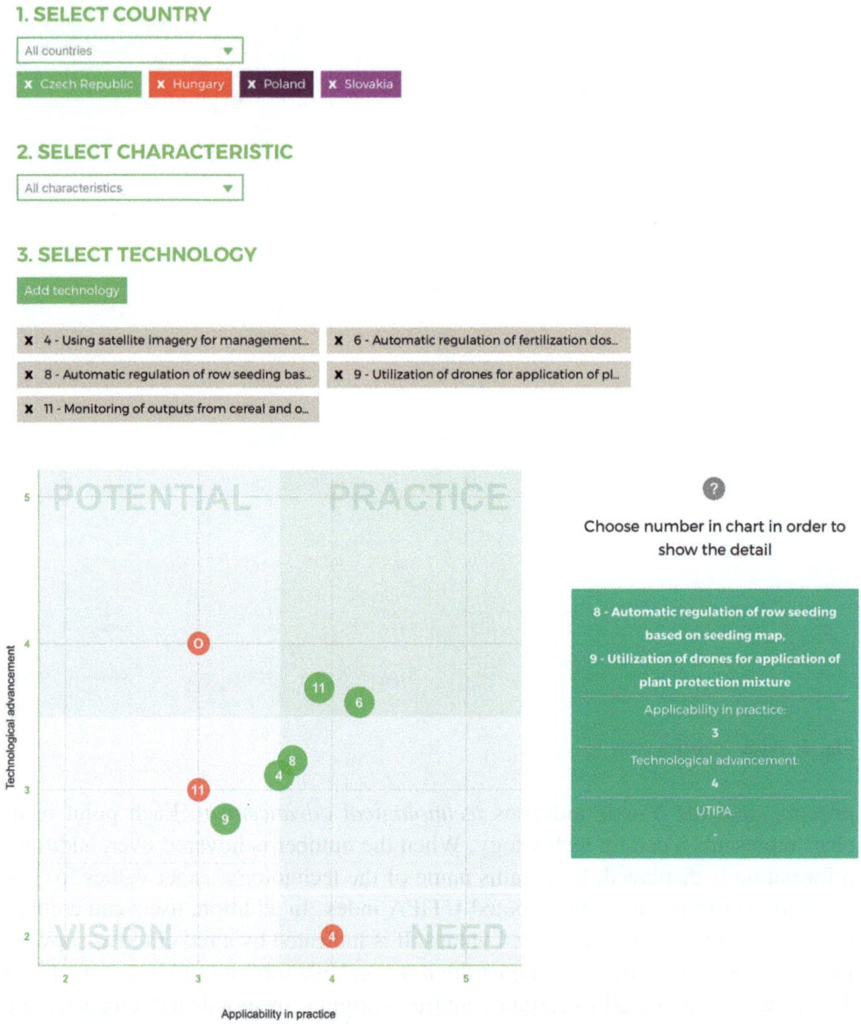

Fig. 5 Technology comparison view

This view is illustrated in the figure. In the chart there is a circle labeled with the G character. It stands for the G-UTIPA index and represents all technologies rated by all respondents from the given country (Fig. 6).

Comparison According to Respondent Characteristics

Another view represents comparison according to respondent characteristics. The main goal of this view is to compare the technologies according to the characteristic

Fig. 6 Country comparison view

of respondents. It is mostly identical to the country comparison mentioned in the previous chapter. As shown in the figure, the difference is the primary viewpoint for the comparison. The G-UTIPA index represents all technologies rated by a certain group of respondents (according to the characteristic) (Fig. 7).

Unfamiliarity of Technologies

Participants of the survey have also an option to indicate the unfamiliarity with a certain technology. It is an important characteristic for the results and represents the lack of knowledge about a certain technology. The output is then a comparison of

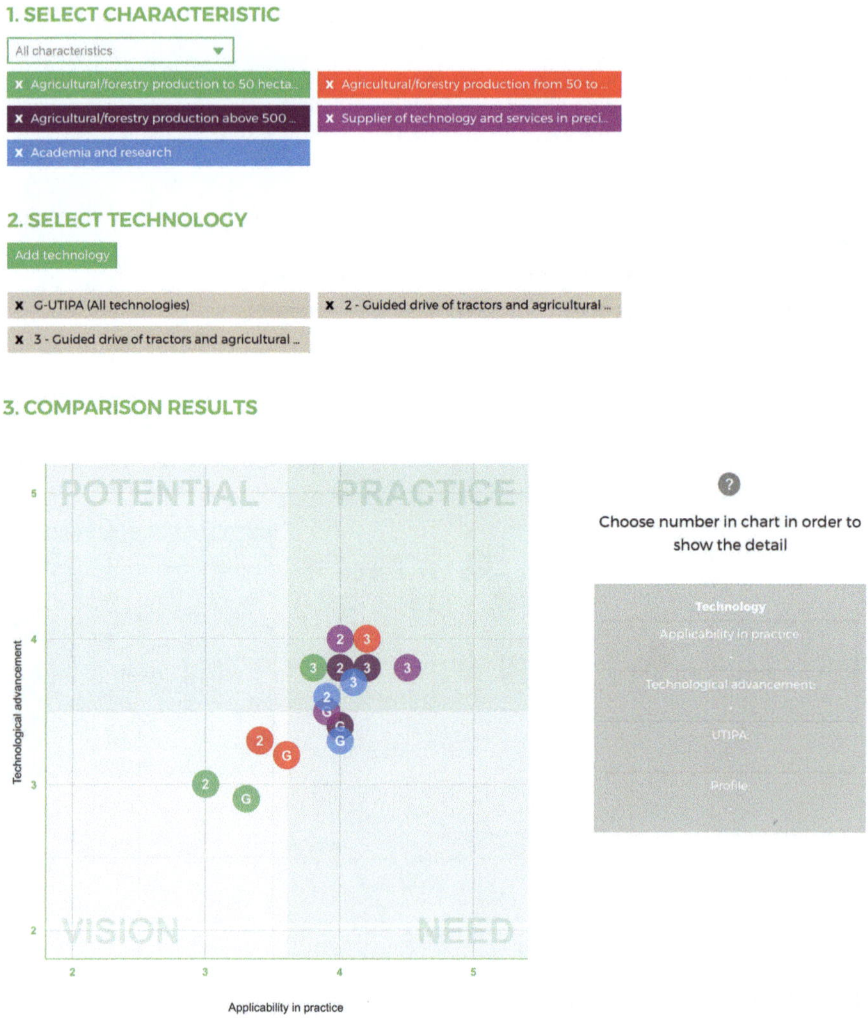

Fig. 7 Comparison according to respondent characteristics view

unfamiliarity of technologies as shown in Fig. 8. The red bars on the chart indicate that the current user marked the “Cannot judge” option for the given technology.

List of Technologies and Technology Overview

The UTIPA web application has a section called Technologies. It displays a simple list of all the technologies with selected results (calculated values) for each one. It shows the UTIPA index, both criteria values, and technology annotation. If the user has not evaluated the technology, the rating option is shown.

Fig. 8 Unfamiliarity of technologies view

Each technology has a detail page with the results' overview. There are tables of basic statistics data, charts for country, characteristics' comparison, and a heat map. The heat map chart, shown in Fig. 9, provides a graphical presentation for scatter of all the ratings. Users have also the possibility to change their ratings when the technology evolves, their opinion changes, or their level of knowledge increases.

4 Conclusion

User-Technological Index of Precision Agriculture is a complex system for the international community of people related to precision agriculture; it is accessible to anyone who respects the terms of use. The fundamental function and contribution of the system is a calculation of a special index called UTIPA. The index serves for various comparisons of precision agriculture technologies.

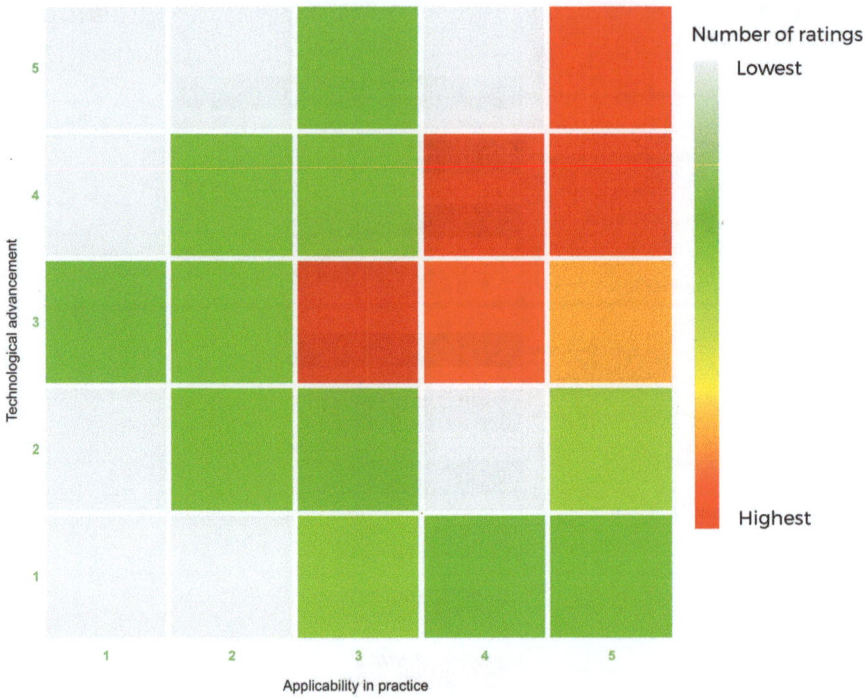

Fig. 9 Heat map visualization of the scatter of ratings by individual respondents

The proposed system for collecting questionnaire data was designed primarily as a web application gathering data online. It is also simultaneously spread in a printed form. It is disseminated in batches at certain occasions. The printed form has significantly higher return rate. On the other side, it requires financial input as well as additional labor to process incoming paper questionnaires. The online form of the survey needs significantly less resources. On the other side, its return rate is on a very low level.

The whole process of the data collection is monitored. For example, data are continuously saved during the filling of the online form. Therefore, users can continue to fill unfinished questionnaires even from a different computer. All data are continuously evaluated, and parts of the system are improved accordingly.

The system provides access to the visualizations of collected data for all participants. It works on the principle of "what data I provide is the type of data I gain access to." There are several views of the results: basic statistics, technology comparison, country comparison, comparison according to respondent characteristics, unfamiliarity of technologies, and particular technology overview.

The UTIPA index benefits all the stakeholders. Farmers can find out whether a given technology is useful and has real importance. Suppliers need to know what their customers (farmers) want or expect but also how they perceive their products.

For academia it can be a source of data for science and research. The importance and significance of the index grows with the number of respondents.

Future research and development is going to focus on several areas. The data about the collection process are going to be continuously monitored and evaluated. The visualization of the obtained data is going to be enhanced according to user needs. New types of views and comparisons will be introduced. To attract new participants, there is an encyclopedia of precision agriculture technologies under active development.

UTIPA system is freely available as a web application at https://www.utipa.info/.

Acknowledgments The results and knowledge included herein have been obtained owing to support from the following institutional grants: Internal Grant Agency of the Faculty of Economics and Management, Czech University of Life Sciences in Prague, grant no. 20181005 –"Processing of precision agriculture and IoT data in regards to big data in agriculture."

References

Andrews D, Nonnecke B, Preece J (2003) Conducting research on the internet: online survey design, development and implementation guidelines. Int J Hum Comput Interact 16:185–210. https://doi.org/10.1207/s15327590ijhc1602_04

Benda P, Smejkalova M, Ulman M (2015) Usability and accessibility analysis of Czech agrarian portals. In: Smutka L, Rezbova H (eds) Agrarian perspectives XXIV: global agribusiness and the rural economy, Agrarian perspectives series, Czech University Life Sciences Prague, Dept Systems Eng, Kamycka 129, Prague 6 165 21, Czech Republic's, pp 47–54

Cambouris AN, Zebarth BJ, Ziadi N, Perron I (2014) Precision agriculture in potato production. Potato Res 57:249–262. https://doi.org/10.1007/s11540-014-9266-0

CEMA (2018) Precision farming: key technologies & amp; concepts [WWW Document]. URL http://cema-agri.org/page/precision-farming-key-technologies-concepts (viděno 3.31.18)

Fan W, Yan Z (2010) Factors affecting response rates of the web survey: a systematic review. Comput Hum Behav 26:132–139. https://doi.org/10.1016/j.chb.2009.10.015

Gordon JS, McNew R (2008) Developing the online survey. Nurs Clin North Am 43:605–619, vii. https://doi.org/10.1016/j.cnur.2008.06.011

Hamilton MB (2009) Online survey response rates and times. Ipathia, Inc, N/A, pp 1–5

Jarolímek J, Stočes M, Masner J, Vaněk J, Šimek P, Pavlík J, Rajtr J (2017) User-technological index of precision agriculture. Agris On-line Pap Econ Informatics 9:69–75. https://doi.org/10.7160/aol.2017.090106

Kai CY, Perumal N (2010) Intelligent informatics platform for precision agriculture. In: Workshop Proceedings of the 6th International Conference on Intelligent Environments, pp 141–147. https://doi.org/10.3233/978-1-60750-639-3-141

Katalin T-G, Rahoveanu T, Magdalena M, István T (2014) Sustainable new agricultural technology – economic aspects of precision crop protection. Procedia Econ Finance 8:729–736. https://doi.org/10.1016/S2212-5671(14)00151-8

Kumhala F, Kroulik M, Masek J, Prosek V (2003) Development and testing of two methods for the measurement of the mowing machine feed rate. Plant Soil Environ 49:519–524

Lindblom J, Lundström C, Ljung M, Jonsson A (2016) Promoting sustainable intensification in precision agriculture: review of decision support systems development and strategies. Precis Agric 18:309–331. https://doi.org/10.1007/s11119-016-9491-4

Lumsden J (2005) Guidelines for the design of online-questionnaire. National Research Council Canada NRC/ERB 11, pp 44–64

Morota G, Ventura RV, Silva FF, Koyama M, Fernando SC (2018) Big Data Analytics and Precision Animal Agriculture Symposium: Machine learning and data mining advance predictive big data analysis in precision animal agriculture. J Anim Sci 96:1540–1550. https://doi.org/10.1093/jas/sky014

Paustian M, Theuvsen L (2017) Adoption of precision agriculture technologies by German crop farmers. Precis Agric 18:701–716. https://doi.org/10.1007/s11119-016-9482-5

Pavlíček J, Jarolímek J, Jarolímek J, Pavlíčková P, Dvořák S, Pavlík J, Hanzlík P (2018) Automated wildlife recognition. Agris On-line Pap Econ Informatics 10:51–60. https://doi.org/10.7160/aol.2018.100105

Schimmelpfennig D, Ebel R (2016) Sequential adoption and cost savings from precision agriculture. J Agric Resour Econ 41:97–115

Šimek P, Stočes M, Vaněk J (2014) Mobile access to information in the agrarian sector. Agris On-line Pap Econ Informatics 6:89–96

Van Selm M, Jankowski NW (2006) Conducting online surveys. Qual Quant 40:435–456. https://doi.org/10.1007/s11135-005-8081-8

Part III
ICT Applications in Agribusiness and New Business Models in the Agri-food Sector

Supplier Selection and Evaluation Using Multicriteria Decision Analysis

Stratos Kartsonakis, Evangelos Grigoroudis, and Michalis Neofytou

Abstract The selection and evaluation of suppliers is a very sensitive and crucial process for all companies and organizations, since the quality of the overall supplier management system may affect the quality of the final product. Suppliers nowadays are considered as partners of any business organization having mutual trust as the core of the relationship with them. In contrast to traditional approaches for selecting/evaluating a supplier, which are mainly based on cost, neglecting other important factors, new research efforts have highlighted the importance of multicriteria decision analysis in this problem. The aim of this paper is to propose an integrated framework for supplier selection and evaluation giving emphasis on the dynamic nature of the problem. The proposed framework is based on the UTA-II method, which is a preference disaggregation approach from the field of multicriteria decision analysis, and its applicability is demonstrated through a real-world application in a food company. The major results of the case study show the importance of quality, cost, delivery, and collaboration in the relationship between companies and suppliers.

Keywords Supplier selection · Supplier evaluation · Multicriteria decision analysis · UTA-II method · Food industry

1 Introduction

One of the most common and at the same time most crucial issue that modern business organizations face is the selection and evaluation of suppliers. This process plays a key role to the supply chain and is related with evaluating, ranking, and selecting supplies over a pool of potential suppliers, taking into account multiple

S. Kartsonakis (✉) · E. Grigoroudis · M. Neofytou
School of Production Engineering and Management, Technical University of Crete, Chania, Greece

A. Theodoridis et al. (eds.), *Innovative Approaches and Applications for Sustainable Rural Development*, Springer Earth System Sciences,
https://doi.org/10.1007/978-3-030-02312-6_11

criteria (Shemshadi et al. 2011). Supplier selection and evaluation can affect the overall performance of any industry, not only from the aspect of cost but also for the final product and the reputation of the company (Koufteros et al. 2012).

The importance of supplier-related decisions has been highlighted by numerous scholars, given their ability to affect the organizations' strategic processes (see, e.g., González et al. 2004). In addition, purchasing decisions are an important determinant of profitability, since in several cases a large part of production costs refers to raw materials. As many industries have established their supplying sources, it would be more efficient to create evaluation systems that could filter the performances of the suppliers (Mummalaneni et al. 1996).

In general, the supplier selection and evaluation problem is part of an overall supplier management process, which mainly involves nine distinct areas (González et al. 2004): (1) supplier selection process, (2) materials quality control, (3) supplier process control, (4) supplier documentation system, (5) supplier management system, (6) supplier availability, (7) supplier quality policies, (8) supplier quality system, and (9) material specification accomplishment.

The importance of supplier management is also emphasized in quality management systems. For example, the ISO 9001 quality management standard considers suppliers as critical partners of the organization. The standard requires that all suppliers should be evaluated, without, however, describing a detailed process or a set of evaluation criteria.

Business organizations with close partnership relationships with important suppliers can produce managerial, technological, and financial benefits (Ellram 1991), while developing a healthy buyer-supplier relationship may, in addition, improve the supplier performance. This relationship must be built with mutual trust and cooperation and demands from the buyer to work with a small number of suppliers. But when it comes to the final decision, the purchase manager has to consider many factors that shape the profile of a proper supplier. Some of the most common criteria, widely used in this respect, include price, quality, and delivery time, fact that justifies the multidimensional nature of the supplier selection and evaluation problem.

In order to have successful buyer-supplier relationships, a regular examination of supplier's performance is necessary. The evaluation of supplier performance has also an essential role in today's industries. It does not involve stakeholders of the supply chain but also improves the supply chain performance (Dey et al. 2015). However, despite the fact that several supplier selection and supplier evaluation systems have been proposed, there is no alignment between them. There is no point in organizing a process for the selection of a supplier without including a well-based evaluation system. This not only helps the buyer to keep track of the suppliers' performance but also motivates the supplier to improve the products and services delivered.

From a decision-making perspective, the supplier selection and the supplier evaluation are different problems, although this is not clear in several studies (see, e.g., Chen 2011). This is because the establishment of a permanent partner relationship between the firm and the supplier is often neglected, while these two problems have similar decision-making characteristics, like similar evaluation criteria or decision-makers (i.e., supply manager). In addition, the terminology used is often rather confusing: supplier selection is sometimes mentioned as supplier evaluation, while supplier evaluation is referred as supplier rating (see, e.g., Agarwal et al. 2011). However, these two decision-making problems have important differences. First, supplier selection is a pre-purchase process, while supplier evaluation is a post-purchase decision-making problem. As a result, although similar, these two problems have, in principle, different evaluation criteria. Moreover, supplier selection aims to identify the best supplier from a set of alternative suppliers, while supplier evaluation is mainly focused in providing a score that quantifies supplier performance. Despite their differences, however, the two problems are not independent; the evaluation of a supplier may affect the supplier's selection in future purchase decisions. On the other hand, the evaluation of a single supply is focused on a specific transaction.

In the aforementioned context, the main aim of the paper is to present a dynamic and holistic supplier selection and evaluation framework that will be able to consider the distinctive characteristics of the problem. Besides its dynamic nature, an additional contribution of the proposed framework is that the different evaluation problems are considered separately while integrated in an overall supplier management system. The proposed approach is based on the UTA-II method, which is a multicriteria decision analysis approach in the area of preference disaggregation models. UTA-II method can overcome several limitations of alternative multicriteria models by handling both quantitative and qualitative attributes, and modeling the preferences and priorities of the business organization regarding suppliers, while requiring limited inputs that can be easily provided by supplier managers. The aforementioned three dynamic systems (i.e., supply evaluation, supplier evaluation, and new supplier evaluation) are integrated together and function as a powerful tool that enables to have up-to-date information over the whole supply base and simultaneously indicate the best suppliers for the next order. The applicability of the proposed model is illustrated through a real-world application in a food industry having a large number of very diverse suppliers.

The paper is organized in five more sections. Section 2 presents briefly a literature review of the supplier selection and evaluation problem with emphasis on the applied quantitative tools. The theoretical background (UTA-II method) is given in Sect. 3, while Sect. 4 describes the proposed evaluation framework. Section 5 presents and discusses the results of a real-world application of the framework in a food industry. Finally, Sect. 6 summarizes concluding remarks and discusses potential extensions of the research.

2 Literature Review

Supplier selection is a process directly linked with the firms' overall performance and competitive advantage. Over the last decades, a shift has been made with respect to the buyer-supplier relationship. It has moved from a simple transaction-oriented relationship to a partnership-oriented relationship. Due to its great importance, the supplier selection problem comes with high complexity, as both qualitative and quantitative factors must be taken into consideration in order to better describe the overall capabilities of the supplier. For each firm, these factors are different, and each factor has a different importance. For this reason, multicriteria decision analysis has become the most popular tool for the supplier selection and evaluation problem.

Ho et al. (2010) reviewed 78 journal papers in 2000–2008, where all of them used multicriteria decision analysis. Other similar researches include Weber et al. (1991) who focused their research on the 23 criteria that Dickson (1966) first introduced and De Boer et al. (2001) who separated the supplier selection process into the pre-qualification of suitable suppliers' methods and the final-phase methods.

Generally, as mentioned by Dotoli et al. (2016), the analytic methods for the supplier selection problem may be categorized into three main domains:

1. Mathematical or operational research models: This category includes approaches like analytical hierarchy process (AHP), linear programming (LP), multi-objective programming (MOP), total cost of ownership (TCO), goal programming (GP), data envelopment analysis (DEA), simulation, and heuristics.
2. Statistical models: It mainly covers statistical-based and data analysis approaches, like cluster analysis, multiple regression, discriminant analysis, conjoint analysis, and principal component analysis.
3. Artificial intelligence models: This category comprises approaches like neural networks, software agents, case-based reasoning, expert systems, and fuzzy set theory (FST).

Dotoli et al. (2016) identify additional approaches that combine some of the aforementioned methods (e.g., AHP and GP, AHP and LP, AHP and FST, DEA and MOP). However, it should be emphasized that the previous approaches have different decision-making objectives, contexts, and orientations.

AHP and its extension ANP are the most widely applied methods in the supplier selection and evaluation problem. AHP was introduced by Saaty (1977, 1980) and uses pairwise comparisons, along with expert judgments to handle the measurement of qualitative or intangible attributes (Chai et al. 2013). The advantage of this method is its ability to structure a complex and multi-attribute problem hierarchically and then examine each level separately as the analysis progresses (Liu and Hai 2005). On the other hand, AHP can be highly subjective as it requires intense involvement from the decision-maker (DM). It can also be time-consuming, especially when there are many levels, due to the pairwise comparisons. Indicative AHP applications in this context include the work of Muraldiharan et al. (2002) who suggest a five-step AHP-based model incorporating nine evaluating criteria and multiple decision-makers involved in the process. Chan and Chan (2004) constructed an AHP model with 6 evaluating factors and 20 sub-factors and tested their model in an advanced technology industry. Other applications include the work of Liu and Hai (2005) who used Noguchi's voting ranking method instead of AHP's pairwise comparisons.

In the same context, the analytic network process (ANP) has been also widely used in the supplier selection and evaluation problem. ANP was developed by Saaty (1996), and its main difference compared to AHP is that it takes into consideration possible dependencies/influences between and within clusters/alternatives, without making any assumption on the interdependencies in the hierarchy (Gencer and Gürpinar 2007). In the examined problem, Sarkis and Talluri (2002) suggest that the evaluating criteria influence each other, and these criteria should consider not only operational attributes but also organizational and performance factors. Moreover, Gencer and Gürpinar (2007) applied ANP to an electronic company, where they classified the evaluating criteria into three clusters and examined their interrelationships. A drawback of this method is that the complexity increases with the increase in the number of factors and their relationships, and as a result more effort is demanded from both the DMs and the analysts (Sarkis and Talluri 2002).

Other approaches based on the multiple attributes of the problem include the Simple Multi-attribute Rating Technique (SMART), which is a simple additive weighting method to obtain total values as a ranking index (Chai et al. 2013) by using both qualitative and quantitative criteria. Indicative applications of this method may be found in Barla (2003) and Huang and Keskar (2007). Other multi-attribute/ multicriteria techniques include the Decision-Making Trial and Evaluation Laboratory (DEMATEL) (Chang et al. 2011) and the ELECTRE method (Sevkli 2010).

Data envelopment analysis (DEA) is a well-known technique that has been widely used for estimating the efficiency of suppliers by considering multiple input and output factors (Talluri et al. 2006). The efficiency of each supplier is expressed as the ratio of the weighted sum of the outputs (i.e., the supplier performance) to the weighted sum of the inputs (i.e., costs of using the supplier) (De Boer et al. 2001). DEA tries to find the best set of weights that maximize the efficiency and then categorizes the supplies to either efficient or inefficient. The main disadvantage of this method is that there is no distinction of the criteria to indicate the different importance of the suppliers. However, DEA has been widely applied in the supplier selection process (Narasimhan et al. 2001; Talluri and Baker 2002; Talluri et al. 2006).

Multiobjective programming (MOP) is another approach that has been used in the supplier selection and evaluation problem. It is a technique that considers multiple objective functions that can be optimized over a set of feasible solutions (Chai et al. 2013). Related research efforts in this field include the works of Narasimhan et al. (2006), Ozkok and Tiryaki (2011), and Wadhwa and Ravindran (2007), who suggested a set of three objective functions referring to the minimization of price, delivery time, and rejects. In the same context, several scholars applied other mathematical programming techniques, like linear programming (Talluri and Narasimhan 2003; Ng 2008), integer linear programming (Talluri 2002), integer nonlinear programming (Ghodsypour and O'Brien 2001), and goal programming (Karpak et al. 2001). Considering the supplier selection problem as a resource allocation problem, the main goal of these studies is to estimate the optimum solution based on a mathematical model under various constraints, which can be linear or nonlinear.

Another important part of the relevant literature refers to supplier selection and evaluation criteria. Identifying the right criteria may give a strategic advantage to the focal firm and should be related to the business process and the stakeholder's requirements (Dey et al. 2015). Dickson (1966) was the first to identify over 20 criteria that have been used in the supplier selection process. Over the years more and more criteria have been considered; still however, quality, delivery, and price/cost remain the most used ones (Ho et al. 2010). Additional criteria that have been widely used include service, flexibility, technology, and finance (see Table 1 for a summary of supplier selection criteria).

Most companies have established supplier bases. What is thus needed is a comprehensive system that allows them to monitor their suppliers in a sufficient way and keep track of their performance. Such a system may also be used to provide feedback to suppliers, thus helping them to improve their capabilities. However the criteria of the performance evaluation should be distinct and focus more on the operational capabilities of the supplier. In the literature there is no such a system that integrates both the supplier selection and performance evaluation. Table 2 presents an indicative list of supplier evaluation criteria.

Table 1 Supplier selection criteria

Criteria	Method	Authors
Quality; responsiveness; discipline; delivery; finance; management; technical capability; facility	AHP	Liu and Hai (2005)
Culture; technology; relationship; cost; quality; time; flexibility	ANP	Sarkis and Talluri (2002)
Outputs (quality, delivery); inputs (price)	DEA	Talluri et al. (2006)
Direct cost; indirect-coordination cost; quality; delivery reliability; complexity of supply arrangement	MOP	Narasimhan et al. (2006)
Supply variety; quality; distance; delivery; price	LP	Ng (2008)

Table 2 Supplier evaluation criteria

Criteria	Authors
On-time delivery; quality; price; professionalism of salesperson; responsiveness to customer needs; quality of relationship with the supplier	Mummalaneni et al. (1996)
Quality; price; delivery; cost reduction performance	Narasimhan et al. (2001)
Quality performance: compliance with quality; quality accreditation and audit; continuous quality improvement Delivery performance: compliance with due date; order to delivery lead time; delivery flexibility Costing performance: cost reduction capability; competitiveness of cost; appropriateness of the material price to the market price	Dey et al. (2015)
Manufacturing capability; supply chain capability; innovation capability; financial capability; human resource capability; service quality capability	Sun (2010)

3 The UTA-II Method

The UTA-II method belongs to the family of UTA methods (UTA, UTASTAR, UTADIS, etc.) that adopt the aggregation-disaggregation principles and are considered as the most representative example of preference disaggregation theory. Contrary to the traditional aggregation paradigm, where the criteria aggregation model is known a priori and the global preference is unknown, the philosophy of disaggregation involves the inference of preference models from given global preferences. The disaggregation-aggregation approach (Jacquet-Lagrèze and Siskos 1982, 2001; Siskos 1980; Siskos and Yannacopoulos 1985; Siskos et al. 1993) aims at analyzing the behavior and the cognitive style of the DM. Special iterative interactive procedures are used, where the components of the problem and the DM's global judgment policy are analyzed and then they are aggregated into a value system (see Fig. 1). The goal of this approach is to aid the DM to improve his/her knowledge about the decision situation and his/her way of preferring that entails a consistent decision to be achieved. A detailed overview of the UTA methods is given by Siskos et al. (2016).

In the UTA method, the criteria aggregation model is assumed to be an additive value function of the following form (Jacquet-Lagrèze and Siskos 1982):

$$u(\mathbf{g}) = \sum_{i=1}^{n} p_i u_i(g_i) \tag{1}$$

subject to normalization constraints:

$$\begin{cases} \sum\limits_{i=1}^{n} p_i = 1 \\ u_i(g_{i*}) = 0, u_i\left(g_i^*\right) = 1 \quad \forall i = 1, 2, \ldots, n \end{cases} \tag{2}$$

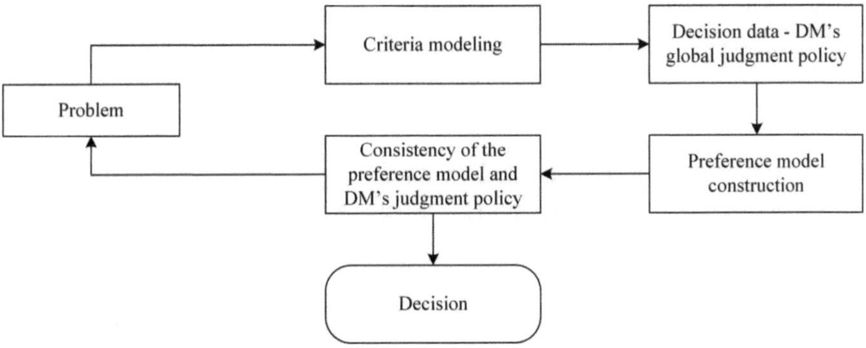

Fig. 1 The disaggregation-aggregation approach (Siskos et al. 2016)

where $u_i(g_i)$ is the marginal value function of criterion g_i, p_i is the weight of criterion g_i, $u(\mathbf{g})$ is the overall value of an alternative, n is the number of criteria, and g_{i*} and g_i^* are the worst and the best value of criterion g_i, respectively. It should be noted that u_i are non-decreasing real valued functions (marginal value or utility functions) normalized in the range [0, 1].

UTA-II is an extension of the UTA method that focuses on the estimation of weighting factors (scaling constants of criteria) p_i, assuming that u_i are known. As mentioned by Siskos et al. (2016), the u_i functions may be constructed using several techniques, such as multi-attribute utility theory (Keeney and Raiffa 1976; Klein et al. 1985), MACBETH method (Bana e Costa and Vansnick 1994, 1997), Quasi-UTA method (Beuthe et al. 2000), or the MIIDAS system (Siskos et al. 1999). The UTA-II algorithm follows a four-step procedure:

Step 1: The DM gives a preference ranking in a set of reference actions $A_R = \{a_1, a_2, \ldots, a_m\}$ where a_1 is the head and a_m is the tail of the ranking. For every pair of consecutive actions (a_m, a_{m+1}), either a preference condition $a_m \succ a_{m+1}$ (a_m is preferred to a_{m+1}) or an indifference condition $a_m \sim a_{m+1}$ (a_m is indifferent to a_{m+1}) holds. Using formula (1), the value of each alternative is given by:

$$u[g(a_m)] = \sum_{i=1}^{n} p_i u_i[g_i(a_m)] \tag{3}$$

where $u_i[g_i(a_m)]$ and $u[g(a_m)]$ are the marginal and the overall value of the alternative a_m, respectively.

Step 2: For each alternative two potential errors are inserted σ_m^+ and σ_m^-, which are the overestimation and the underestimation errors of $u[g(a_m)]$, respectively. Using these errors, the following differences are assessed for each pair of consecutive alternatives:

$$\Delta(a_m, a_{m+1}) = u[g(a_m)] - \sigma_m^+ + \sigma_m^- - u[g(a_{m+1})] + \sigma_{m+1}^+ - \sigma_{m+1}^- \tag{4}$$

Step 3: The following LP model is formulated and solved:

$$\begin{cases} [\min]F = \sum_{a_k \in A_R} \left[\sigma_k^+ + \sigma_k^-\right] \\ \text{subject to} \\ \Delta(a_m, a_{m+1}) \geq \delta \quad \text{if} \quad a_m \succ a_{m+1} \\ \Delta(a_m, a_{m+1}) = 0 \quad \text{if} \quad a_m \sim a_{m+1} \\ \sum_{i=1}^{n} p_i = 1 \\ p_i, \sigma_k^+, \sigma_k^- \geq 0 \quad \forall i, k \end{cases} \tag{5}$$

where δ is a very small positive number.

Step 4: A stability analysis is performed for the weights calculated from the solution of the LP model of step 3. This step can be described as a post-optimality analysis problem. Possible other optimal or near-optimal solutions can be found in order to lead to a perfect representation of the ranking of alternatives. The post-optimal analysis is described by a polyhedron (Fig. 2), which is defined by the constraints of the LP in step 3 and the following additional constraint:

$$\sum_{a_k \in A_R} \left[\sigma_k^+ + \sigma_k^-\right] \leq F^* + \varepsilon \tag{6}$$

where F^* is the optimal solution of the LP model solved in step 3 and ε is a very small positive number.

The final solution is calculated as the average of $2n$ solutions, provided by a set of linear programs, which maximize/minimize p_i subject to the constraints (5)–(6). In case of instability, the variation of the provided solutions is large, indicating that the average solution is less representative. In any case, the solutions of the above LPs give the internal variation of the weight of all criteria and consequently give an idea of the importance of these criteria in the DM's preference system.

The original development of the UTA-II method is presented in Siskos (1980), while applications of the UTA-II method may be found in Spyridakos et al. (2001) who combined the UTA-II method with the Cook and Seiford voting technique in order to handle the collaborative decision-making process with the construction of an additive value preference model for job evaluation and Androulaki and Psarras (2016) who developed a framework for assessing a large number of onshore and

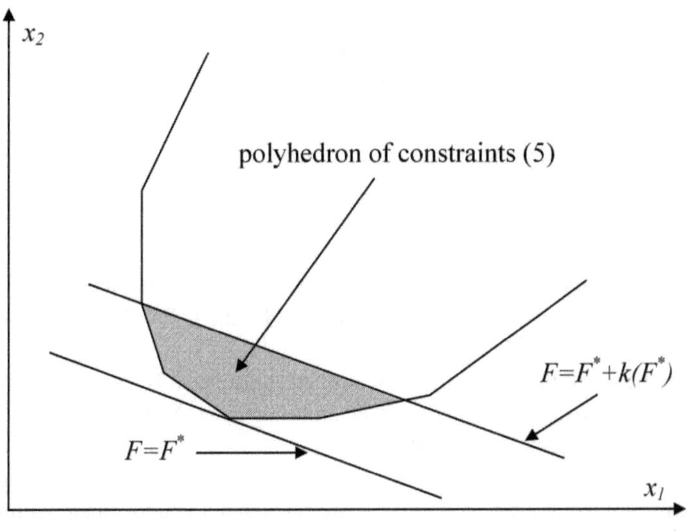

Fig. 2 Post-optimality analysis (Siskos et al. 2016)

offshore gas supply pipeline corridors for Greece. The application of the UTA-II method in collaborative group decision-making problems is also discussed in Spyridakos and Yannacopoulos (2018).

4 Proposed Approach

The first step to develop the proposed model was to categorize all the products and services that the company is being supplied. Given that in several cases the number of different supplies is huge, such an approach may decrease the complexity of the evaluation problem, as different sets of evaluation criteria may be assessed for a limited number of supply categories. The next step concerns the development of three discrete but related evaluation systems.

The first system refers to the evaluation of a single supply. For every product/service that the company buys, the supplier manager should rate the supply based on a predefined set of criteria. By implementing the UTA-II method, the weights of each criterion are estimated; therefore the total score of a supply is estimated using a simple weighted sum formula as follows:

$$U_{ik} = \sum_{j=1}^{n} p_j u_{ikj} \tag{7}$$

where U_{ik} is the score of the i-th supply of the k-th supplier, p_j is the weight of the j-th supply evaluation criterion ($j = 1, 2, \ldots, n$) as estimated by the UTA-II algorithm, and u_{ikj} is the score of the i-th supply of the k-th supplier on the j-th criterion.

The second evaluation system refers to the annual supplier evaluation. In this context, every supplier is rated by the supplier manager using a predefined set of criteria. The annual supplier evaluation score is calculated similarly to the previous case:

$$U_k^{year} = \sum_{l=1}^{m} p_l u_{kl} \tag{8}$$

where U_k^{year} is the annual score of the k-th supplier, p_l is the weight of the l-th supplier evaluation criterion ($l = 1, 2, \ldots, m$) as estimated by the UTA-II algorithm, and u_{kl} is the score of the k-th supplier on the l-th criterion.

It should be noted that in the aforementioned formulas, a 10-point scale is used for u_{ikj} and u_{kl}. Since p_j andp_l are normalized so that $\sum p_j = \sum p_l = 1$, the resulting U_{ik} and U_k^{year} are also normalized in [0, 10].

Based on (7) and (8), the overall supplier evaluation may be calculated as the weighted sum of the average score of the supplies that a vendor made with the score of the vendor's annual evaluation:

$$U_k^{global} = a\overline{U_k} + (1 - a)U_k^{year} \tag{9}$$

where U_k^{global} is the overall score of the k-th supplier, $\overline{U_k}$ is the average score of the annual supplies made by the k-th supplier, U_k^{year} is the annual score of the k-th supplier, and a is a weighting factor that aggregates the supplier annual evaluation and the average scores of supplies.

Depending on the previous overall score U_k^{global}, each supplier is enlisted in a different category. Although alternative categorizations may be adopted, for the purpose of this study, three categories are used:

- Authorized suppliers
- Suppliers under examination
- Unauthorized suppliers

As the score is dynamically updated, it is possible for a supplier to enter in different categories during such time. However, if it becomes an unauthorized supplier, then it will not be selected again.

The final evaluation system refers to new suppliers. Again the supplier manager rates the new supplier in each of the evaluation criteria defined for this case. This time also a threshold is assigned to each criterion. If the score exceeds the respective threshold, then the examined supplier is approved; otherwise it is not selected and becomes inadequate. In other words, a new supplier is approved only if $u_{kr} \geq b_r \, \forall \, r$, where u_{kr} is the score of the k-th supplier on the r-th new supplier evaluation criterion and b_r is the minimum acceptance score (threshold) of the r-th criterion ($r = 1,2,\ldots, t$). As it can be observed, the new supplier evaluation system does not apply the UTA-II algorithm.

Figure 3 summarizes how the aforementioned three evaluation systems are integrated together, offering a dynamic tool that may be used in the overall supplier management system of a business organization.

For each one of the previous evaluation systems, a set of criteria has been selected. The assessment of these evaluation criteria has been based on the literature review presented in Sect. 2 and the collaboration with the quality manager of the examined food industry. Table 3 presents the criteria for the evaluation of supplies, the annual evaluation of suppliers, and the evaluation of new suppliers.

The criteria were selected in a way that could express the strengths and weaknesses of each supplier. As a result, the company would have an overall better view of their performance, given that the evaluation criteria are important to fulfill company's requirements. Some of the selected criteria such as cost, quality, and delivery time have been applied in numerous previous studies, while others like cooperation (Chan et al. 2007) and quality accreditation and audit (Karpak et al. 2001) have been found in more specialized works. Furthermore, some criteria proposed in this study have not been considered in previous researches, but they are very important for the examined food industry. These criteria include guarantee that expresses the prerequisites that a supplier should have if a problem with the order occurs, as well as reliability that expresses the image of the supplier in the market.

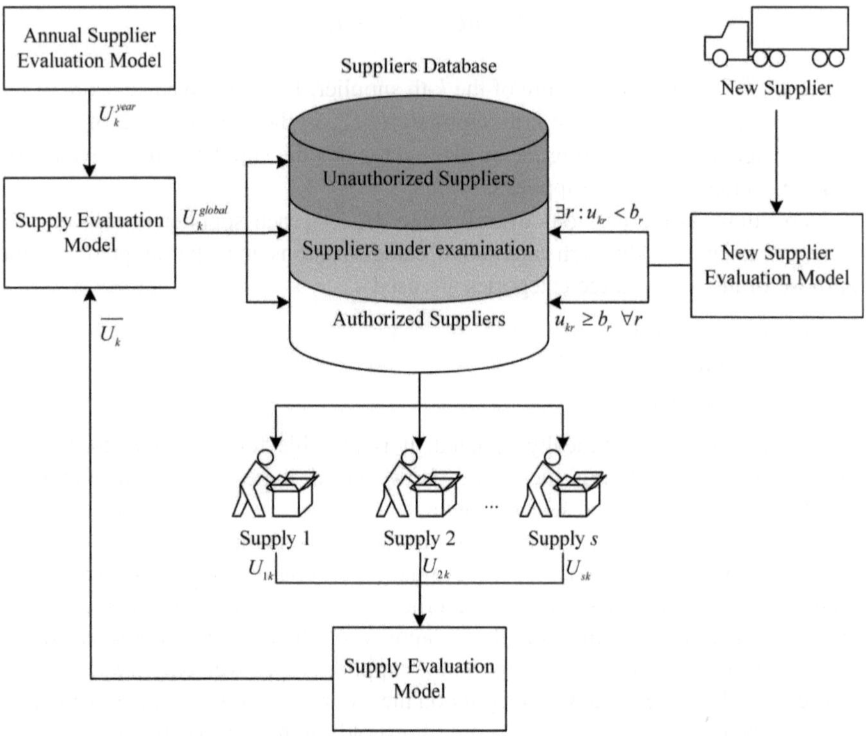

Fig. 3 Integrated supplier evaluation approach

Table 3 Evaluation criteria in the proposed approach

Supply evaluation	Annual supplier evaluation	New supplier
Price Quality of product/service Cooperation Responsiveness to customer needs Delivery time	Cost Cost reduction performance Quality Cooperation Reliability Delivery time Guarantee	Reliability Cooperation Free products for trial Quality accreditation and audit Payment Guarantees

5 Application to a Food Industry

The proposed framework has been applied in a food industry which needed an improved supplier evaluation system. The company is a mill industry certified according to the ISO 9001 standard for quality management. The company produces a wide range of flour products for home use, professional, as well as for animal feeding. The supplies include a very large number of different materials: raw materials for flour and animal feed, auxiliary materials for flour and animal feed, packaging materials, services, consumables, etc.

Having chosen the evaluation criteria, the proposed approach was applied and tested considering all the supplies that the company made within a 1-year period. In order to facilitate the whole process, a database using MS Excel was developed. Table 4 presents some indicative results for the evaluation of a supply where the company cooperated with three different suppliers. More specifically, the evaluation of a supply takes place in the following way:

- The assessor (person that made the order) opens the database where all the suppliers and the products that the company purchased are recorded.
- As soon as the product and the supplier that he/she wants to evaluate is chosen, he/she gets immediately information of the supplier's overall score, its annual evaluation score, and the average score of its past supplies.
- Then he/she simply rates the product currently received in the set of defined evaluation criteria, in order to automatically evaluate the score and also the state of the supply.

As shown in Table 4, the food industry is not satisfied with the price it got for the purchased laptops from all the three suppliers. It is also dissatisfied by supplier 3 who got a poor evaluation in all criteria. Moreover, supplier 2 did not receive a good score at the criteria of cooperation and delivery time. At this point it should be mentioned that if a supply receives a score 5 or lower, the company may not accept it and ask for a refund.

The results of the UTA-II method reveal that quality, collaboration, and delivery are the most important criteria for the examined food industry. More specifically, the estimated criteria weights are the following: price (0.1462), quality (0.2134), cooperation (0.2590), responsiveness (0.1221), and delivery time (0.2592).

Table 4 Example for the evaluation of different supplies of the same product

	Supply 1	Supply 2	Supply 3
Category	Office supplies	Office supplies	Office supplies
Product	Laptops	Laptops	Laptops
Supplier	Supplier 1	Supplier 2	Supplier 3
Overall score	8.07	7.17	5.02
Annual evaluation	8.12	8.08	5.04
Average score of supplies	8.02	6.26	5.00
Score of the supply	8.02	6.26	5.00
State of the supply	Authorized	Under examination	Unauthorized
Price	6	4	5
Quality	9	9	5
Cooperation	9	6	5
Responsiveness	9	9	5
Delivery time	9	5	5

As already described, the annual supplier evaluation process is held once a year. The supplier manager evaluates all the suppliers that cooperated with the company over the past year, giving a score based on the assessed evaluation criteria. As soon as he/she gives these individual ratings, the annual evaluation score is calculated, and the overall supplier score is estimated accordingly. In all of these cases, the supplier database is updated. In the example of Table 5, two suppliers received good scores in all the criteria, while the other two underperformed. Their low score classify them as unauthorized suppliers. More importantly, this classification is the result of both the annual supplier evaluation and the supply evaluation. If the company did not use an integrated supplier evaluation system and was based only on the annual supplier evaluation, then supplier 4 would not be characterized as an anauthorized supplier. Instead he would be considered as an under examination supplier which could lead to an undesirable collaboration in the future.

Similar to the previous case, the results of the UTA-II regarding the annual supplier evaluation show the importance of cost and quality. More specifically, the estimated criteria weights are the following: cost (0.2684), cost reduction (0.1365), quality (0.1318), cooperation (0.12), reliability (0.1167), delivery time (0.1156), and guarantee (0.1111).

Regarding the new supplier selection process, when there is a new supplier that wants to cooperate with the firm (or the opposite), the supplier manager evaluates the candidate based on the previously described set of criteria. In the proposed approach, a threshold is set for each criterion, and therefore if the candidate supplier receives a score equal or higher of the threshold, then it is classified as approved; otherwise it is rejected. Based on the examples of Table 6, supplier 1 receives a score lower than the threshold to the criteria of reliability and cooperation and thus is categorized as rejected. On the other hand, supplier 2 receives scores above the threshold in all of the criteria, and for that reason he is being approved. The approved suppliers of the new supplier selection model are then being classified as suppliers under examination.

Table 5 Example of annual supplier evaluation

	Supplier 1	Supplier 2	Supplier 3	Supplier 4
Category	Animal feed	Animal feed	Animal feed	Animal feed
Product	Corn	Corn	Corn	Corn
Overall score	8.34	8.80	7.81	4.93
State	Authorized	Authorized	Unauthorized	Unauthorized
Annual evaluation	8.41	8.80	4.81	5.76
Cost	9	9	5	5
Cost reduction	7	7	5	5
Quality	9	9	3	4
Cooperation	9	9	5	5
Reliability	9	9	5	5
Delivery time	7	9	5	5
Guarantee	7	9	5	5

Table 6 Example of new supplier selection of the same product

	Threshold	Supplier 1	Supplier 2
Category		Animal feed	Animal feed
Product		Soda	Soda
Reliability	7	6	8
Cooperation	7	6	9
Free products for trial	7	8	8
Quality accreditation and audit	7	7	9
Payment	8	7	8
Guarantee	7	7	8
Approval		Rejected	Approved

Concluding, the process described in the previous examples may be applied in all products. The food industry in this way can have all the necessary and constantly updated information regarding suppliers and keep track of their detailed performances.

6 Conclusions and Future Research

The dynamic nature of the proposed framework is its main advantage compared to other alternative approaches, given its ability to provide a valuable tool to business organizations of all types, where supplier evaluation information is being constantly updated. The process of a continuous evaluation of a supply offers to the supplier manager the capability to keep track of a vendor's performance through the year. In addition, the proposed model functions not only as a performance measurement tool for the supplier, but it helps the supply manager to find the best supplier for the next order. As the scores are updated, he/she can check who has the best overall score at a specific time and pick the best one. Moreover, the assessment of an appropriate set of evaluation criteria gives the ability to the buyer to track the weaknesses of a supplier. For instance, a vendor may have a good score in delivery time and a poor one in quality. So the company will have the ability to take the right measures next time; it will trust the same supplier.

The presented case study shows the applicability of the proposed approach in the case of a food company. The major results of the UTA-II method show that the most important evaluation criteria are related to quality, cost, delivery, and collaboration, although all criteria are somehow important to the food industry. In general, this is consistent with several studies in the field (see, e.g., Ho et al. 2010; Dickson 1966). However, some differences may be found comparing the criteria weights between the evaluation of supplies and the evaluation of suppliers. In the first case, greater importance is given to fundamental attributes or attributes that affect a particular transaction (e.g., cost, quality, and cooperation), while in the second case, cost-related criteria seem to have the most important role.

Although the proposed supplier evaluation/selection framework uses the UTA-II method, other alternative preference disaggregation approaches may also be applied in this problem. UTA-II has been chosen mainly because of its simplicity and our intention to focus on the weighting factors (scaling constants of criteria) of the evaluation process. In addition, as noted in Sect. 3, several techniques for constructing the criteria value functions may be incorporated in the proposed approach (multi-attribute utility theory, MACBETH MIIDAS, etc.).

The proposed framework has some limitations, which may provide opportunities for future research. One such issue is who will be responsible for the supply and the supplier evaluation. Optimally, the evaluations should be given by the department that made the order. For instance, for raw materials the responsible personnel for the evaluation is the production manager. However, due to the fact that multiple criteria have been applied and the raw materials may affect different aspects of the company, more departments should be involved in the evaluation process. The production manager should first be informed by the quality department that the product meets the requirements. In this context, the quality department is responsible for evaluating the supply in the criterion of quality. Such issues can be time-consuming and demand a very good coordination.

Furthermore, the database developed for the purpose of the study is relatively friendly, but it comes also with some limitations. As the supplier selection and evaluation systems are already complex, it should be adopted to more advanced software or even be incorporated to other software that business organizations already use, like SAP. This way, the supplier management processes can be incorporated in the company's procedures, affecting the overall organizational culture.

Finally, a limitation of the proposed approach refers to criteria weights. The products are classified into eight large categories, so that the appropriate criteria may be identified more easily. However, for each product category, the same set of criteria weights is applied. Although this may increase the complexity of the problem, future research may examine different weighting schemes for the product categories.

Conclusively, the supplier selection and evaluation problem are one of the most critical strategic processes of modern business organizations, offering numerous research opportunities for developing and implementing new multicriteria decision analysis methods.

References

Agarwal P, Sahai M, Mishra V, Bag M, Singh V (2011) A review of multi-criteria decision making techniques for supplier evaluation and selection. Int J Ind Eng Comput 2(4):801–810

Androulaki S, Psarras J (2016) Multicriteria decision support to evaluate potential long-term natural gas supply alternatives: the case of Greece. Eur J Oper Res 253(3):791–810

Bana e Costa CA, Vansnick JC (1994) MACBETH: an interactive path towards the construction of cardinal value functions. Int Trans Oper Res 1(4):489–500

Bana e Costa CA, Vansnick JC (1997) Applications of the MACBETH approach in the framework of an additive aggregation model. J Multi-Criteria Decis Anal 6(2):107–114

Barla SB (2003) A case study of supplier selection for lean supply by using a mathematical model. Logist Inf Manag 16(6):451–459

Beuthe M, Eeckhoudt L, Scanella G (2000) A practical multicriteria methodology for assessing risky public investments. Socio Econ Plan Sci 34(2):121–139

Chai J, Liu JNK, Ngai EWT (2013) Application of decision-making techniques in supplier selection: a systematic review of literature. Expert Syst Appl 40(10):3872–3885

Chan FTS, Chan HK (2004) Development of the supplier selection model: a case study in the advanced technology industry. Proc Inst Mech Eng B J Eng Manuf 218(12):1807–1824

Chan FTS, Chan HK, Ip RWL, Lau HCW (2007) A decision support system for supplier selection in the airline industry. Proc Inst Mech Eng B J Eng Manuf 221(4):741–758

Chang B, Chang CW, Wu CH (2011) Fuzzy DEMATEL method for developing supplier selection criteria. Expert Syst Appl 38(3):1850–1858

Chen YJ (2011) Structured methodology for supplier selection and evaluation in a supply chain. Inf Sci 181(9):1651–1670

De Boer L, Labro E, Morlacchi P (2001) A review of methods supporting supplier selection. Eur J Purch Supply Manag 7(2):75–89

Dey PK, Bhattacharya A, Ho W (2015) Strategic supplier performance evaluation: a case-based action research of a UK manufacturing organization. Int J Prod Econ 166:192–214

Dickson GW (1966) An analysis of vendor selection systems and decisions. J Purch 2(1):5–17

Dotoli M, Epicoco N, Falagario M, Sciancalepore F (2016) A stochastic cross-efficiency data envelopment analysis approach for supplier selection under uncertainty. Int Trans Oper Res 23 (4):725–748

Ellram LM (1991) A managerial guideline for the development and implementation of purchasing partnerships. Int J Purch Mater Manag 27(3):2–8

Gencer C, Gürpinar D (2007) Analytic network process in supplier selection: a case study in an electronic firm. Appl Math Model 31(11):2475–2486

Ghodsypour SH, O'Brien C (2001) The total cost of logistics in supplier selection, under conditions of multiple sourcing, multiple criteria and capacity constraint. Int J Prod Econ 73(1):15–27

González ME, Quesada G, Mora Monge CA (2004) Determining the importance of the supplier selection process in manufacturing: a case study. Int J Phys Distrib Logist Manag 34 (6):492–504

Ho W, Xu X, Dey PK (2010) Multi-criteria decision making approaches for supplier evaluation and selection: a literature review. Eur J Oper Res 202(1):16–24

Huang SH, Keskar H (2007) Comprehensive and configurable metrics for supplier selection. Int J Prod Econ 105(2):510–523

Jacquet-Lagrèze E, Siskos Y (1982) Assessing a set of additive utility functions for multicriteria decision making: the UTA method. Eur J Oper Res 10(2):151–164

Jacquet-Lagrèze E, Siskos Y (2001) Preference disaggregation: 20 years of MCDA experience. Eur J Oper Res 130(2):233–245

Karpak B, Kumcu E, Kasuganti RR (2001) Purchasing materials in the supply chain: managing a multi-objective task. Eur J Purch Supply Manag 7(3):209–216

Keeney RL, Raiffa H (1976) Decisions with multiple objectives: preferences and value tradeoffs. Wiley, New York

Klein G, Moskowitz H, Mahesh S, Ravindran A (1985) Assessment of multiattribute measurable value and utility functions via mathematical programming. Decis Sci 16(3):309–324

Koufteros X, Vickery SK, Dröge C (2012) The effects of strategic supplier selection on buyer competitive performance in matched domains: does supplier integration mediate the relationships? J Supply Chain Manag 48(2):93–115

Liu FHF, Hai HL (2005) The voting analytic hierarchy process method for selecting supplier. Int J Prod Econ 97(3):308–317

Mummalaneni V, Dubas KM, Chao C (1996) Chinese purchasing managers' preferences and trade-off in supplier selection and performance evaluation. Ind Mark Manag 25(2):115–124

Muraldiharan C, Anantharaman N, Deshmukh SG (2002) A multi-criteria group decision-making model for supplier rating. J Supply Chain Manag 38(3):22–33

Narasimhan R, Talluri S, Méndez D (2001) Supplier evaluation and rationalization via data envelopment analysis: an empirical examination. J Supply Chain Manag 37(2):28–37

Narasimhan R, Talluri S, Mahapatra SK (2006) Multiproduct, multicriteria model for supplier selection with product life-cycle considerations. Decis Sci 37(4):577–603

Ng WL (2008) An efficient and simple model for multiple criteria supplier selection problem. Eur J Oper Res 186(3):1059–1067

Ozkok BA, Tiryaki F (2011) A compensatory fuzzy approach to multi-objective linear supplier selection problem with multiple-item. Expert Syst Appl 38(9):11363–11368

Saaty TL (1977) A scaling method for priorities in hierarchical structures. J Math Psychol 15 (3):234–281

Saaty TL (1980) The analytic hierarchy process. McGraw-Hill, New York

Saaty TL (1996) Decision making with dependence and feedback: the analytic network process. RWS Publications, Pittsburgh

Sarkis J, Talluri S (2002) A model for strategic supplier selection. J Supply Chain Manag 38 (1):18–28

Sevkli M (2010) An application of the fuzzy ELECTRE method for supplier selection. Int J Prod Res 48(12):3393–3405

Shemshadi A, Shirazi H, Toreihi M, Tarokh MJ (2011) A fuzzy VIKOR method for supplier selection based on entropy measure for objective weighting. Expert Syst Appl 38 (10):12160–12167

Siskos J (1980) Comment modéliser les préférences au moyen de fonctionsd'utilité additives. RAIRO RechercheOpérationelle 14:53–82

Siskos Y, Yannacopoulos D (1985) UTASTAR: an ordinal regression method for building additive value functions. Investigação Operacional 5(1):39–53

Siskos J, Spyridakos A, Yannacopoulos D (1993) MINORA: a multicriteria decision aiding system for discrete alternatives. J Inf Sci Technol 2(2):136–149

Siskos Y, Spiridakos A, Yannacopoulos D (1999) Using artificial intelligence and visual techniques into preference disaggregation analysis: the MIIDAS system. Eur J Oper Res 113(2):236–246

Siskos Y, Grigoroudis E, Matsatsinis NF (2016) UTA methods. In: Greco S, Ehrgott M, Figueira J (eds) Multiple criteria analysis: state of the art surveys, 2nd edn. Springer, New York, pp 315–362

Spyridakos A, Yannacopoulos D (2018) Collaborative decision making for small groups utilizing UTA methods. In: Matsatsinis NF, Grigoroudis E (eds) Preference disaggregation in multiple criteria decision analysis. Springer, New York, pp 59–76

Spyridakos A, Siskos Y, Yannacopoulos D, Skouris A (2001) Multicriteria Job evaluation for large organizations. Eur J Oper Res 130(2):375–387

Sun CC (2010) A performance evaluation model by integrating fuzzy AHP and fuzzy TOPSIS methods. Expert Syst Appl 37(12):7745–7754

Talluri S (2002) A buyer-seller game model for selection and negotiation of purchasing bids. Eur J Oper Res 143(33):171–180

Talluri S, Baker RC (2002) A multi-phase mathematical programming approach for effective supply chain design. Eur J Oper Res 141(3):544–558

Talluri S, Narasimhan R (2003) Vendor evaluation with performance variability: a max-min approach. Eur J Oper Res 146(3):543–552

Talluri S, Narasimhan R, Nair A (2006) Vendor performance with supply risk: a chance-constrained DEA approach. Int J Prod Econ 100(2):212–222

Wadhwa V, Ravindran AR (2007) Vendor selection in outsourcing. Comput Oper Res 34 (12):3725–3737

Weber CA, Current JR, Benton WC (1991) Vendor selection criteria and methods. Eur J Oper Res 50(1):2–18

Competitive Advantage Establishment Through Sustainable Environmental Management and Green Entrepreneurship: A Proposed Differential Equations Framework

Michalis Skordoulis, Miltiadis Chalikias, Spyros Galatsidas, and Garyfallos Arabatzis

Abstract Worrying phenomena connected to climate change need to be treated effectively as they can lead to negative consequences in peoples' lives. At the same time, as businesses are dominant forms of social organization, it can contribute to the forming of the natural environment. This interaction has led businesses to face a set of new challenges concerning environmental management. However, the effective response to these challenges is not just only an obligation for businesses; it can also lead them to the development of a competitive advantage as many researchers state. The aim of this paper is to propose the framework of applying operational research models in the creation of competitive advantage based on green entrepreneurship principles and the sustainable management of the environment. Such mathematical models are based on differential equations and were originally used to predict the outcome of battles. However, the similarity of business competition to battlefields has led to several applications of such models in business cases.

Keywords Environmental sustainability · Competitive advantage · Competitive strategy · Operations research · Green entrepreneurship

1 Introduction

The main goal of any kind of business is to obtain a sustainable competitive advantage (Wheelen and Hunger 2012). There are two main trends: the first one defines competitive advantage as a business's ability to impose conditions favorable

M. Skordoulis (✉) · S. Galatsidas · G. Arabatzis
Department of Forestry and Management of the Environment and Natural Resources, Democritus University of Thrace, Nea Orestiada, Greece

M. Chalikias
Department of Tourism Management, University of West Attica, Egaleo, Greece

A. Theodoridis et al. (eds.), *Innovative Approaches and Applications for Sustainable Rural Development*, Springer Earth System Sciences,
https://doi.org/10.1007/978-3-030-02312-6_12

in commercial transactions, while the second trend defines competitive advantage as any cause or factor leading a business to superior performance compared with its competitors (Sigalas et al. 2013).

In this context, various theoretical models have been developed over time. According to Black and Boal (1994), searching for business opportunities, focusing on strengths, improving weaknesses, and avoiding risks have been a key framework for developing a competitive advantage. Porter (1980, 1985) states that businesses can gain a competitive edge either by recognizing opportunities in existing markets or by recognizing new markets. Furthermore, he developed a framework for examining competition in a market, whereby the environment of each business is determined by the threat of new businesses entry, the negotiating power of the business's suppliers, the negotiating power of the business's customers, the threat of substitute products, as well as the intensification of competition among existing businesses (Porter 1991).

All the traditional theoretical models for business competition analysis like the aforementioned ones have a static character. On the contrary, the current environment in which businesses operate is very dynamic, similar to the environment during a war clash (Chalikias and Skordoulis 2014, 2016). For the analysis of such an environment, operations research models have been developed and can be applied to business competition cases (Chalikias and Skordoulis 2014, 2016).

At the same time, businesses have been faced with a set of new challenges in environmental management. Worrying phenomena such as the climate change, the increase in average temperatures, and the greenhouse gas emissions lead to increasingly negative environmental and economic impacts, which need to be treated effectively (Melville 2010; Papageorgiou et al. 2015; Ntanos et al. 2018a, b). In addition, the need for environmentally friendly products is increasingly imperative (Chalikias et al. 2010a, b; Kolovos et al. 2011; Kyriakopoulos et al. 2010a, b).

Thus, since the early 1990s, the goal is to create a sustainable future (Hutchinson 1992). A number of researchers (Nidumolu et al. 2009; Arabatzis et al. 2015; Ojo et al. 2015) note that the integration of sustainable environmental management into a business strategy can be a new important source for developing a competitive advantage.

The aim of this paper is to propose a theoretical framework to establish competitive advantage based on differential equations models. This will be achieved by using operations research models in order to analyze business competition based on sustainable environmental management and green entrepreneurship principles.

2 Sustainable Environmental Management and Competitive Advantage Establishment

In recent years, more and more businesses are turning to practices that take into account environmental protection and sustainable management. On the one hand, this trend stems from increasing pressure from so-called business stakeholders such as environmental organizations, consumers, and regulators (Orlitzky et al. 2011). In this sense, sustainability of the environment could be the most important challenge of corporate social responsibility for business (DesJardins 2007). On the other hand, the trend toward integrating sustainable environmental management into business strategy has to do with contributing to the creation of a competitive advantage.

Corporate social responsibility is inseparably linked to environmental protection. The main benefits a business can gain from the adoption and implementation of corporate social responsibility practices linked to environmental sustainability deal primarily with the enhancement of corporate reputation, improved taxation, improved relations with the local community, and better risk management and security (Urip 2010). Consequently, actions related to environmental protection which can be expressed through a framework of corporate social responsibility have the potential to lead to higher economic return and profitability; thus, they can lead to the establishment of a competitive advantage (Urip 2010; Skordoulis et al. 2013; Ntanos et al. 2014).

According to Danjelico and Pujari (2010), sustainable environmental management is linked to a business's competitive advantage by reducing energy and raw material usage as well as pollutant production. Furthermore, they state that the production of green products, i.e., products produced by methods that minimize the burden on the environment (Ottman et al. 2006), leads to a competitive advantage through the agreement with the established legal framework for the environment, corporate social responsibility, and competitiveness.

Wu and Lin (2016) report that the production of green products contributes to minimizing the environmental impact of products throughout their lifetime, including the purchase of raw materials, manufacture, sale, consumption, and disposal after use.

According to Rohm and Montgomery (2010), revenue growth, risk reduction, and cost reduction will increase a business's profitability and lead to an advantage over its competitors.

In particular, a strong emphasis on environmental sustainability leads to the development of more eco-efficient products, strengthens cooperation with environmental regulators more effectively, and reduces the environmental impact of business operations (Rohm and Montgomery 2010).

In addition, new technology capabilities are created to support more effectively environmental impact analysis. By producing more eco-efficient products, a business can provide value to an increasing number of green consumers, which will lead to increased sales. It will also be possible for better and more efficient cooperation with regulators. This will reduce business risk, since eco-friendly and safe products will reduce the potential risk of environmental liability and consumers. This reduced risk will have a positive impact on the cost of capital. In turn, actions to reduce the impact of productive activity on the environment will in turn lead to immediate cost savings for fuel, water, electricity, and waste disposal (Rohm and Montgomery 2010).

Similarly, Ambec and Lanoie (2008) connect environmental sustainability with the competitive advantage by quoting the following four main points:

- Lower cost of raw materials, energy, and services
- Lower production costs through improved efficiency
- Lower capital costs
- Improved risk management and stakeholder relations

Based on the aforementioned data, a successful environmental management can be closely linked to the development of a competitive advantage. This concerns both large businesses and small- and medium-sized enterprises (SMEs). Many researchers have dealt with the correlation between environmental management and SMEs' financial performance using variables such as return on equity (ROE), return on assets (ROA) and earnings before interest, tax, depreciation, and amortization (EBITDA) (King and Lenox 2001; Ong et al. 2014; Jayeola 2015; Malesios et al. 2018).

A useful guide for any businesses on the road to reorganization aiming to focus on environmental sustainability is the certification through appropriate standards. In particular, certification with an environmental standard such as ISO 14001 clearly defines business procedures and processes that lead to cost savings, improved and more efficient management, customer satisfaction and regulatory requirements, and improved environmental performance (Rondinelli and Vastag 2000; Zhang et al. 2000; Lawrence et al. 2002; Soerger Zaro et al. 2015). ISO implementation, in most of the cases, may need bilateral agreements between customers and suppliers in B2B transactions. These agreements may lead to a wider implementation of sustainable environmental management principles.

Finally, we cannot forget that that the cost of organizing a business focusing on sustainable environmental management can be considered quite high. However, businesses should not have to perceive the adoption of sustainable environmental management practices as unnecessary costs but as long-term investments that eventually lead to the development of a competitive advantage (Becker-Olsen et al. 2006; Porter and Kromer 2006).

3 Business Competition Analysis and Strategy Development Based on Differential Equations Research Models

3.1 Differential Equations Model Development

During the two world wars, mathematical models were widely used in various combat cases. These models were designed to solve various optimization problems related to the ongoing war operations. By the end of these wars, it was argued that after the appropriate modifications, such models could be applied to business cases. The main concept was that business competition is very similar to armies' operation in the battlefields (Chalikias and Skordoulis 2014, 2016). The developement and use of optimization problems' solving models are one of the greatest advances in mathematics and science from the 1930s (Naidu et al. 2007).

Optimization models which are basically based on differential equations have been used in a wide range of business-related issues such as marketing strategy analysis (Taoka 1997), oligopoly competition strategies (Fruchter and Kalish 1997), product development based on customers' needs (Fehlmann 2008), advertising costs (Chintagunta and Vilcassim 1992; Chalikias and Skordoulis 2014; Chalikias et al. 2016a, b), and supply chains (Chalikias and Skordoulis 2016).

Unlike traditional static models that can only explain competition in a market, combat models based on differential equations such as these of Lanchester (1916, 1956) can not only be applied in a market's multidimensional analysis but also are able to predict the future sizes of it based on the established competition (Fehlmann 2008).

In this case, the competition between two businesses could be analyzed by a system of differential equations as the following one:

$$\begin{cases} \dfrac{dx}{dt} = -ay + f(t) \\ \dfrac{dy}{dt} = -bx + g(t) \end{cases} \tag{1}$$

where $x(t)$ and $y(t)$ would be the operating revenues of the two competitors and $f(t)$ and $g(t)$ the rate at which they are increased or decreased. This rate is affected by one or more variables in each model. Such variables could be advertising costs (Chalikias and Skordoulis 2014), sales at points of sale (Chalikias and Skordoulis 2016), or environmental investments.

Operational research models like the above one have been successfully implemented in various business cases. Chalikias and Skordoulis (2016) applied Lanchester's combat models in the oligopoly of Coca-Cola and Pepsi Cola concerning their logistics. Specifically, they have used a differential equations model where $x(t)$ and $y(t)$ are the number of available product units for sale at a specified time (t). During the operation of the two competitors, the rate of change of the quantities $x(t)$ and $y(t)$ is equal to the rate of growth of refueling at the points of sale, minus the rate of sales.

If $f(t)$ and $g(t)$ are the rates of product units sold which are increasing and decreasing and $ay(t)$ and $bx(t)$ are the rates of the available for sale product (a and b positive constants), the differential equations model, according Lanchester's combat model, is this of Eq. 1.

The solution of the above system for $x = 0$ and $y = 0$ is:

$$x(t) = -\sqrt{\frac{a}{b}} y_0 \frac{e^x + e^{-x}}{2} \left(\sqrt{ab} t\right) + \int_0^t \frac{e^x + e^{-x}}{2} \left(\sqrt{ab}(t-s)\right) f(s) ds \tag{2}$$

$$y(t) = -\sqrt{\frac{b}{a}}x_0 \frac{e^x + e^{-x}}{2}\left(\sqrt{ab}t\right) - \int_0^t \frac{e^x + e^{-x}}{2}\left(\sqrt{ab}(t-s)\right)g(s)ds \quad (3)$$

In the case where $f(t) = 0$ and $g(t) = 0$, which means that sales don't change, the differential equations system (1) is now:

$$\begin{cases} \frac{dx}{dt} = -ay \\ \frac{dy}{dt} = -bx \end{cases} \quad (4)$$

The above model's solution is:

$$x(t) = \frac{1}{2}c_1 e^{-\sqrt{a}\sqrt{b}t}\left(e^{2\sqrt{a}\sqrt{b}t} + 1\right) + \frac{\sqrt{a}c_2 e^{-\sqrt{a}\sqrt{b}t}\left(e^{2\sqrt{a}\sqrt{b}t} - 1\right)}{2\sqrt{\lambda}} \quad (5)$$

$$y(t) = \frac{1}{2}c_2 e^{-\sqrt{a}\sqrt{b}t}\left(e^{2\sqrt{a}\sqrt{b}t} + 1\right) - \frac{\sqrt{a}c_1 e^{-\sqrt{a}\sqrt{b}t}\left(e^{2\sqrt{a}\sqrt{b}t} - 1\right)}{2\sqrt{a}} \quad (6)$$

The solution results in c_1 and c_2. These coefficients are equal to $x(t)$ and $y(t)$. When $t = 0$ then $c_1 = x(0)$ and $c_2 = y(0)$.

The differential equation system's (4) trajectories are the solutions of the following differential equation:

$$\frac{dy}{dx} = \frac{bx}{ay}$$

From the above equation, we can have the following relationship which is called Lanchester's square type, because of squares that appear in this.

$$ay^2 - bx^2 = ay_0^2 - bx_0^2 = k \quad (7)$$

Chintagunta and Vilcassim (1992) have used Lanchester's combat model to examine the investments in advertising of a duopoly. A similar model using data of Coca-Cola and Pepsi advertising expenditures was used by Erickson (1992), in order to analyze the advertising strategies of the two firms. Fruchter and Kalish (1997) have used Lanchester's model to describe the dynamics of a duopoly market. They proved that the model solution is a global Nash equilibrium. In their work, Chalikias and Scordoulis (2016) use data concerning the abovementioned firms' revenues from sales between 2003 and 2007, and they estimated the model's coefficients. The statistical analyses they carried showed a good fitting of the model's predictions compared to the real data.

In order to expand the usage of Lanchester's combat model, Chalikias et al. (2016a, b) applied a 3×3 differential equations model in order to describe the advertising expenditures in the mobile communications market in Greece. This market consists of three firms, namely, Cosmote, Vodafone, and Wind. The model was:

$$\begin{cases} \dfrac{dx}{dt} = \dfrac{b}{a}y + \dfrac{c}{a}z \\ \dfrac{dy}{dt} = \dfrac{a}{b}x + \dfrac{c}{b}z \\ \dfrac{dz}{dt} = \dfrac{a}{c}x + \dfrac{b}{c}y \end{cases} \tag{8}$$

The variables x, y, and z represent the advertising expenditures of the three firms, and the parameters a, b, and c are the quotas of the three firms in Greek market.

In the following solution, parameter t represents the time:

$$\begin{cases} x(t) = -c_1\dfrac{c}{a}e^{-t} - c_2\dfrac{b}{a}e^{-t} + c_3\dfrac{c}{a}e^{2t} \\ y(t) = c_2e^{-t} + c_3e^{2t}\dfrac{c}{b} \\ z(t) = c_1e^{-t} + c_3e^{2t} \end{cases} \tag{9}$$

In order to measure the time with a good fitting and representative way, a function $t(i)$ where $i\epsilon[0,1]$ was assumed. At the beginning $i = 0$, and at the end of the examined time ($i \in [0,1]$), $i = 1$. It was made the correspondence $i = 0$ for 2000, $i = 1/6$ for 2001, $i = 2/6$ for 2002, etc. After the iterative method, where all the functions of the form $t(i) = ic^i$ $c \in [0,1]$ were examined, it was concluded that the $t(i) = i0.4^i$ gives very good fitting results.

For the examination of the model's good fit on the real data, proper statistical analyses have been carried out, such as parametric (paired sample T-test) and nonparametric (Wilcoxon signed-rank test) tests.

The Lanchester models are not limited here. Generalized Lanchester model describes the evolution of the battle conducted between homogeneous regular military forces (they hold the same weapons such as planes, tanks, etc.), when there is no aid or operating loss, and:

- Each side sprinkles targeted fires.
- The rate of loss of each side is proportional to the number of remaining units of weapon of the opponent side.

An example of such a battle is the battle between scheduled conventional (not rebel) forces targeted fire. For the definition of the model, we have the following (Daras 2001).

Let us consider two rival armed military forces, 1 and 2. We assume that the force 1 has M_1 different types of weapons:

$$O\Sigma_1^{(1)},\ O\Sigma_1^{(2)},\ O\Sigma_1^{(3)}, \ldots, O\Sigma_1^{(M_1)},$$

with respective cardinality (congeners) weapon units for each type of weapon system at the time t from the start of the battle:

$$\mathrm{X}_1^{(1)} = \mathrm{X}_1^{(1)}(t), \mathrm{X}_1^{(2)} = \mathrm{X}_1^{(2)}(t), \ldots, \mathrm{X}_1^{(M_1)} = \mathrm{X}_1^{(M_1)}(t) \tag{10}$$

Similarly, we assume that force 2 has M_2 different types of weapons:

$$O\Sigma_2^{(1)},\ O\Sigma_2^{(2)},\ O\Sigma_2^{(3)}, \ldots, O\Sigma_2^{(M_2)},$$

with respective cardinality (congeners) weapon units for each type of weapon system at the time t from the start of the battle:

$$\mathrm{X}_2^{(1)} = \mathrm{X}_2^{(1)}(t), \mathrm{X}_2^{(2)} = \mathrm{X}_2^{(2)}(t), \ldots, \mathrm{X}_2^{(M_2)} = \mathrm{X}_2^{(M_2)}(t) \tag{11}$$

We denote as $\mathrm{X}_1 = \mathrm{X}_1(t)$ the matrix – column:

$$\mathrm{X}_1 = \mathrm{X}_1(t) = \left[\mathrm{X}_1^{(1)}\ \mathrm{X}_1^{(2)} \ldots\ \mathrm{X}_1^{(M_1)}\right]^{\mathrm{T}} = \left[\mathrm{X}_1^{(1)}(t)\ \mathrm{X}_1^{(2)}(t) \ldots\ \mathrm{X}_1^{(M_1)}(t)\right]^{T}$$

and $X_2 = X_2(t)$ the matrix – column:

$$\mathrm{X}_2 = \mathrm{X}_2(t) = \left[\mathrm{X}_2^{(1)}\ \mathrm{X}_2^{(2)} \ldots\ \mathrm{X}_2^{(M_1)}\right]^{\mathrm{T}} = \left[\mathrm{X}_2^{(1)}(t)\ \mathrm{X}_2^{(2)}(t) \ldots\ \mathrm{X}_2^{(M_1)}(t)\right]^{T}$$

For every $i = 1,2,\ldots,M_1$ and $j = 1,2,\ldots,M_2$, we define: $(i \to j)$– Coefficient Military Efficacy of force 1 against the force 2:

$$\theta_1^{(i \to j)} = \theta_1^{(i \to j)}(t, X_1, X_2\ ) \geq 0$$

as the rate of the overall wear causing all (similar) weapon units of the weapon system force $1O\Sigma_1^{(i)}$ against (congeners) weapon units of the power $2O\Sigma_2^{(j)}$.

Similarly, for every $j = 1,2,\ldots,M_2$ and $i = 1,2,\ldots,M_1$, we define: $(j \to i)$– Coefficient Military Efficacy of force 2 against the force 1:

$$\theta_2^{(j \to i)} = \theta_2^{(j \to i)}(t, X_1, X_2\ ) \geq 0$$

as the rate of the overall wear causing all (similar) weapon units of the weapon system force 2 $O\Sigma_2^{(j)}$ against (congeners) weapon units of the power 1 $O\Sigma_1^{(i)}$.

The coefficient of Total Military Efficacy of force 1 against force 2 is defined as the $(M_2 \times M_1)$ matrix:

$$\Theta_1 := \left(\theta_1^{(i\to j)}\right)_{j,i} = \begin{pmatrix} \theta_1^{(1\to 1)} & \theta_1^{(2\to 1)} & \cdots & \theta_1^{(M_1\to 1)} \\ \theta_1^{(1\to 2)} & \theta_1^{(2\to 2)} & \cdots & \theta_1^{(M_1\to 2)} \\ \vdots & \vdots & \ddots & \vdots \\ \theta_1^{(1\to M_2)} & \theta_1^{(2\to M_2)} & \cdots & \theta_1^{(M_1\to M_2)} \end{pmatrix}$$

Similarly, the coefficient of Total Military Efficacy of force 2 against force 1 is defined as the $(M_1 \times M_2)$ matrix:

$$\Theta_2 := \left(\theta_2^{(j\to i)}\right)_{i,j} = \begin{pmatrix} \theta_2^{(1\to 1)} & \theta_2^{(2\to 1)} & \cdots & \theta_2^{(M_2\to 1)} \\ \theta_2^{(1\to 2)} & \theta_2^{(2\to 2)} & \cdots & \theta_2^{(M_2\to 2)} \\ \vdots & \vdots & \ddots & \vdots \\ \theta_2^{(1\to M_1)} & \theta_2^{(2\to M_1)} & \cdots & \theta_2^{(M_2\to M_1)} \end{pmatrix}$$

We are now able to give the deterministic Lanchester model that describes every battle between the armed military forces with targeted shots:

$$\begin{cases} \dot{X}_1 = -\Theta_2 \mathrm{X}_2 \\ \dot{X}_2 = -\Theta_1 \mathrm{X}_1 \end{cases}$$

More analytical, we may have:

$$\frac{dX_1^{(1)}}{dt} = -\theta_2^{(1\to 1)} X_2^{(1)} - \theta_2^{(2\to 1)} X_2^{(2)} - \ldots - \theta_2^{(M_2\to 1)} X_2^{(M_2)}$$
$$\frac{dX_1^{(2)}}{dt} = -\theta_2^{(1\to 1)} X_2^{(1)} - \theta_2^{(2\to 1)} X_2^{(2)} - \ldots - \theta_2^{(M_2\to 1)} X_2^{(M_2)}$$
$$\vdots$$
$$\frac{dX_1^{(M_1)}}{dt} = -\theta_2^{(1\to M_1)} X_2^{(1)} - \theta_2^{(2\to M_1)} X_2^{(2)} - \ldots - \theta_2^{(M_2\to M_1)} X_2^{(M_2)}$$

$$\frac{dX_2^{(1)}}{dt} = -\theta_1^{(1\to 1)} X_1^{(1)} - \theta_1^{(2\to 1)} X_1^{(2)} - \ldots - \theta_1^{(M_1\to 1)} X_1^{(M_1)}$$
$$\frac{dX_2^{(2)}}{dt} = -\theta_1^{(1\to 2)} X_1^{(1)} - \theta_1^{(2\to 2)} X_1^{(2)} - \ldots - \theta_1^{(M_1\to 2)} X_1^{(M_1)}$$
$$\vdots$$
$$\frac{dX_2^{(M_2)}}{dt} = -\theta_1^{(1\to M_2)} X_1^{(1)} - \theta_1^{(2\to M_2)} X_1^{(2)} - \ldots - \theta_1^{(M_1\to M_2)} X_1^{(M_1)}$$

Therefore, we have the following:

$$\dot{X} = \Theta X \tag{12}$$

where:

$$X{:=}[X_1X_2]^T$$

and Θ is the coefficient of Total Military Efficacy of the two forces, hence the $(M_1{+}M_2) \times (M_1{+}M_2)$ matrix:

$$\Theta{:=}\begin{bmatrix} O & -\Theta_2 \\ -\Theta_1 & O \end{bmatrix}$$

For the numerical solution of differential system (9), there are various numerical methods known. The simplest is derived from the formula:

$$\frac{1}{\Delta t}(X_{n+1} - X_n) = \Theta\, X_n$$

which leads directly to the following linear system:

$$X_{n+1} = (I_n + \Delta t_n\, \Theta)X_n \quad (n = 0, 1, 2, \ldots, N) \tag{13}$$

where:

- X_n and X_{n+1} are the values of the matrix – column: $X := [\,X_1X_2]^{\mathrm{T}}$ at times t_n and t_{n+1}, respectively.
- $\Delta t_n := t_{n+1} - t_n$ is the length of time of the numerical integration.
- N is the number of repetitions.

Furthermore, it is worthy investigating the case where one of the armies accepts enhancement. In this case, it is very difficult to find the analytical solution, but the numerical solution is possible. If during the time $[t_n, t_{n+1}]$, one of the forces receives aid or withdrawal of some weapon units from battle, then relation 10 takes the form:

$$X_{n+1} = (I_n + \Delta t_n\, \Theta)X_n + \mathrm{E}_n \quad (n = 0, 1, 2, \ldots, N) \tag{14}$$

where E_n is the matrix:

$$E_n = \begin{pmatrix} E_1^{(1)}(\Delta t_n) \\ E_1^{(2)}(\Delta t_n) \\ \vdots \\ E_1^{(\mathrm{M}_1)}(\Delta t_n) \\ E_2^{(1)}(\Delta t_n) \\ E_2^{(2)}(\Delta t_n) \\ \vdots \\ E_2^{(\mathrm{M}_2)}(\Delta t_n) \end{pmatrix}$$

which at the first M_1 rows represent the unit number of enhancements or withdrawal for every $O\Sigma_1^{(i)}$ $i = 1, 2, 3, \ldots, M_1$ of force 1 during time $[t_n, t_{n+1}]$, and at next M_2 rows represent the units number of enhancements or withdrawal for every $O\Sigma_2^{(j)}$ $i = 1, 2, 3, \ldots, M_2$ of force 1 during time $[t_n, t_{n+1}]$.

In the above differential equations models, the regular variables related to military operations can be replaced by business variables as in all the examined cases. Such variables could concern green entrepreneurship components such as investments for environmental sustainability, corporate social responsibility investments, energy consumption reduction, gas emissions reduction, or raw materials usage reduction.

3.2 Business Strategy Development

Using differential equations models like the above ones, the competition between two or more businesses can be analyzed based on specified variables. In this way, it is possible to capture the strong and weak points of each business in the competition analysis; in this way, the dimensions that need to be improved can be pointed out (Fehlmann 2008).

Once the strong and weak points have been recorded, the next step will be to formulate the appropriate business strategy. For this purpose, tools of modern business management can be used.

Such a tool is the quality function deployment (QFD) which aims to integrate customer needs and requirements at all stages of the development of a business strategy (Felhmann 2008; Chaudha et al. 2011). The QFD method can indicate a business where and how to focus in order to gain a competitive advantage (Hwarng and Teo 2001).

Based on the QFD framework, an environmentally friendly business should operate in environmentally responsible way (Sakao 2007; Lam and Lai 2015). Therefore, it should translate environmental requirements into processes resulting in the possible development of a competitive advantage as already discussed (Oudrihiri, 2005; Yazdani et al., 2016).

Another tool that could be used to achieve environmental sustainability is the balanced scorecard (BSC). BSC concept is that the effective use of investment funds is no longer the sole determinant of developing a competitive advantage. On the contrary, there are other components such as human capital, knowledge creation, and customer orientation that could contribute to a competitive advantage development (Kaplan and Norton 2005).

In order to apply BSC to a business's environmental sustainability strategy, the proper business units should be selected, the environmental aspects should be identified, and the relationship between the environmental aspects and the business's strategy should be analyzed (Figgie et al. 2002). By adding the environmental dimension to BSC, in addition to its default dimensions, a business can translate its environmental strategy into a set of performance and business requirements criteria (Sidiropoulos et al. 2004). Based on this framework, the environmental dimension of BSC can be used to model different events and provide the ability to develop and evaluate different scenarios (Sidiropoulos et al. 2004).

A very useful method to support the above tools' implementation is this of 5 R's concept, namely, refuse, reduce, reuse, repurpose, and recycle. "Refuse" refers to stop using products or ways of production harmful to the environment; "reduce" means less usage of raw materials, energy, and services; "reuse" is about to use materials over and over again; "repurpose" is about taking materials and using them for something else; and "recycle" is the final step and a way to support the abovementioned dimensions (De Souza et al. 2016; Sear 2016).

In any case, government policies could strengthen and support business initiatives for adopting sustainable environmental management. Such policies could concern the increase of green business practices and initiative funding; the provision of proper infrastructure, information and technical assistance; as well as the popularization of sustainable environmental management (Lee 2008).

4 Conclusions

Developing an environmental sustainability strategy through operations research models and business management tools is more effective than traditional ways, since it takes into account all possible variables and it is possible to predict the results by creating, running, and evaluating different scenarios (White and Lee 2009; Paucar-Caceres and Espinosa 2011).

From the above analysis, it is obvious that integrating environmental requirements into business strategy leads to a sustainability framework that can be reflected by the development of a competitive advantage (Skordoulis et al., 2017). This can be achieved by improving efficiency, reducing costs and risks, and increasing compatibility with customers' and regulations' requirements.

Therefore, a business strategy that includes environmental sustainability will be more effective when it is based on operations research model and business tools such as the QFD method, as this will provide a better understanding of the competition and the requirements (Skordoulis et al., 2017). Moreover, the analysis of as many scenarios as possible including different variables will become feasible.

A strategic approach as described above will result in significant changes in the structure, the mission, the vision, and the purpose of any business that will adopt it. Such a strategy can ultimately enhance the reputation and results of a business in a constantly changing market that creates new challenges (Skordoulis et al., 2017).

References

Ambec S, Lanoie P (2008) Does it pay to be green? A systematic overview. Acad Manag Perspect 22(4):45–62

Arabatzis G, Galatsidas S, Intze C, Chalikias M, Tsiantikoudis S, Mamalis S (2015) Green entrepreneurship and green products: consumers' views and attitudes in regional unit of Evros. In: Proceedings of the 7th International Conference on ICT in Agriculture, Food and Environment (HAICTA 2015). Kavala, September 2015. Athens: HAICTA, pp 291–297

Becker-Olsen L, Cudmore A, Hill P (2006) The impact of perceived corporate social responsibility on consumer behaviour. J Bus Res 59(1):46–53

Black JA, Boal KB (1994) Strategic resources: traits, configurations and paths to sustainable competitive advantage. Strateg Manag J 15:131–148

Chalikias M, Kalaitzidis I, Karasavvidis G, Pechlivanis EF (2010a) Relationship between sustainable farming and agricultural training: The case of Pella prefecture (Northern Greece). J Food Agric Environ 8(3–4 Part 2):1388–1393

Chalikias M, Kyriakopoulos G, Kolovos KG (2010b) Environmental sustainability and financial feasibility evaluation of wood fuel biomass used for a potential replacement of conventional space heating sources. Part I: A Greek case study. Oper Res 10(1):43–56

Chalikias M, Skordoulis M (2014) Implementation of Richardson's arms race model in advertising expenditure of two competitive firms. Appl Math Sci 8(81):4013–4023

Chalikias M, Skordoulis M (2016) Implementation of FW Lanchester's combat model in a supply chain in duopoly: the case of Coca-Cola and Pepsi in Greece. Oper Res Int J 17:737–745. https://doi.org/10.1007/s12351-016-0226-0

Chalikias M, Lalou P, Skordoulis M (2016a) Modeling advertising expenditures using differential equations: the case of an oligopoly data set. Int J Appl Math Stat 55(2):23–31

Chalikias M, Lalou P, Skordoulis M (2016b) Modeling a bank data set using differential equations: the case of the Greek banking sector. In: Proceedings of 5th International Symposium and 27th National Conference on Operation Research. Piraeus, June 2016. Piraeus: Hellenic Operation Research Society, pp 113–116

Chaudha A, Jain R, Singh AR, Mishra PK (2011) Integration of Kano's Model into quality function deployment (QFD). Int J Adv Manuf Technol 53(5–8):689–698

Chintagunta PK, Vilcassim NJ (1992) An empirical investigation of advertising strategies in a dynamic duopoly. Manag Sci 38(9):1230–1244

Dangelico RM, Pujari D (2010) Mainstreaming green product innovation: Why and how companies integrate environmental sustainability. J Bus Ethics 95(3):471–486

Daras, N. (2001). Differential equations and mathematical theories of war. Volume II. Athens: Evelpidon Military Academy.

DesJardins JR (2007) Business, ethics, and the environment: Imagining a sustainable future. Pearson Prentice Hall, Upper Saddle River, NJ

De Souza RG, Clímaco JCN, Sant' Anna AP, Rocha TB, Do Valle RDAB, Quelhas OLG (2016) Sustainability assessment and prioritisation of e-waste management options in Brazil. Waste Manag 57:46–56

Erickson, G. (1992) Empirical analysis of closed-loop duopoly advertising strategies. Management Science. 38(12):1732–1749

Fehlmann T (2008) New Lanchester theory for requirements prioritization. In: Proceedings of the Second International Workshop on Software Product Management. Barcelona, September 2008. Bacelona: I.E.E.E, pp 35–40

Figge F, Hahn T, Schaltegger S, Wagner M (2002) The sustainability balanced scorecard–linking sustainability management to business strategy. Bus Strateg Environ 11(5):269–284

Fruchter G, Kalish S (1997) Closed-loop advertising strategies in a duopoly. Manag Sci 43 (1):54–63

Hutchinson C (1992) Corporate strategy and the environment. Long Range Plan 25(4):9–21

Hwarng HB, Teo C (2001) Translating customers' voices into operations requirements - A QFD application in higher education. Int J Qual Reliab Manag 18(2):195–226

Jayeola O (2015) The impact of environmental sustainability practice on the financial performance of SMEs: A study of some selected SMEs in Sussex. Int J Bus Manag Econ Res 6(4):214–230

Kaplan RS, Norton DP (2005) The balanced scorecard: measures that drive performance. Harv Bus Rev 83(7):172

King A, Lenox M (2001) Does it really pay to be green? An empirical study of firm environmental and financial performance. J Ind Ecol 5(1):105–116

Kolovos KG, Kyriakopoulos GL, Chalikias M (2011) Co-evaluation of basic wood fuel types used as alternative heating sources to existing energy network. J Environ Prot Ecol 12(2):733–742

Kyriakopoulos G, Kolovos KG, Chalikias MS (2010a) Environmental sustainability and financial feasibility evaluation of wood fuel biomass used for a potential replacement of conventional space heating sources. Part II: A combined Greek and the nearby Balkan Countries case study. Oper Res 10(1):57–69

Kyriakopoulos GL, Kolovos KG, Chalikias M (2010b) Wood fuels prosperity towards a more sustainable energy production. In: World Summit on Knowledge Society: 19–25. Springer, Berlin Heidelberg

Lam JSL, Lai KH (2015) Developing environmental sustainability by ANP-QFD approach: the case of shipping operations. J Clean Prod 105:275–284

Lanchester FW (1916) Aircraft in warfare: the dawn of the fourth arm. Constable Limited, London

Lanchester FW (1956) Mathematics in warfare. The World of Mathematics 4:2138–2157

Lawrence L, Andrews D, Ralph B, France C (2002) Identifying and assessing environmental impacts: investigating ISO 14001 approaches. TQM Mag 14(1):43–50

Lee SY (2008) Drivers for the participation of small and medium-sized suppliers in green supply chain initiatives. Supply Chain Manag Int J 13(3):185–198

Lin Y, Cheng HP, Tseng ML, Tsai JC (2010) Using QFD and ANP to analyze the environmental production requirements in linguistic preferences. Expert Syst Appl 37(3):2186–2196

Malesios C, Skouloudis A, Dey PK, Abdelaziz FB, Kantartzis A, Evangelinos K (2018) The impact of SME sustainability practices and performance on economic growth from a managerial perspective: Some modeling considerations and empirical analysis results. Bus Strategy Environ, In Press

Melville NP (2010) Information systems innovation for environmental sustainability. MIS Q 34 (1):1–21

Naidu N, Badu K, Rajendra G (2007) Operations research: question bank. I.K. International Publishing House, New Delhi

Nidumolu R, Prahalad CK, Rangaswami MR (2009) Why sustainability is now the key driver of innovation. Harv Bus Rev 87(9):56–64

Ntanos S, Kyriakopoulos G, Chalikias M, Arabatzis G, Skordoulis M (2018a) Public perceptions and willingness to pay for renewable energy: a case study from Greece. Sustainability 10(3):687

Ntanos S, Kyriakopoulos G, Chalikias M, Arabatzis G, Skordoulis M, Galatsidas S, Drosos D (2018b) Social assessment of renewable energy sources usage and contribution to life quality: the case of an Attica Urban Area in Greece. Sustainability 10(5):1414

Ntanos A, Skordoulis M, Ntanos S (2014) Greek millennial consumers' perceptions on organic products. In: Proceedings of eRA-9 International Scientific Conference. Economy II Session. Piraeus, September 2014. Piraeus: T.E.I. of Piraeus, pp 26–35

Ojo E, Mbohwa C, Akinlabi E (2015) Sustainability - competitive advantage? In: Proceedings of the 2015 International Conference on Operations Excellence and Service Engineering. Florida, September 2015. Canton, MI: IEOM Society, pp 592–600

Ong TS, Teh BH, Ang YW (2014) The impact of environmental improvements on the financial performance of leading companies listed in Bursa Malaysia. Int J Trade Econ Finance 5 (5):386–391

Orlitzky M, Siegel DS, Waldman DA (2011) Strategic corporate social responsibility and environmental sustainability. Business & Society 50(1):6–27

Ottman JA, Stafford ER, Hartman CL (2006) Green marketing myopia. Environment 48(5):22–36
Oudrihiri R (2005) Six Sigma and DFSS for IT and Software Engineering. The Quarterly Journal of the TickIT Software Quality Certification Scheme 4:7–9
Papageorgiou A, Skordoulis M, Trichias C, Georgakellos D, Koniordos M (2015) Emissions trading scheme: evidence from the European Union countries. In: Kravets et al. (ed) Communications in Computer and Information Science. 535: Proceedings of creativity in intelligent technologies & data science conference, Volgograd, September 2015. Switzerland: Springer International Publishing, pp 222–233
Paucar-Caceres A, Espinosa A (2011) Management science methodologies in environmental management and sustainability: discourses and applications. J Oper Res Soc 62(9):1601–1620
Pitman G, Motwani J, Kumar, A, Cheng, CH (1996). QFD application in an educational setting: A pilot field study. International Journal of Quality & Reliability Management 13(4):99–108
Porter M (1980) Competitive strategy. The Free Press, New York
Porter M (1985) Competitive advantage: creating and sustaining superior performance. The Free Press, New York
Porter M (1991) Towards a dynamic theory of strategy. Strateg Manag J 12:95–117
Porter M, Kromer M (2006) Strategy and society: the link between competitive advantage and corporate social responsibility. Harv Bus Rev 84(12):78–92
Rohm H, Montgomery D (2010) Link sustainability to corporate strategy using the balanced scorecard. Balanced Scorecard Institute, Cary, NC
Rondinelli D, Vastag G (2000) Panacea, common sense or just a label? The value of ISO 14001 environmental management systems. Eur Manag J 18(5):499–510
Sakao T (2007) A QFD-centred design methodology for environmentally conscious product design. Int J Prod Res 45(18–19):4143–4162
Sear M (2016) Why loose parts? Their relationship with sustainable practice, children's agency, creative thinking and learning outcomes. Educating Young Children: Learning and Teaching in the Early Childhood Years 22(2):16
Sidiropoulos M, Mouzakitis Y, Adamides E, Goutsos S (2004) Applying sustainable indicators to corporate strategy: the eco-balanced scorecard. Environ Res Eng Manag 1(27):28–33
Sigalas C, Pekka-Economou V, Georgopoulos NB (2013) Developing a measure of competitive advantage. J Strateg Manag 6(4):320–342
Skordoulis M, Tsoulfas A, Kornelaki E, Samanta I (2013) The effect of corporate social responsibility (CSR) actions on consumers' behaviour. In: Proceedings of eRA-8 International Scientific Conference. Economy Session. Piraeus, September 2013. Piraeus: T.E.I. of Piraeus, pp 47–58
Skordoulis M, Galatsidas S, Arabatzis G (2017) Business strategies and competitive advantage through green entrepreneurship and sustainable environmental management. In: Proceedings of the 8th International Conference on ICT in Agriculture, Food and Environment (HAICTA 2017). Chania, September 2017. Aachen: CEUR Workshop Proceedings, pp 205–213
Soerger Zaro E, Soerger Zaro C, Richartz F, Borgert A, Van Bellen HM (2015) The impact of ISO 14001 certification on cost behavior of petrochemical companies listed in the BM&FBOVESPA Stock Exchange. Environ Qual Manag 24(3):57–70
Taoka N (1997) Lanchester strategy: an introduction. Lanchester Press, London
Urip S (2010) CSR strategies for a competitive edge in emerging markets. Wiley, Toronto
Wheelen LT, Hunger JD (2012) Strategic management and business policy toward global sustainability, 13th edn. Prentice Hall, Upper Saddle River, NJ
White L, Lee GJ (2009) Operational research and sustainable development: Tackling the social dimension. Eur J Oper Res 193(3):683–692
Wu SI, Lin SR (2016) The effect of green marketing strategy on business performance: a study of organic farms in Taiwan. Total Qual Manag Bus Excell 27(1–2):141–156
Yazdani M, Hashemkhani Zolfani S, Zavadskas EK (2016) New integration of MCDM methods and QFD in the selection of green suppliers. J Bus Econ Manag 17(6):1–17
Zhang Z, Shon L, Love P, Treloar G (2000) A framework for implementing ISO 14000 in construction. Environ Manag Health 11(2):139–149

Food System and Sustainability: The Role of Crowd-Funding

Rosa Misso, Monica Varlese, and Gian Paolo Cesaretti

Abstract Financial support for companies that pursue ethically and responsibly goals that we could call "off-market" is very important for the company's sustainability. On the other side, food and sustainability are now an inseparable binomial for the development of our society or rather an inevitable strategic priority in the pursuit of a socio-economic and environmental well-being, both individually and collectively, to be projected in time and space. In this context, support for entrepreneurial projects aimed at achieving this goal represents a crucial leverage not only for the enterprise but for the entire society. In the light of these considerations, this paper aims to clarify the important role that crowd-funding platforms can play because not only they allow the sustainability-oriented agrifood enterprise to "finance its vocation" but above all to potentiate in the society the culture of sustainability in the food field.

Keywords Food sustainability · Sustainable finance · Financing instrument · Well-being sustainability · Sustainable Development Goals

1 Introduction

The starting point of this paper is the recognition of sustainability as an inevitable perspective for the development of the agrifood system. The development strategies of this last are in fact correlated to every form and size of sustainability and are evaluated with respect to the construction and maintenance of well-being in time and

The chapter is the result of the joint commitment of the authors; in particular, paragraph nos. 1 and 5 have been edited by all the authors, paragraph no. 2 by Professor Gian Paolo Cesaretti, paragraph no. 3 by Dr. Monica Varlese, and Prof. Rosa Misso has edited paragraph no. 4.

R. Misso (✉) · M. Varlese · G. P. Cesaretti
Department of Economic and Legal Studies, University of Naples "Parthenope", Naples, Italy
e-mail: misso@uniparthenope.it

A. Theodoridis et al. (eds.), *Innovative Approaches and Applications for Sustainable Rural Development*, Springer Earth System Sciences,
https://doi.org/10.1007/978-3-030-02312-6_13

space. This orientation, today a compulsory goal for the food system, is the result of internal and external variables to the system itself, and, at institutional level, it is due to the indications of Agenda 2030, to the programming guidelines of the European Union, and to the actions of different states aimed at implementing economic, social, and environmental sustainability in each territory. At the sectoral level, instead, it is due to the propensity of businesses to pursue their mission to meet a need for well-being in the broadest sense, to an increasingly informed, skilled, and aware consumer of food choice importance.

More precisely, in September 2015, the 193 United Nations member countries adopted a new global development framework, the "2030 Agenda for Sustainable Development." Among the 17 Sustainable Development Goals (SDGs), Goal 2 made a step forward in fighting hunger by adopting a "zero hunger" target to "achieve food security and improved nutrition and promote sustainable agriculture" by 2030.

To meet this goal, agriculture is the key element even if the new faces of food insecurity, such as the urban malnutrition, must be addressed (Sassi 2018).

In this context, agrifood businesses become part of a larger socioeconomic and environmental project that goes beyond the core business of the company itself.

Specifically, they are an active part of this project for well-being sustainability and use various tools in full respect of this role. In financial terms, in particular, the crowd-funding is widely used, and it is made possible by platforms on the Internet, which, on the one hand, allow the company to find an economic response to its design interest; on the other, they trigger a process informative directed to the growth of the culture of sustainability.

Taken into consideration the results of the project funded by the University of Naples "Parthenope" (Italy) on the theme "Sustainability Practices in Food system and Financial Instruments" (Scientific Director, Prof. Rosa Misso), this chapter provides, first of all, an overview on the sustainability goals in the food sector, and then it focuses on the role that crowd-funding platforms can play as a multiplicator of the vocation of the food system to the individual and collective well-being sustainability.

2 Food System and Sustainability

The pursuit of well-being sustainability is now a universal aim or, rather, the goal toward which all businesses operating in the agrifood sector should aim in order to give new value to their core business. This theme, more precisely, represents a topic of great interest for each productive sector as well as for any level of society.

Analyzed no longer exclusively in its environmental dimension, sustainability is increasingly understood as an extension in the future, or as a strengthening, of the economic, social, and cultural (in addition to environmental) patrimony currently held by an individual, community, or territory (Misso et al. 2013). After all the action plan signed in September 2015 by the Governments of the 193 UN member countries,

"Agenda 2030 for Sustainable Development," is a clear signal that no one is excluded, nor should it be left behind on the road to bring the world on a sustainable path. The signatory countries have pledged to achieve by 2030 the Development Goals (which act on the results of the Millennium Development Goals that preceded them) representing common goals on a set of important issues for development: the fight against poverty, the elimination of hunger, and fighting climate change, to name just a few. In agriculture, of course, the goal of greater environmental sustainability led to the definition of the concept of integrated control of harmful organisms in the principles of integrated production, the concepts and practices of which were the subject of general and specific documents for individual crops. On this basis, numerous private and public guidelines have been proposed on the integrated production in order to meet the growing demand for healthy products by producers and consumers, products that require processing processes with minimal environmental impact (FAO 2013, 2017; FAO, IFAD, UNICEF, WFP, WHO 2017).

While at an international level, Agenda 2030 is marking the transition to a new sustainable society; at European level, undoubtedly, the principles of sustainable development are an integral part of the Common Agricultural Policy (CAP). From the Treaty of Rome to date, this policy has evolved in line with the needs that emerged on the European Horizon from time to time in order to be always current and meeting the expectations of its members' agriculture. Today, the CAP, in more apparent way than in the past, has to overcome the existing conflicts between the market, society, and the environment and to synthesize in its action the needs arising from these three spheres in order to ensure the sustainability of its agriculture, both at global and at European level (Table 1).

So, if at institutional level, international and European programming has outlined new pathways for sustainable development; at the sectoral level, an increasingly informed consumer stimulates businesses to pursue their mission to meet a need for well-being in the broadest sense.

Consumers are, in fact, increasingly sensitive to issues related to the improvement of information on agrifood products but also to the production processes used. New consumer trends are heavily influenced by changes in social work organization and demographic changes (macroeconomic variables). Such dynamics of consumption greatly affect the behavior of the operators and the economic system, and, on the other hand, this then affects the consumption of food, in a logic of circular evolution of the phenomenon (Arcese et al. 2015; Sidali et al. 2016).

In this context, the ability of a food system to develop over the long term depends on the ability to set strategies that ensure the sustainability of production factors. The presence of an intact environment and social capital and, more generally, the existence of all those factors that guarantee economic, social, environmental, territorial, and generational sustainability are, in fact, necessary elements for the survival of the sector. Therefore, an enterprise strategy oriented to ensure the sustainability of factors and production processes focuses on a responsible and equitable management of the workforce, the preservation and enhancement of environment, and the introduction of sustainable innovations capable of balancing sustainability and competitiveness goals (Weitz et al. 2014).

Table 1 The CAP between market, society, and environment

	Market	Society	Environment
Internal dimension	Simplify access to credit and favor a widespread diffusion	Increase farmers' income	Management of landscaping and natural resources (biodiversity)
	Growth of the sector	Fight the aging of the rural population	Attenuating vulnerability to climate change and greenhouse gas emissions
	Increase the contractual power of farmers	Repopulate the country sides	
	Quality standards	Occupation	
		Investments in human capital	
International dimension	Sustainable competitiveness on markets	Attract young people toward rural areas and activities	Preserving the identities of European territories
	Volatility of prices	Support the role of farmers as providers of public goods	Promote eco-friendly practices and green energy production
	Quick action tools for any crisis	Guaranteeing healthy products	Environmental standards
	Food security of emerging countries		Adapt the CAP to the global fight against climate change
	Food standards		

Source: Misso (2010)

3 Crowd-Funding

With the evolution of social media, crowd-funding has been increasingly emerging as a new funding method for entrepreneurial projects where the investors may support an idea and contribute to its realization. Crowd-funding fundamentally affects how our economic and social system functions as it changes how, why, and which products and services are brought into existence (Gerber and Hui 2013; Agrawal et al. 2014).

Financial support for companies that pursue ethically and responsibly goals that we could call "off-market" is very important for the company's sustainability. A bottom-up microfinance practice that mobilizes people and resources to support sustainability-oriented businesses is crowd-funding. It is a relatively new and evolving method of using the Internet to raise capital to support a wide range of ideas and ventures. An enterprise raising funds through crowd-funding typically seeks small individual contributions from a large number of people. Individuals interested in the crowd-funding campaign – members of the "crowd" – may share information about the project, cause, idea, or business with each other and use the information to decide whether to fund the campaign based on the collective "wisdom of the crowd" (Bartenberger et al. 2013).

The recent rise in crowd-funding activity is due to different reasons: firstly, matching funders with creators is now more efficient and effective due to lower search costs online. Secondly, risk exposure is reduced because funding in small increments is economically feasible online. Finally, low communication costs facilitate better information gathering and progress monitoring for distant funders and also better enable funders to participate in the development of the idea.

Indeed, there are several crowd-funding models that differ for the reward that is recognized in favor of the supporters. A type of crowd-funding is based on donation, so, it does not offer any reward to its funders. But there are also crowd-funding schemes that could recognize economic rewards to the funders or various other kinds of not economic reward as participation to the results of the project or simply symbolic acknowledgments (Stemler 2013; Giudici et al. 2017; OECD 2015).

More in particular, Agrawal et al. (2014) make a difference between equity and non-equity crowd-funding. The most critical differences between equity and non-equity crowd-funding will arise due to the amplification of information asymmetries. Whereas the asymmetry problem currently concerns the feasibility of the creator's ability to deliver the product, in the equity setting, the asymmetry problem includes the above as well as the creator's ability to generate equity value by building a company rather than just delivering a product (Agrawal et al. 2014).

Hemer (2011) identifies five types of crowd-funding according to the type of reward offered, crowd donations, crowd sponsoring, crowd pre-selling, crowd lending, and crowd equity, and ranked by process complexity:

Crowd donations: Although a donation is – in essence – an altruistic act without any obligation for the recipient to give the donor anything in return, one feature of crowd-funding is for donors to be given some "reward" for their support.

Crowd sponsoring: In the case of sponsoring, the project initiator and the sponsor agree on a defined reward which the initiator is obligated to give. Often these rewards take the form of services like PR or marketing for the sponsor.

Crowd pre-selling: Very often the donation takes the form of pre-selling or pre-ordering. The donation is meant to help produce something (a book, a film, a music album, a theater performance, software, some new technical product, an agricultural product, a service concept, etc.), and the promised return is the delivery of an early version of the product or service. In such a case, crowd-funding is basically an advance order of a product and represents a purchasing act which is subject to turnover tax.

Crowd lending: Here the rewards are normally the interest and the payback after the lending period. One alternative to this is long-term lending based on the revenue sharing principle. Here, the creditor gives a risk-bearing loan. He does not get interest but receives, at the defined end of the lending period, an amount including an agreed share of the earnings of the venture, which could be a multiple of the original loan but could – in the case of bad performance – also be nothing.

Crowd equity: This variant of micro-investments is – in administrative terms – the most complicated alternative in the spectrum of crowd-funding instruments. Crowd funders invest equity; the rewards are either shares of the venture, dividends, or voting rights.

The different forms of crowd-funding referred to above show differences in user groups, risks, complexity, and purpose, which warrant a distinction among these various forms and, importantly, a distinction between financial and nonfinancial return models. Crowd-funding with financial returns is slightly less well-known and is considered to carry higher risks for contributors who take the position of investors. The main issues EU legislation addresses with regard to all types of crowd-funding include anti-money laundering, advertising, consumer protection, and – where relevant – intellectual property protection. Financial returns campaigns and platforms may be subject to further rules at both EU and national level, again depending on the specific business model used (European Commission 2014).

4 Crowd-Funding and Food Sustainability

Most crowd-funding platforms allow you to invest in projects related to different sectors: art, comics, crafts, dance, design, fashion, film and video, music, food, etc. This is the case of the most renowned international platforms, Kick starter and Indiegogo, which include a generic food category. Many other platforms do not distinguish projects by sector, but they include food projects with big platforms as AngelList, Crunchbase, or Crowdcube.

Instead, the most important crowd-funding platforms entirely dedicated to the food industry are Barnaiser, WoopFood, and *PieShell*. The first two platform are dedicated to sustainable agriculture and are aimed at promoting small, organic, sustainable farms and projects related to proper and healthy nutrition. In fact, there is a specific section dedicated to sustainability. In particular, on the Barnaiser platform, created in the United States in 2014, the categories of projects that are sponsored are very different from one another, with a single denominator: the one of quality food production. In "farm," from the urban farming, you go to young peasants, from bio-farms to exchange markets; in "community" we find initiatives related to food-related justice, community gardens, cooperative kitchens, and more. Food includes projects related to accommodation facilities (restaurants, bars, and clubs), handicrafts, and distribution; the section "education" deals with workshops and didactic farms; food media focus on books, TV shows, cellular applications, and creative projects dedicated to sustainable food.

WoopFood is the first reward microfinance platform for small businesses in the Italian agricultural industry. Companies that aim to expand their business or to new agrifood products can launch a bottom-up funding campaign, collecting sums of between 5 and 2 thousand euros. The prize varies according to the generosity of the donors, according to the scheme already experienced by giants like Indiegogo and Kickstarter. The selection filter for the projects is twofold: the eco-sustainable impact of the activity or a certain rate of innovation, in the form of technologies applied to the production.

PieShell, instead, is more than just a crowd-funding platform. It can be defined as a community of people passionate about food, innovation, and donations. Based on the incubation approach, this crowd-funding platform prepares projects for the launch and increases their chances of success and through the stepping-stone model lets set realistic and obtainable goals on your way to funding the complete project. In this way it gives the projects a better chance to achieve their goals.

In general, crowd-funding in agrifood field appears in many different shapes. There are many different cases of projects such as funding for making a community farm, foundation of small business, presale for something (cooking equipment, farm products, processed farm products, garden equipment, and so on), farm operation cost, facility cost, farming education management cost, supporting events (market, party), and publishing a cookbook. We can classify crowd-funding projects on agrifood into three types: presale, event, and funds for operation cost. Presale type is funding projects before the production stage of food. Event type is a funding for the events related to agrifood, such as opening a market, launching a festival, or making educational event. Two cases succeeded in funding by providing enjoyable opportunities which people can't normally experience. Funds for operation cost are supporting essential operation expenses such as cost for setting up company, equipment, research, and development. Mainly these projects have social benefit or can help to produce healthy food (Yoo et al. 2014).

In particular, numerous sustainable agrifood projects, internationally, have been financed through the platforms mentioned above, to preserve endangered food products, to produce health food, and to encourage food education. For example, through the Barnraiser platform, in California, an organic farmer has financed the project "Eureka! Help Save Heirloom Fruit & Nut Trees of California's Gold Rush" in order to plant an "orchard" of organic fruit relics and walnut trees from specimens that had survived the era of gold fever in the mountains of the Sierra Nevada; three young women who have studied the healing properties of bone broths have raised funds for the project "Kitchen Witch Broth – Nourish Your Soul" in order to buy a 60-gallon kettle and other equipment to increase production in their small startup "Kitchen Witch Bone Broth," which sells six different types of broth, made from different types of animal bones and packed in glass jars; in New York, an educator for the project "The Green Bronx Machine CAN!" has raised funds to convert an old school library into a national health center, an indoor urban farm, and a cooking and education program for school children in the Bronx (www.barnraiser.us). Instead, through the PieShell platform, a family-owned producer of high-quality plant powders that use organic, biodynamic, and wild vegetables has raised funds in order to make biodiversity more accessible to the average consumer and therefore in order to expand in retail stores so that more people can appreciate their products. The main objective of this project, which is titled "Launch Dr. Cowan's Garden into Retail!!," is to transform agriculture by creating a better demand for organic products and by improving the environment (www.pieshell.com).

From this point of view, these projects are relevant for sustainability, as well as for the pursuit of the objectives of Agenda 2030, in particular of Goal 2 that places food and nutrition issues at center stage (the correlations between targets of Goal 2 and crowd-funding projects mentioned above are illustrated in Table 2).

Table 2 Targets of Goal 2 of the SDGs and crowd-funding projects

Targets	Crowd-funding projects
2.1. By 2030, end hunger and ensure access by all people, particularly the poor and people in vulnerable situations, including infants, to safe, nutritious and sufficient food all year round	"Kitchen Witch Broth – Nourish Your Soul"
2.2. By 2030, end all forms of malnutrition, including achieving, by 2025, the internationally agreed targets on stunting and wasting in children under 5 years of age, and address the nutritional needs of adolescent girls, pregnant and lactating women, and older persons	"The Green Bronx Machine CAN!"
2.3. By 2030, double the agricultural productivity and incomes of small-scale food producers, in particular women, indigenous peoples, family farmers, pastoralists, and fishers, including through secure and equal access to land, other productive resources and inputs, knowledge, financial services, markets, and opportunities for value addition and nonfarm employment	"Kitchen Witch Broth – Nourish Your Soul" "Launch Dr. Cowan's Garden into Retail!!"
2.4. By 2030, ensure sustainable food production systems, and implement resilient agricultural practices that increase productivity and production; that help maintain ecosystems; that strengthen capacity for adaptation to climate change, extreme weather, drought, flooding, and other disasters; and that progressively improve land and soil quality	"Eureka! Help Save Heirloom Fruit & Nut Trees of California's Gold Rush" "Launch Dr. Cowan's Garden into Retail!!"
2.5. By 2030, maintain the genetic diversity of seeds, cultivated plants, and farmed and domesticated animals and their related wild species, for instance, through soundly managed and diversified seed and plant banks at the national, regional, and international levels, and promote access to and fair and equitable sharing of benefits arising from the utilization of genetic resources and associated traditional knowledge per international agreements	"Eureka! Help Save Heirloom Fruit & Nut Trees of California's Gold Rush" "Launch Dr. Cowan's Garden into Retail!!"
2.A. Increase investment, for instance, through enhanced international cooperation, in rural infrastructure, agricultural research and extension services, technology development, and plant and livestock gene banks to enhance agricultural productive capacity in developing countries, particularly the least developed countries	"The Green Bronx Machine CAN!"
2.B. Correct and prevent trade restrictions and distortions in world agricultural markets, for instance, through the parallel elimination of all forms of agricultural export subsidies and all export measures with equivalent effects, in accordance with the mandate of the Doha development round	
2.C. Adopt measures to ensure the proper functioning of food commodity markets and their derivatives and facilitate timely access to market information, including on food reserves, to help limit extreme food price volatility	

Source: Our elaboration on https://sustainabledevelopment.un.org/sdg2

The push from the Internet can lead both to supporting small business and to providing a more showcase for niche productions, fixed on zero KM or local distribution. Like all reward-based crowd-funding systems, the company generates its revenue with the collected fees at the end of the campaign. The sum is borne by the project proposer, with a share that may vary depending on the amount reached (Magnani 2016).

The benefits of this crowd-funding web-based platform are:

- The creation of a new demand
- The reduction in investment risk for producers
- The establishment of direct relations with consumers
- The sponsorship of a product
- The valorization and protection of the traditional ecosystem made in Italy
- The promotion and dissemination of the concept of environmental sustainability
- The strengthening of food security and quality assurance

In a food system, a variety of fairly large-scale ICT-enabled projects demonstrate economic viability and provide significant social and economic value. Such projects are directly linked to income-generating activities (e.g., providing better selling opportunities for agro-products), making their value easily visible for end users. At the same time, these projects collect individuals around a topic that allow them to improve their knowledge on a product or on an activity in terms of sustainability (Bargain et al. 2016; Carvalho et al. 2012).

Surely, crowd-funding can be a good tool to connect the producers and consumers of food. But above all it can promote the achievement of the company's sustainability goals through the launch of projects geared toward this (European Commission 2014).

Application of crowd-funding in agrifood sector can provide a huge help in foundation of small businesses and stabilizing cash flow of small farmers. Because it reduces pressure on financing and sales, the producer can make their products in the way he wants without anxiety. Also, unlike other agrifood transactions, relationship between producers and consumers grows up during the funding process. This leads to interest and active participation of consumer which can result additional purchases and popularity. Through the proceeding of crowd-funding, project creator can get not only funds but also various helpful effects for the success of the project. The project creator can check point for improvement through this. Also, participants sometimes form a supporting community and play a role as a co-producer. In case of presale type, it can help the pricing of products or services. For the success of crowd-funding, active participation of friends and fans in early funding, promotion through social media, and communication effort of a creator such as information provision are important.

In some ways, crowd-funding almost always benefits the public, and in this case, it has constructed rollover bars provided to poverty-stricken farmers through the local agricultural education program. The public is able to view campaigns and donate a small amount of money to be used toward the project.

Oftentimes, projects with clear goals, specific needs, and a defined end date are more successful at raising money than projects without clear objectives. Most often, family members and friends of project creators who are involved in the project are the most likely to donate. Crowd-funding is less about raising large amounts of money from a few investors and, rather, more about raising small donations from a large group of people, "the crowd." One reason crowd-funding is so successful is due to the "feel good" or philanthropy attached to donating (Morgan et al. 2016; Servato et al. 2013; Lehner 2013). Belleflamme et al. (2014) states that crowd-funding produces the same amount of funds as if you were to seek funds from a bank, but what compels the public to donate to a crowd funded project is the perceived benefit to what is being created.

The financial support for entrepreneurial projects aimed to achieving the sustainability goal represents a crucial leverage not only for the enterprise but for the entire society. In this context, crowd-funding web-based platforms can play an important role because they not only allow the sustainability-oriented agrifood enterprise to "finance its vocation" but above all to potentiate in the society the culture of sustainability in the food field.

Moreover it needs to consider that *sustainability-oriented* ventures are especially constrained as their social and environmental goals cause them to be perceived as less attractive investments compared to traditional entrepreneurial ventures focused on marked needs and economic returns (Choi and Gray 2008).

However, academic contributions on the role of crowd-funding in moving toward a sustainable society are mainly conceptual. As underlined by Nielsen and Reisch (2016), in the current literature remain arguably difficult to detangle cause and effect relationships between crowd-funding and orientation to sustainability.

Some conceptual and empirical works like Vasileiadou et al. (2015) or Calic and Mosakowxki (2016) suggest a positive relationship between environmental or sustainability orientation and the likelihood of success of crowd-funding projects (Lam and Law 2016; Mollick 2014).

Conversely, in line with part of the crowd-funding literature claiming that crowd funders are likely to act similar to conventional financiers (Moss et al. 2015), Hörisch (2015) does not observe in his investigation any positive connection between environmental sustainability and crowd-funding success. For example, as stated by Ordanini et al. (2011), other elements than sustainability orientation are determinant for the projects' funding success, such as prospect of financial return.

From the point of view of its concrete application, crowd-funding transcends simple financial support to business projects trough credit and is a powerful business accelerator capable of enhancing sustainable projects by halting the weaker ones.

This is because crowd-funding platforms – from the past – use a new paradigm to select the worthy financing projects, which consists of an estimate "market test" that can provide valuable indications and projections of business success.

Crowd-funding platforms initially show the idea-project and its goal to reach: afterward, they give users the ability to decide whether and how to finance the projects and also identify the economic commitment they intend to sign up. Projects that reach the finish line will be realized, and lenders will get the agreed interest.

Projects that otherwise will not reach the minimum amount of funding will not be realized, and the sums eventually paid will be fully refunded to the lender, without any loss. This process allows an extremely effective selection of "profitable" projects, starting with the design phase of the business, thanks to the impact that the project produces on consumer-investors (Servato et al. 2013; Manning and Bejarano 2016).

Crowd-funding platforms also have an incentive to attract projects that can generate a disproportionate share of media attention because they both expand the existing community of funders (further increasing network effects) and allow the platform to expand into new categories (Agrawal et al. 2014).

5 Conclusion

In recent years, the integration of sustainability criteria by investors is constantly increasing all over the world. In this regard, crowd-funding is considered today a financing option particularly suited to entrepreneurs of sustainability projects because the lenders themselves are attracted by the ideas and fundamental values of projects funded by crowd-funding (Lehner 2013). In other words, lenders, even if they do not derive economic benefits, have prosocial reasons to support projects funded by crowd-funding (Hörisch 2015).

In this context, crowd-funding platforms specialized in social and environmental projects contribute to increasing the financing of sustainable projects. In particular, in the agrifood sector, there were numerous projects funded through crowd-funding and based on a sustainable way of eating, farming, and distributing food products, respecting the people and the planet we live on. With this in mind, the crowd-funding provides an important contribution to sustainability.

References

Agrawal A, Catalini C, Goldfarb A (2014) Some simple economics of crowd funding. Innov Policy Econ 14(1):63–97

Arcese G, Flammini S, Lucchetti MC, Martucci O (2015) Evidence and experience of open sustainability innovation practices in the food sector. Sustainability 7(7):8067–8090

Bargain O, Cardebat JM, Vignolles A (2016) Crowd funding in wine (no. 234638). Aawe working paper N. 196, 1–21

Bartenberger M, Leitner P (2013) Crowd sourcing and crowd funding: approaches to Foster social innovation. In Proceedings of the IADIS international conference web based communities and social media 2013, pp 81–85

Belleflamme P, Lambert T, Schwienbacher A (2014) Crowd funding: tapping the right crowd. J Bus Ventur 29(5):585–609

Calic G, Mosakowski E (2016) Kicking off social entrepreneurship: how a sustainability orientation influences crowd funding success. J Manag Stud 53(5):738–767

Carvalho A, Feinberg C, Klarsfeld L, Lepicard F, Posthumus B (2012) Leveraging information and communication Technology for the Base of the pyramid-innovative business models in education, health, agriculture and financial services. Proceedings of M4D New Delhi, India

Choi DY, Gray ER (2008) The venture development processes of "sustainable" entrepreneurs. Manage Res News 31(8):558–569

European Commission (2014) Communication from the Commission to the European Parliament, the Council, the European Economic and Social Committee and the Committee of the Regions unleashing the potential of crowd funding in the European Union

FAO (2013) Coping with the food and agriculture challenge: smallholders' agenda. Preparation and outcomes of the 2012 United Nations conference on sustainable development

FAO, IFAD, UNICEF, WFP, WHO (2017) The state of food security and nutrition in the world

FAO (2017) The future of food and agriculture

Gerber EM, Hui J (2013) Crowd funding: motivations and deterrents for participation. ACM Trans Computer-Human Interaction (TOCHI) 20(6):34

Giudici G, Guerini M, Rossi-Lamastra C (2017) Reward-based crowd funding of entrepreneurial projects: the effect of local altruism and localized social capital on proponents' success. Small Bus Econ:1–18

Hemer J (2011) A snapshot on crowd funding. Working paper firms and region R2/2011

Hörisch J (2015) Crowd funding for environmental ventures: an empirical analysis of the influence of environmental orientation on the success of crowd funding initiatives. J Clean Prod 107:636–645

Lam PT, Law AO (2016) Crowd funding for renewable and sustainable energy projects: an exploratory case study approach. Renew Sust Energ Rev 60:11–20

Lehner OM (2013) Crowd funding social ventures: a model and research agenda. Ventur Cap 15 (4):289–311

Magnani A (2016) Woopfood, il crowdfunding finanzia la filiera agricola. www.ilsole24ore.it

Manning S, Bejarano TA (2016) Convincing the crowd: entrepreneurial storytelling in crowd funding campaigns. Strateg Organ 15(2):194–219

Misso R, Cesaretti GP, Viola I (2013) Sostenibilità del benessere e responsabilità, Rivista di studi sulla sostenibilità, N.1, pp 77–92

Misso R (2010) L'uomo, il clima e l'agricoltura – Verso nuovi strumenti e politiche. FrancoAngeli, Milano, pp 1–144

Mollick E (2014) The dynamics of crowd funding: an exploratory study. J Bus Ventur 29(1):1–16

Morgan L, Vincent SK, Mazur J (2016) Establishing a crowd funding sustainability project for resource – depleted secondary agricultural education programs. Southern region conference of the American Association for Agricultural Education February 7–9, 2016 San Antonio, pp 34–37

Moss TW, Neubaum DO, Meyskens M (2015) The effect of virtuous and entrepreneurial orientations on microfinance lending and repayment: a signaling theory perspective. Entrep Theory Pract 39(1):27–52

Nielsen KR, Reisch LA (2016) Crowd funding for sustainability – The role of value orientation frames in guiding individual investment behavior. Paper presented at The International Conference on Business, Policy and Sustainability. Copenhagen Business School, 16–17 June, Copenhagen, Denmark

OECD (2015) New approaches to SME and entrepreneurship financing: broadening the range of instruments. OECD, Paris

Ordanini A, Miceli L, Pizzetti M, Parasuraman A (2011) Crowd-funding: transforming customers into investors through innovative service platforms. J Serv Manag 22(4):443–470

Sassi M (2018) The history of food security: approaches and policies. In: Understanding food insecurity. Springer, Cham, pp 89–120

Servato F, Miglietta A, Parisi E, Pessione M (2013) Gli strumenti finanziari a supporto delle start-up innovative: le opportunità offerte dal CrowdFunding. XXV Convegno annuale di Sinergie

Referred Electronic Conference Proceeding. L'innovazione per la competitività delle imprese, Ottobre 2013 – Università Politecnica delle Marche, pp 491–506

Sidali KL, Spiller A, von Meyer-Hofer M (2016) Consumer expectations regarding sustainable food: insights from developed and emerging markets, international food and agribusiness. Manag Rev 19(3):141–170

Stemler AR (2013) The JOBS Act and crowdfunding: Harnessing the power-and money-of the masses, in Business Horizons 56(3):271–275

Business Horizons 56(3):271–275

Vasileiadou E, Huijben JCCM, Raven RPJM (2015) Three is a crowd? Exploring the potential of crowd funding for renewable energy in The Netherlands. J Clean Prod 30:1–14

Yoo YG, Choe YC (2014) Current status of Korean crowd funding industry and its applicability to agrifood sector. Adv Sci Tech 49(SoftTech 2014):213–221

Weitz N, Nilsson M, Davis M (2014) A nexus approach to the post-2015 agenda: formulating integrated water, energy, and food SDGs. SAIS Rev Int Aff 34(2):37–50

Evaluation of Innovative Tools for the Trade Enhancement of Fresh Agrifood Products

Luisa Sturiale and Alessandro Scuderi

Abstract The e-Marketplaces have become the global platform for exchange, which have modified social and economic behavior. The digital revolution has opened the doors of the global village to industries, services, cities, and territories. Information and communications technology (ICT) can be a tool for the development of rural and agricultural areas, as demonstrated by real experiences reported in the literature. The advent of new technologies has also led to major changes in consumer buying processes and market relationships. The interest in shopping places is shifting from large shopping malls to "zero-km" shopping and e-commerce.

The new e-commerce frontier is represented by logistics adaptations for zero-km sale of fresh produce and same-day delivery.

The objectives of this study were to contribute to the knowledge of the phenomenon of e-marketplaces in the Italian agrifood economy and to analyze the new e-commerce models for fresh fruit and vegetables in Italy.

Keywords Digital economy, e-marketplaces · Trading platform · e-commerce models · Zero-km sale · Online-to-offline commerce (O2O)

1 Introduction

The information society is changing every aspect of daily life, and the digital revolution has radically changed the concepts of space, time, and mass (Kotler et al. 2010). Cyberspace allows us to transform purchases and sales into ever faster,

L. Sturiale (✉)
Department of Civil Engineering and Architecture (DICAR), University of Catania, Catania, Italy
e-mail: luisa.sturiale@unict.it

A. Scuderi
Department of Agrifood and Environmental Systems and Management (Di3A), University of Catania, Catania, Italy

A. Theodoridis et al. (eds.), *Innovative Approaches and Applications for Sustainable Rural Development*, Springer Earth System Sciences,
https://doi.org/10.1007/978-3-030-02312-6_14

more automated operations without costs or barriers related to space and time limits. The digital revolution has opened the doors of the global village to newly established companies, to niche companies, and to small or very small businesses. The strong development of information and communications technology (ICT) is at the base of what has been defined for many years as the "net economy," an economic system based on online activities through the use of networks (Sturiale 2000),

In the ICT sector, innovations follow each other swiftly, as it is a very dynamic market, where Web 4.0 represents the latest phenomenon after Web 3.0, which followed Web 2.0, which revolutionized the global world of information, following the initial stage of Web 1.0[1] (Pitt et al. 2012; Neilson et al. 2010; Sturiale and Scuderi 2015). There has been an evolution of sites and services on the web, such as wiki sites and social networks, where interaction and sharing of content by users are fundamental. The web has become a "web ecosystem" in which users create value by sharing and creating experiences on the web (through e-mail, blogs, networking, forums, communities, chat, etc.). The internet has become the global platform for exchange—the space where an economic force of global dimensions operates, which has radically modified social and economic behavior (Brush and McIntosh 2010; Kumar and Rajan 2012; Sturiale 2000; Wang and Zhang 2012).

The development of digital technology and spreading of computer networks have transformed production processes and access to, transfer of, and use of information. Communication technologies allow the easiest access to knowledge and the easiest way to create it by simple sharing of information, from e-mails to forums to social networks, with the consequent reduction of space–time barriers (Busalim and Che Hussin, 2016; Sturiale and Scuderi 2011).

The digital economy is growing at a rate seven times faster than the real economy. Much of this growth has been fueled by broadband internet. Today, high-speed broadband networks are having an impact equally important to those of electricity and transport a century ago. They are also paving the way for innovative services such as online healthcare (e-health), smart cities, and data-based production.

The *2030 Agenda for Sustainable Development* recognizes that the spread of ICT and global interconnectedness have great potential to accelerate human progress, to bridge the digital divide, and to develop knowledge societies.

The new technologies are offering a platform supporting businesses in rural regions, in both the agricultural sector and other sectors, to innovate production processes and management functions, to support direct online interaction with customers, to search for new market opportunities, etc. The adoption of digital

[1]In particular, the initial stage of the internet, Web 1.0, was represented by static sites; in 2004 we moved on to Web 2.0, with the introduction of dynamic programming languages that allowed users to interact with the content of websites, with the possibility of using online applications such as blogs, wikis, forums, and social networks. In 2006 we moved on to Web 3.0, characterized by the possibility of using the network as huge *web data*—semantic, enhanced, and three-dimensional. The next step, already under way, is Web 4.0, a network of very intelligent interactions where the interconnection and communication between things and people will generate the decision-making process.

models is increasing the potential competitiveness of rural areas in the new framework of the net economy. Research in the literature on applications of ICT in rural and agricultural areas, in Italy and several other European countries, highlights continuous developments and innovative approaches (Bruckmaier and Tovey 2009; Fountas et al. 2015; FAO 2005; Galati et al. 2013; Hammond and Paul 2006; Stratigea 2011; Warren 2002). The most important applications for rural regions in the business sector are e-commerce, e-marketing, e-training, and e-marketplaces.

ICT has opened up new lands to economic competition, which is deeply affecting trade in different sectors and in some cases also altering the very rules of competition themselves. Cultural aspects, safety, and the health of products represent obstacles to the diffusion of ICT and web tools within the agrifood sectors, as do the lack of quality of websites and the scarce attention paid to interactivity with web consumers, which is fundamental for web marketing. Besides, there is a poor level of standardization of these sector products and several difficulties in managing the quality of e-commerce. Among the relevant works of literature, we must mention (in order of the year of publication) a few concerning farms and agrifood small-to-medium enterprises (SMEs) in Italy and elsewhere: Sturiale (2000), Sturiale and Scuderi (2001, 2011, 2013, 2015), Fritz et al. (2004), Bucca et al. (2006), Brush and McIntosh (2010), Neilson et al. (2010), Lehmann et al. (2012), Rapisarda et al. (2015), and Scuderi et al. (2015).

The advent of new technologies has also led to major changes in consumer buying processes and market relationships. In fact, the interest in shopping places is shifting from large shopping malls to "zero-km" shopping and to online shopping.

ICT provides the consumer with more information, compares products and prices across sites, and verifies greater convenience, meeting the need to know the global scenario. However, there are several critical issues in the new online shopping models, primarily the inability to examine the purchased product—an action that still remains primary for consumers, through use of the five senses. Virtual purchasing also limits opportunities for conviviality and socialization, which are characteristic of offline purchases.

These limits are felt to be most important in the case of fresh agricultural produce, for which e-commerce in Italy still lags behind other commodity sectors and other countries, although there have recently been some interesting developments in e-food. In fact, there is interest in both international marketplaces (AmazonFresh) and retailers who have developed new online sales channels for fresh produce, creating online-to-offline (O2O) models, virtual platforms, marketplaces, etc., and improving logistics services to facilitate zero-km delivery and same-day delivery (or even delivery within an hour in some cases).

These new dynamics in the agrifood sector in general—and in fresh produce in particular—still appear to be underestimated in the literature. This study aimed to build a framework, even if concise, of the evolution of the digital economy and of the innovations that characterize the new e-commerce models for fresh agrifood products, typical of rural areas and strategic for their sustainable development. Given the growing importance of e-marketplaces, the phenomenon of their diffusion and the opportunities offered for agrifood products will be broadened.

In addition, a specific survey on Italian web consumers of fresh agrifood products, structured as two complementary phases, allowed us to highlight willingness to buy these products online and the variables that affected this choice.

2 The Global Evolution of the Digital Economy and the New Models of e-Business for Agrifood Products

Society is experiencing the "fourth industrial revolution," characterized by a fusion of technologies that blurs the boundaries between the physical, digital, and biological spheres. Further innovations in areas such as artificial intelligence, robotics, the "internet of things," three-dimensional (3D) printing, and autonomous vehicles will further accelerate this technological revolution.

Increasing connectivity has led to a fundamental change in the business world: digital has enabled it to expand both online and offline, and is already influencing various industries. The current consumer seeks and buys goods and services differently from previous generations.

In the world, 1.3 billion people buy online (with a total of 2.8 billion people connected to the internet), making 7.0% of all retail purchases (8.0% in Europe, 5.0% in Italy) (Fig. 1).

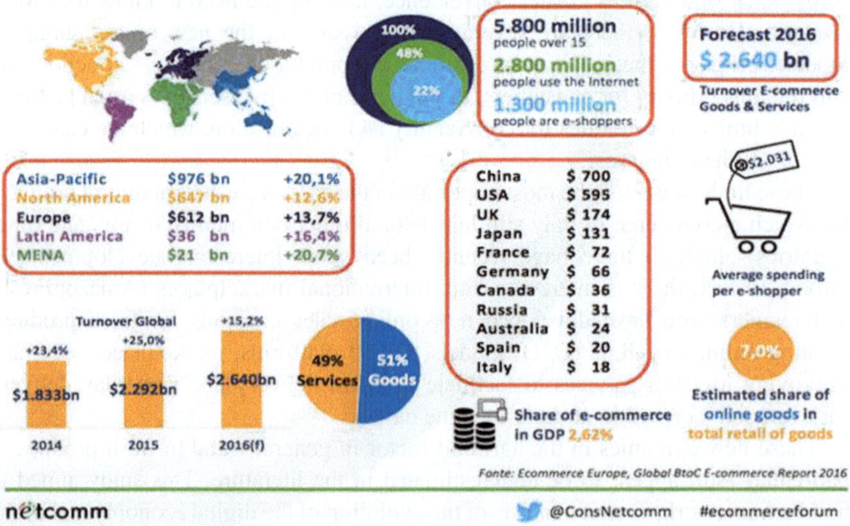

Fig. 1 Stastistics on e-commercee-commerce in the world (in 2016)

In 2016, online sales of products and services accounted for 7.0% of the total retail market on a global level—a total near US$2.7 billion. By 2019, this value will have more than doubled to reach US$3.578 billion, while continuing to represent only a fraction (13.0%) of total retail purchases. In the business-to-consumer (B2C) market, the USA and China remain the most important markets; the major players in the market are global giants such as Amazon and, for the Chinese market, Alibaba.

Considering the European Union (EU), ICT accounts for almost 5.0% of the EU economy and generates a quarter of total business spending. Investments in this sector are responsible for 50.0% of productivity growth across the EU.

In 2016 the value of e-commerce in Europe (Fig. 2) was estimated to be US$600-billion, with the UK, Germany, and France accounting for around two thirds of total sales. Italy ranked seventh after Russia, Spain, and the Netherlands. Online sales represented 8.0% of all European retail sales, exceeding (albeit only slightly) the US percentage of 7.0%, but in southern Europe, online sales barely reached 2.0% (Ecommerce Europe 2016).

ICT has profoundly changed consumers' attitudes toward purchasing. In fact, consumers now browse the web to compare offers and services, moving toward the so-called info-commerce, and 65.0% of web surfers show a multichannel-purchase attitude, where social media play a more and more important role.

There has been an evolution of sites and services on the web, such as wiki sites and social networks, where interaction and sharing of content by users are fundamental. The web has become a web ecosystem in which users create value by sharing and creating experiences on the web (through e-mail, blogs, networking, forums, communities, chat, etc.) (Hanna et al. 2011; Sturiale and Scuderi 2015). The internet has become the global platform for exchange—the space where an economic force of global dimensions operates—which has radically modified social and economic

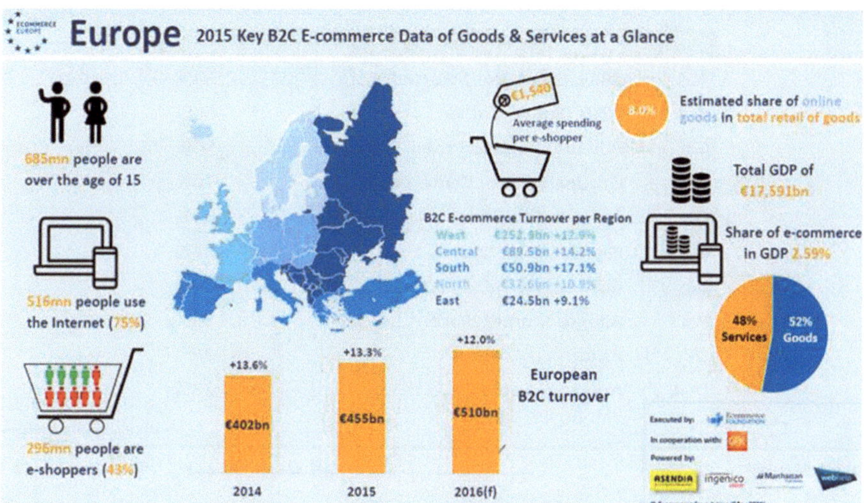

Fig. 2 Statistics on e-commerce in Europe (in 2015 and forecast for 2016)

behavior and models of online sales (Brush and McIntosh 2010; Fritz et al. 2004; Kumar and Rajan 2012; Sturiale and Scuderi 2013, 2016; Sturiale et al. 2017; Wang and Zhang 2012).

Shoppers are certainly changing, but some aspects of shopping remain the same. People enjoy it as a social activity, and this is particularly true of the younger age group. The social nature of shopping has changed and is linked to the phenomenon of online socialization (Verdouw et al. 2013; Ecommerce Europe 2016).

In a 2017 survey of over 1200 US internet users who qualified as primary household shoppers, they preferred to do most of their shopping in-store or via digital channels; only the oldest respondents in the study—those aged 60 years and older—said they preferred in-store shopping. For all other age groups, digital channels were preferred for shopping, and for those aged 21–29 years, mobile was the preferred shopping channel.

As Amazon and other e-commerce players increasingly cut into the market share of traditional retailers, now seems like the perfect time to give the O2O model and pickup services a second look.

The grocery landscape is changing quickly. Amazon has opened physical supermarkets, Walmart is heavily investing in e-commerce, and many regional players now offer either curbside pickup or delivery.

The internet has significantly changed the commercial playing field for B2C sales. Many retailers and product brands already sell their products through (international) online marketplaces, present in different typologies (Groothuist and Gabriner 2016). International examples of these e-marketplaces are Amazon, Zalando, Alibaba's TMall, and Mercado Libre. Most countries also have strong national players such as Bol.com (in the Netherlands) and Allegro (in Poland).

	Players	Share
Aus	eBay, Coles, Woolworths	34%
Bra	B2W Digital, Cnova	36%
Chi	Alibaba, Jingdong Mall	80%
Fra	Groupe Casino, Vente-Privee	20%
Ger	Amazon DE, Otto	55%
Ind	Flipkart, Snapdeal	25%
Jap	Rakuten Ichiba, Amazon JP	40%
Pol	Allegro Market Place	50%
UK	Amazon UK, Argos	28%
USA	Amazon, Apple	28%

Fonte: elaborazioni Netcomm

netcomm @ConsNetcomm #ecommerceforum

Fig. 3 The most important e-marketplaces and e-retailers in world e-commerce in 2016 (in single major countries)

These marketplaces are growing very rapidly, with Amazon and Alibaba taking the lead (Fig. 3).

There are different definitions of e-marketplaces, such as that provided by Wikipedia:

> An online marketplace (or online e-commerce marketplace) is a type of e-commerce site where product or service information is provided by multiple third parties, whereas transactions are processed by the marketplace operator. Online marketplaces are the primary type of multichannel e-commerce and can be a way to streamline the production process.

The terms "trading platform" and "marketplace" are used interchangeably. According to another definition, a trading platform or marketplace can be defined as a website where multiple B2C sellers can sell their products and where the commercial transaction (with the consumer) takes place on the platform. It can also be a website on which the consumer is directly linked to the product sales page of the related webshop (Groothuist and Gabriner 2016).

There are different types of trading platforms, varying from platforms that started as a webshop and later opened up to other sellers, to platforms that have been initiated to bundle the forces of local players in the online market. Besides comparison sites, search engines and social media platforms such as Pinterest and Facebook increasingly offer the possibility to instantly buy products. These websites can also be considered marketplaces. The following types of different trading platforms can be distinguished:

- *Comparison websites (with a buy button):* Examples are Kelkoo (in France), Kieskeurig.nl (in the Netherlands), and Allegro (in Poland).
- *Search engines with a buy option:* Google Shopping or Bing Shopping are examples of these.
- *Open platforms:* These are platforms that used to be webshops but nowadays also allow third-party sellers to offer their products through their webshop. Examples of these are Bol.com, Zalando, and Amazon.
- *Local platforms:* There are more and more platforms that let a local neighborhood form their own online community. Nine Streets in Amsterdam is an example, where visitors can virtually shop on the platform. Also, smaller players such as boutiques, which would be too small to open their own webshops, can bundle their forces this way. Examples are Locals United and Miinto.
- *Marketplaces:* Two examples are eBay and AliExpress.
- *Auctions:* Usually, these are websites that often change their assortment, depending on remaining parties of vendors. Examples are iBood and Groupon.
- *Theme portals:* More and more consumers and small businesses are setting up sites offering product reviews (e.g., of clothes), where the consumer is forwarded to a vendor's webshop. Fashionchik and Asos are prime examples of this type of marketplace (Groothuist and Gabriner 2016).

Marketplaces and aggregators of various kinds carry about 30.0% of online transactions and will play a key role in the e-commerce market in the near future.

The planning and operation of marketplaces can also be found in rural areas, as marketing and transaction platforms that are based on local enterprises and local products. Several interesting examples of marketplaces have been developed in rural regions, both in advanced economies and in developing countries.[2]

With regard to the fresh agrifood sector, there has been increasing interest in the most important e-marketplaces in recent years to widen the supply range with fresh produce with very short delivery times. First of all, Amazon launched AmazonFresh, but others are now being added, as well as special platforms being created for the marketing of fresh agrifood products.

According to specific research, it is possible to identify different business models of fresh produce e-commerce (Lo and Lam 2015):

- *Integrated B2C e-commerce marketplace:* Food brands and fresh produce manufacturers are invited to be online sellers in the marketplace (Amazon); the categories of products are fresh food (seafood, and fruit and vegetables), grains and oils, and organic food; the organization of the logistics is based on providing minimal logistics support, or logistics and distribution are handled by the online sellers; the expected revenue model is the registration fee or the commission.
- *Vertical B2C e-commerce platforms:* The focus is on online selling of food and fresh produce, with full control of the entire cold-chain logistics from farm to fork; the categories of products are fresh food (seafood, and fruit and vegetables) and organic food; the logistics involve a hybrid model of self-operation and use of third-party logistics companies; the expected revenue model is the product markup.
- *e-Commerce platform operated by a logistics company:* The features of this model are strength in entire cold-chain logistics and last-mile delivery; the categories of products are fresh food (seafood, and fruit and vegetables); it is expected self operated logistics and last-mile delivery of the goods sold by the platform; the expected revenue model is the product markup.
- *O2O model operated by a supermarket:* This is a new model of e-commerce for fresh produce, with O2O integration; it includes an increasing number of offline stores that are creating a specific network in different countries in the world and development of the O2O model of e-commerce; the categories of products are fresh food (seafood, and fruit and vegetables) and grains and oils; the logistics include a hybrid model of self-operation and use of express delivery companies but also a self-pickup service at offline stores; the expected revenue model is the product markup.
- *Community O2O model:* The features of this model, based on the creation of a community of offline stores and supermarkets, are specific targets (the white-collar market or university students), direct sourcing, pickup of goods from the nearest supermarket/convenience store/mom-and-pop store on the basis of

[2]Examples in developing countries are B2Bpricenow.com in the Philippines and Agriwatch in India (Hammond and Paul 2006). Such markets also exist in small urban settlements; an example is a business-to-consumer marketplace in Trikala, Greece (Stratigea 2011).

location-based service (LBS) positioning of the online shopper, and delivery of the goods to the designated location within 1 hour; the categories of fresh produce are cooked food, fresh food, and fresh fruit; the expected revenue model is the product markup or the commission.

Growing connectivity is transforming the business experience, and firms have also introduced new and engaging ways of marketing that have changed consumer behavior and broken long-established business models. Marketplaces are claiming to be the dominant players in e-commerce.

In more mature markets such as the UK, the first three players (Amazon, eBay, and Tesco) manage one third of the turnover of physical products sold online. In the market that is positioning itself as the main market in the world—China—marketplaces already manage over half of all online transactions. The reasons for this market dominance lie in their ability to invest in service and in the relationship with customers (who always look for an infinite range to avoid having to search on multiple sites), and to continuously innovate customer service, thanks to global economies of scale. The formula of service on e-commerce sites means that there is also a propensity on the part of merchants to exploit the visibility and services provided by the marketplaces (Eng 2004; Rask and Kragh 2004).

It is foreseeable that in Italy we will also witness sustained growth of these actors over the next few years. The marketplaces that are placed in the Italian market are mainly Amazon (used by 63.0% of dealers in the marketplace) and eBay (used by 57.0%) (Casaleggio Associati 2016). There are, however, new players such as Pixplace (used by 4.0%), which is part of Pixmania and allows sales throughout Europe; Etsy (used by 2.0%), which is perhaps the easiest way to enter the US market for craft products; Alibaba (used by 2.0%), which is now a global initiative aimed not only at business-to-business (B2B); and local initiatives such as BuyMe (used by 4.0%). Time-based sales sites are also entering this sector; in the future it may also include comparator and ad sites, which currently do not allow conclusion of purchases, as customers are sent back to the merchants' sites.

Operators who use marketplaces to sell their products are mainly interested in integrating the sales made by direct e-commerce through their presence in the marketplace and use of the related services offered by the marketplaces themselves (payments and profiled advertising), and in developing their own brands and their own online sales, since in any case these sites are also a showcase with millions of potential customers. In addition, the marketplaces make it possible to sell internationally, which is mainly offered to other European countries but in some cases also to countries such as the USA (e.g., Etsy), China (e.g., Tmall), South America (e.g., Mercado Libre), Africa (e.g., Jumia), or worldwide, with the best-known competitors being Amazon, eBay, and Rakuten (in Japan) (Casaleggio Associati 2016).

3 The New Frontier for e-Commerce of Fresh Agrifood Products in Italy: e-Marketplaces and Other Physical Versus Virtual Models

The value of Italian e-commerce had an average annual growth rate of 14.0% between 2010 and 2016, but in the same period the value more than doubled (from €8.012 billion in 2010 to €19.282 billion in 2016). The digital economy thus grew at a considerably faster rate than the real one. Italian internet users seem to be web users with maturity and more confidence in the characteristics of the channel. In 2016, the number of Italian web shoppers (i.e., consumers who made at least one online purchase during the year) grew by 7.0% and reached 19 million, equal to about 60.0% of internet users.

There are differences between different product sectors. In Italian e-commerce, tourism and consumer products have been predesigned, respectively, on services and products. But now there are many product categories that make up the online shopping basket, and in 2016 the composition was as follows: tourism 44.0%, computers and electronics 14.0%, clothing 9.0%, insurance 7.0%, publishing 4.0%, furnishings and home living 3.0%, food and groceries 3.0%, and other purchases 16.0%.

The contribution of emerging sectors (food and groceries, furniture and home living, beauty, and toys) is increasingly important, with over €1.5 billion in value and growth rates ranging between 30.0% and 50.0%. These sectors are expected to record the highest growth rates in the future.

In 2016, online sales in Italy represented 5.0% of total retail; in the services category they represented 9.0%, while in the products category they accounted for 3.0%. If we consider the online penetration rates of the different product sectors, unfortunately the food and grocery trade is still lagging behind with 0.3%, versus 30.0% in the tourism sector.

The food and grocery trade is one of the sectors that today arouses more interest in the B2C e-commerce landscape in Italy, both because of the enormous potential that is still unfulfilled and because of the vivacity shown by the offer in the last 2 years.

In 2016 the e-commerce market in food and groceries purchased by Italian consumers on both Italian and foreign sites was worth €575 million, an increase of 30.0% in comparison with 2015. Although food and groceries represent one of the main areas of Italian spending, their online distribution has been limited so far; their share of online purchases was only 0.35% of total retail purchases.

Food products represent 90.0% of the market. Health and care products account for 10.0% but show a significantly higher growth rate. In the food category, over 90.0% of purchases are food products and fewer than 10.0% are wine.[3]

[3]In particular, the breakdown of expenditure for online food in Italy is as follows: dry food 55.0%, fresh produce 29.0%, wine 8.0%, drinks 6.0%, and frozen foods 2.0% (www.osservatori.net).

The most important limitation in the development of the fresh meat market is the lack of opportunity for customers to see and choose the product before buying (according to 49.0% of the sample), followed by inadequate quality and freshness (according to 32.0%) and lack of confidence in the mode of transport (according to 25.0%) (Monitor Ortofrutta 2016; Ecommerce Europe 2016).

On the one hand, these data point to the still-limited use of e-commerce in the Italian agrifood sector; on the other hand, are well-deserved, as broad growth areas are foreseen on the basis of what is already happening in other European countries (such as France, the UK, and Germany), in the USA (where it is a fast-growing phenomenon), and in Japan.

In summary, consumers prefer to resort to e-commerce (through the use of computers, tablets, and smartphones) for the following advantages: convenience, availability of information in real time, availability of products and services, lower cost, and time efficiency. On the other hand, the main reasons for not using e-commerce are unsecured payment, slowness of transportation, receipt of unwanted products, circulation of spam or viruses, receipt of boring e-mails, and annoying technological issues.

Traditional operators (retailers and producers) are the protagonists of the sector, but dot-coms are growing at a faster rate. Amazon's entry into the sector, the growing contribution of flash-sales players, and the development of the ready-to-eat food segment—dominated by aggregators—are the main causes of this trend.

Italy is trying to fill the gap in the agrifood sector and especially in the fresh produce sector. Indeed, in recent years there has been a proliferation of online initiatives. The offerings have increased in all areas, from supermarket products to gastronomic products, and from food and wine to fresh agricultural produce, for which innovative marketing methods are also developing, very often linked to the initiative of producers and territories.[4]

Italy is trying to bridge the gap in the agrifood sector, especially in fresh produce. Very active platforms set up by large distribution chains (Esselunga, Carrefour, Unes, Naturasì, etc.) provide access to specific online supermarkets offering the same assortment as retail outlets, with customer advantages of a loyalty card scheme and a choice of delivery times (a commission is provided).

[4]One of various initiatives is "The Beehive That Says Yes!" (alvearechedicesi.it), an online network of farmers and consumers who periodically meet at a set point to exchange local and seasonal agricultural produce. The project, launched in 2015, has more than 100 collection sites in Italy and over 600 in France (where the initiative originated in 2011 as "La Ruche qui Dit Oui") .The website serves as a meeting tool between supply and demand. Consumers can register for free on the web and discover the produce (fruit, vegetables, bread, meat, beer, etc.) made available by small and medium-sized regional companies (currently around 400 in Italy). They are always online and the products are available for pickup at a collection point, which is a real temporary market with a short supply chain. Consumers can consume local foods and support the economy in their territory, while the producers perceive that their remuneration is "equitable" and share emotions and values (www. italiafruit.net, 02/12/2016).

"Subscription models" are being launched. Subscription sales models are based on repeated purchases and low risk perception, according to a logic that transforms customers into subscribers. In Italy, Cortilia is an example of a marketplace based on the subscription model in the food sector (like Good Eggs in the USA). Even single producers and distributors (Agrispesa, Zolle, Almaverde Bio, and others) are making use of this new e-commerce model.

The new frontier of e-commerce for agrifood products, and especially for fresh produce, is represented by integration between the physical and the virtual, with marketplace examples that in recent years have been developing new forms of integration and multichannels, where physical stores are integrated into the digital ecosystem.

According to Netcomm (2016), physical stores have entered the digital market; in fact, according to a survey involving 30,000 Italian direct sales points, in 8.6% of them it is possible to pick up online purchases in the shop; moreover, in half of them, online booking is available to reserve products that will be bought in the shop.

In fact, the presence of a physical store offers value in terms of the ability for the customer to touch the product, or collect an order placed online, complementing the online experience. Logistics has become more efficient in terms of the services offered, with shorter delivery times; consequently, the value of deliveries of products purchased online doubled in value from €6 million to €12 million in the period from 2012 to 2016.

According to a survey conducted by Milan Polytechnic (2016), the investment priorities in online sales channels for Italian retailers are as follows (in descending order of importance): an e-commerce website controlled by the company, mobile commerce through a mobile app controlled by the company, an e-commerce website controlled by an intermediary/retailer, and mobile commerce through a mobile app controlled by an intermediary. The first two channels listed above register the greatest preferences on the part of Italian companies.

The traditional physical sales space and the space dedicated to customer care (which represent most of the current cases) are expected to shrink, to the advantage of physical space dedicated to delivering products purchased online (according to the "pick and collect" model) and the "experimental" physical space, aimed at creating a relationship with the client (such as an "experiential room"). In the near future, they will also be able to take on other physical space functions, including the creation of online showroom products/services and "temporary experimental spaces" (Spiller 2016).

The delivery of products purchased online is one of the most interesting challenges for the future of e-commerce, especially for fresh agricultural produce: the possibility of pickup from collection points (which may be automated, such as lockers, or manned by human personnel) that are open 24 hours a day and where customers can go when the e-commerce store advises them that their goods are ready for collection. At present, 94.0% of deliveries take place at home or in the office, while only 6.0% are collected from a designated place.

The new e-commerce frontier is represented by logistics adaptations for zero-km sale of fresh produce and same-day delivery. Amazon Prime Now, which provides delivery within an hour in some urban areas of the USA, is the most prominent example, but there are other examples of such marketplaces in several European countries. In France and Spain, Amazon has launched the grocery version of its service, AmazonFresh—a division that markets and delivers even fresh food such as fruit and vegetables. The website offers many items (30,000), divided into 25 categories, including known brands, local products, and specific lines such as gluten-free products. Also, in Madrid, the capital of Spain, Now, in Madrid, capital of Spain, the Prime service distributes fresh products, with online purchase and home delivery of fruit and vegetables (UNATA 2017).

In Italy, 16,000 companies participated in e-commerce in 2016, and this number is projected to reach 50,000 by 2025. With regard to food, e-commerce in Italy is slowly beginning to close the gap with other European countries, thanks to the growth of local players, but especially the arrival of international players that have entered the market through acquisitions; for example, Just Eat (a London-based company) has acquired Click and Eat in Milan, DeliveRex in Rome, HelloFood, and PizzaBo (Casaleggio Associati 2016)[5].

Also, in the agrifood industry, 2015 was the year in which Amazon Prime Now was launched for delivery of shopping in Milan and in 34 municipalities in its hinterland. Launched in Italy in November 2015, it allows purchase of more than 20,000 grocery items and other types of products; in February 2016 the service was also expanded to include more than 30 types of fresh fruit and vegetables. Purchase is possible only via the mobile app, with home delivery from 8 am to midnight every day of the week, within an hour of purchase in certain areas and in the customer's preferred time slot in other areas (Sturiale et al. 2017).

Coop Italia has responded to Amazon with the launch of the Easy Coop service for online shopping and home delivery of food products. Customers can place orders on a dedicated website (www.easycoop.com) for 10,000 different foods, of which about 3000 are different types of fresh fruit and vegetables, meat, fish, salami, dairy products, bread, and pastry. Delivery takes place at home the day after the order date, at the customer's preferred time.

So, we are developing e-commerce models in Italy where virtual and physical space is integrated, even for fresh agricultural produce. In particular, it is about the

[5]Even for wholesale fruit and vegetable markets, the web offers interesting online sales potential. Cesena's Fruit Market has been operating for a couple of years and has reached an agreement with the Tippest platform for sale on the web; also, the one in Florence operates online through MaxFruit.

O2O (online-to-offline)[6] channel, alongside the well-known "click-and-mortar" model.[7]

In these multichannel and customer-centric approaches, mobile will become increasingly crucial. Mobile has generated an ecosystem that enables increasingly large portions of the economy and helps to break down the digital divide. The funds available from companies and Public Administration (PA) for the development of mobile solutions to support business processes grew by 24.0% and those oriented to marketing and communication on smartphones and tablet grew by 41.0% in the last five years. The development of mobile commerce and mobile payment brought the total value of the mobile economy to over € 37 billion in 2017 (2.3% of the gross domestic product (GDP)).

Agrifood enterprises will have to consider the mobile channel as a potentiator and amplifier at other points of contact throughout all stages of the buying process. All of this requires a transformation of business models into a logic of multichannel transformation, mobile marketing and mobile service.

4 Methodology

As was confirmed in the previous chapter, studies of e-commerce have shown the influence of the interaction between consumers in the buying process, especially if the latter has to buy a new product. Therefore, it is interesting to highlight in this survey the salient differences between traditional and online purchases—as the difference in the impact is not well distinguished in the literature—but also to clarify the transmission of messages from the consumer to other consumers in the two contexts, both traditional and virtual. We interviewed 1000 consumers in 2017,

[6]Online-to-offline (O2O) commerce is a business strategy that draws potential customers from online channels to physical stores. It identifies customers in the online space (e.g., through e-mails and internet advertising) and then uses a variety of tools and approaches to entice the customer to leave the online space. This type of strategy incorporates techniques used in online marketing with those used in brick-and-mortar marketing. Some companies that have both an online presence and an offline presence (physical stores) treat the two different channels as complements rather than competitors. The goal of O2O commerce is to create product and service awareness online, allowing potential customers to research different offerings and then visit a local brick-and-mortar store to make a purchase (www.investopedia.com).

[7]"Click-and-mortar" is a type of business model that includes both online and offline operations, which typically include a website and a physical store. A click-and-mortar company can offer customers the benefits of fast online transactions or traditional face-to-face service. This model is also referred to as "clicks-and-bricks." An increasing number of big-brand retail stores such as Walmart, Best Buy, and Nordstrom follow the click-and-mortar business model. Also referred to as an omnichannel strategy, the merging of online and offline channels provides customers with an enhanced shopping experience with more choices, greater flexibility, more convenience, and more services (www.investopedia.com).

through the ModulesGoogle.com platform, through a semistructured questionnaire with multiple-choice and Likert scale questions.

The first phase of the research aimed to evaluate, through a consumer preference scale[8], what the strengths and weaknesses of online purchase of fruit and vegetables were, and the role of a physical point of contact for the logistics service of an online store to orient the consumer of fruit and vegetables to e-commerce.

The analysis was continued to assess the consumer's intention to report to third parties their intention to purchase a product on the basis of suggestions from other consumers (such as the phenomenon of "e-wom" (e-word of mouth) (Brown et al. 2007; Scuderi and Sturiale 2014)). In particular, the study—called "demand side"—was analyzed to focus attention on the figure of the consumer as a recipient of a message and as a decision maker in the buying process. The survey therefore aimed to identify the structure of buying intent as a result of influences in the two environments: offline and online. The main purpose of the second phase was therefore to precisely determine the influence of inputs from the two different environments, and the assumption on which the research was based was the belief that the individual experiences different purchasing intentions in relation to different sources of information (offline or online). We investigated the possibility that this intention to buy a given product (which, in our experiment, was 1 kg of oranges) may differ depending on whether the comment comes from a friend in a traditional/offline environment rather than from a user in a virtual environment such as an online community.

The parameter estimation process was done by a process of minimizing the distances between the data produced by the model and the observed data. The available estimation functions were different, and we used a linear model (LM) to study the relationship between a dependent variable (y) and a set of independent variables (x1, x2, ... xn) To understand the impact these variables had on the object of study, we used the following equation:

$$\begin{aligned} \text{buy\_probi} = {} & \beta 0 + \beta 1\text{BUY\_off} + \beta 2\text{BUY\_on} + \beta 3\text{POSTi} + \beta 4\text{WHYi} \\ & + \beta 5\text{GENDER} + \beta 6\text{AGEi} + \beta 7\text{EDU} + \varepsilon t \end{aligned}$$

5 Results

This research analyzed and aimed to understand, for the consumers who were interviewed, the evolving trends in the purchase of online agrifood products and the decisive factors influencing the buying process.

[8]In particular, a Likert scale (1 = not influential, 2 = not very influential, 3 = moderately influential, 4 = very influential, 5 = determinant) was used.

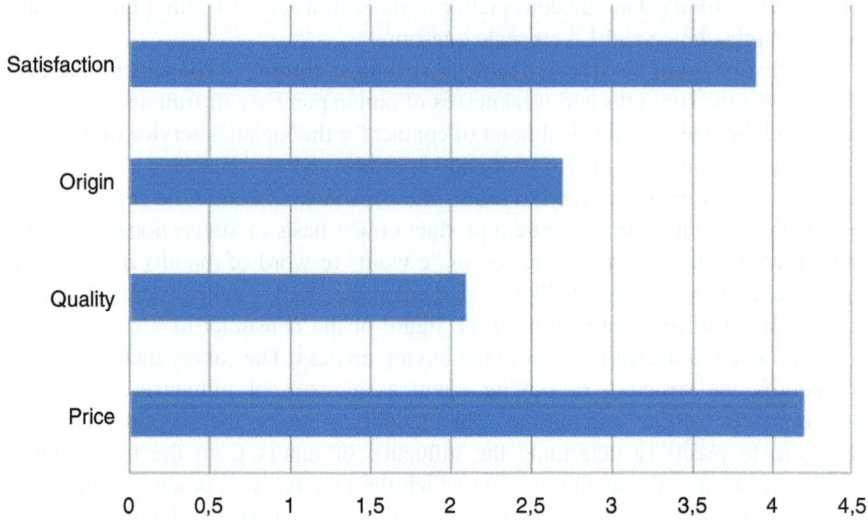

Fig. 4 The driving factors consumers to online shopping

The results of the first part showed that an element that could lead to the online channel (Fig. 4) was the current price, which was a discriminating element in the choice of purchase. However, another discriminating element was the ease of purchase (defined as "satisfaction") as a sum of services related to the comfort of payment, purchase, and delivery of the product. Less influence was attributed to quality and origin, as the first factor was assumed to be in compliance with the standards, and secondly, it was a discriminating factor in the choice factors from the beginning.

We asked consumers which shopping channel they preferred for the same quality, price, origin, and packaging of fruit. It turned out that 84.0% of consumers we interviewed always preferred the traditional channel; in relation to their normal shopping habits, they saw buying of fruit and vegetables as being included in the act of daily expenditure for food through the traditional channels (supermarkets, local shops, and local markets). However, the percentage of consumers who did not always prefer the traditional channel (16%) was significant rate, which showed that current behavioral barriers are being pulled down and confirmed that there are growing spaces for there are very interesting growth spaces for fruit and vegetables.

Finally, consumers were asked whether they preferred an online store to also have a physical point of sale in their area, to orient their purchase and what their reasons were. The results (Fig. 5) showed that a physical point of sale and collection for the e-commerce channel is a major strength for the consumer, as it represents a transition from offline to online, a material element of the purchase act, and the possibility to collect the product, as well as a contact point to enable customers to verify the quality of products and to have access to the products of a farmers market (direct sale, zero-km).

Fig 5 The presence of a traditional shop alongside an online site can guide consumer choices

Table 1 Estimated structural parameters of opinions derived from the traditional environment (of_med, of_pos) and the virtual environment (on_med, on_pos) and how they affect consumers in their intention to buy (buy_prob)

	Estimate	Std. Error	*t* value	Pr (>|*t*|)
(Intercept)	18.352	9.125	2.087	0,039258
med off	3.987	4.687	1.025	0,325874
pos off	14.259	3.257	2.253	0,016587
med_on	5.312	4.851	1.214	0,278957
pos_on	11.257	3.458	3.508	0,000258
no.rev	6.981	3.625	1.258	0,235784
post	6.235	1.259	3.398	0,000768

The series of regression models inherent in the part of the experiment called the demand side focused on some models (Table 1) that relate views from the traditional offline environment and from the online context and the potential impact those views have on the intention to buy of the consumer who receives such information. Table 1 shows the coefficients of the different variables chosen to observe a possible influence of the same on the dependent variable (i.e., the willingness to purchase a product, which in our experiment was 1 kg of oranges) measured in percentage terms.

The opinions of other consumers regarding a product have a different impact on the consumer's intention to buy, depending on whether they come from the online environment or the offline environment. The impact of information from the online and offline environments on the intention to buy varies according to the value of such opinions (positive, negative, neutral) and is significant only if both opinions (online and offline) about a product are positive. In the case of negative opinions, the offline channel prevails, even in the presence of positive online opinions. The results show that the impact on purchase intent comes mainly from the offline channel, but online comments have meaning as a source of information although they are not decisive in the final purchase choices.

On the basis of the results that were collected and the recent actions of international players (Amazon, Alibaba, etc.), it is reasonable to assume that in the coming years there will be a growing online demand for fresh fruit and vegetables.

According to some research, the success factors are:

- *Shopper convenience:* Online shopping saves time and reduces the need to make a trip to the store and carry heavy loads.
- *Growth in click-and-collect:* This will increase consumer participation in the online sales channel.
- *Omnichannel growth:* This is the complete integration of stores, e-commerce, mobile apps, and social media, which will deepen consumer experiences.
- *Increased internet usage:* This also includes increased usage of mobile and tablet devices for online shopping.
- *Changes in purchase behavior:* This refers to consumers making bulk purchases online and using offline stores to "top up" on a daily basis.
- *Demand for organic products through online channels* (Sturiale et al. 2017).

6 Discussion and Conclusion

ICT is at the base of the net economy, an economic web system based on online activities through the use of networks that have changed the way consumers shop and the way consumers wish to receive their purchases. The adoption of digital models is increasing the potential competitiveness of rural areas in the new framework of the net economy.

The current consumer seeks and buys goods and services differently from previous generations: consumers may start to shop for an item online, but they may well finish their purchase in-store, deciding they want to touch and feel the item just to be sure or because they want immediate gratification. But they could also spot something in-store and then go online to complete the transaction. The experience needs to be seamless and connected. Consumers need to be able to see where their purchases are and when they can get their hands on them.

The internet has significantly changed the commercial playing field for B2C sales. Many retailers and product brands already sell their products through (international) e-marketplaces, present in different typologies, which will play a key role in the e-commerce market in the near future.

With regard to the fresh agrifood sector, there has been increasing interest in the most important e-marketplaces in recent years to widen the supply range with fresh produce with very short delivery times. The new e-commerce frontier is represented by logistics adaptations for zero-km sale of fresh produce and same-day delivery. Stores are creating an opportunity to enhance the customer experience and delight shoppers into coming back time and time again (Ecommerce Europe 2016).

Food is a key sector for e-commerce, and the market in fruit and vegetables has the potential to grow. According to various surveys, food is expected to become the most important part of e-commerce worldwide over the next 10 years. Italy has become an interesting laboratory for alternative food experiences, with the spread of phenomena such as delivery of food ordered online.

There is, however, a delay with regard to e-grocery, although almost all large retailers are developing e-commerce projects, which see a synergy between offline and online.

Our research aimed to evaluate what the strengths and weaknesses of online purchase of fresh fruit and vegetables were, and the role of a physical point in the service of logistics of an online store to orient the consumer of fresh fruit and vegetables to e-commerce. In the agrifood economy there are new e-commerce models for fresh agrifood products; the presence of e-marketplaces is increasingly important, and their numbers continues to grow in Italy and in other European countries. On the basis of the results that were collected and the recent actions of international players (Amazon, Alibaba, etc.), it is reasonable to assume that in the coming years there will be a growing online demand for fresh fruit and vegetables. The success factors are shopper convenience, growth in click-and-collect, omnichannel growth, increased internet usage, changes in purchase behavior, and growth of demand for organic products through online channels.

The research shows that consumers of fruit and vegetables still base their acquisitions on the values of the traditional market but observe with interest what is happening in the online market. The role of fresh agrifood products in the online channel will depend on the evolution of ICT consumption models and technology platforms in the years to come.

References

Brown J, Broderick AJ, Lee N (2007) Word of mouth communication within online communities: conceptualizing the online social network. J Interact Mark 21(3):4

Bruckmaier K, Tovey H (2009) Rural sustainable development in the knowledge society—perspectives on rural policy and planning. Ashgate, Surrey.

Brush GJ, McIntosh D (2010) Factors influencing e-marketplace adoption in agricultural micro-enterprises. Int J Electron Business 8(4/5):405–432

Bucca M, Scuderi A, Sturiale L (2006) Metodologie di analisi delle strategie di web marketing delle imprese agroalimentari nelle Regioni dell'Obiettivo 1. Rivista di Economia Agro-Alimentare 1:101–125

Busalim AH, Hussin CHE, Razak AB (2016) Understanding social commerce: a systematic literature review and directions for further research. Int J Inf Manag 36(6):1075–1088

Casaleggio Associati (2016) E-commerce in Italia 2016. https://www.casaleggio.it/focus/rapporto-e-commerce-in-italia-2016/

Ecommerce Europe (2016) Southern Europe B2C E-commerce report 2016. www.ecommerceeurope.eu (on 30 may 2018)

Eng T-Y (2004) The role of e-marketplaces in supply chain management. Ind Mark Manag 33:97–105

Food and Agriculture Organization, United Nations (2005) World Summit on the Information Society: Geneva 2003–Tunis 2005. Plan of action, paragraph 21. Document WSIS-03/GENEVA/DOC/5-E. http://www.itu.int/dms_pub/itu-s/md/03/wsis/doc/S03-WSIS-DOC-0005!!PDF-E.pdf

Fountas S, Carli G, Sørensen CG, Tsiropoulos Z, Cavalaris C, Vatsanidou A, Liakos B, Canavari M, Wiebensohn J, Tisserye B (2015) Farm management information systems: current situation and future perspectives. Comput Electron Agric 115:40–50

Fritz M, Hausen T, Schiefer G (2004) Development and development directions of electronic trade platforms in US and European Agrifood market: impact on sector organization. Int Food Agribus Manag Rev 7(1):1–20

Galati A, Schimmenti E, Asciuto A, Borsellino V (2013) The role of information and communication technologies and logistics organisation in the economic performance of Sicilian fruit and vegetable enterprises. Int J Business Globalisation 10(2):185–193

Groothuist A, Gabriner R (2016) Marketplaces. www.eucommercewiki.org

Hammond A, Paul J (2006) A new model for rural connectivity. World Resources Institute – Development Thought Enterprise, Washington DC.

Hanna R, Rohm A, Crittenden VL (2011) We're all connected: the power of the social media ecosystem. Bus Horiz 54(3):265–273

Kotler P, Kartajaya H, Setiawan I (2010) Marketing 3.0, dal prodotto al cliente, all'anima. Gruppo 24 Ore, Milan

Kumar V, Rajan B (2012) Social coupons as a marketing strategy: a multifaceted perspective. J Acad Mark Sci 40(1):120–136

Lehmann RJ, Reiche R, Schiefer G (2012) Future internet and the agrifood sector: state of the art in literature research. Comput Electron Agric 89:158–174

Lo W, Lam G (2015) E-commerce business models. Manag Theor Bus Models 3:8–12

Monitor Ortofrutta (2016) Speciale Frutta & Verdura 2016. www.italiafruit.net (accessed 10 may 2018)

Neilson LC, Madill J, Haines GH Jr (2010) The development of eBusiness in wine industry SMEs: an international perspective. Int J Electron Bus 8(2):126–147

Netcomm (2016). Netcomm global B2C Ecommerce report 2016. www.consorzionetcomm.it (accessed 15 june 2018)

Pitt L, Berthon P, Plangger K, Shapiro A (2012) Marketing meets Web 2.0, social media and creative consumers implications for international marketing strategy. Bus Horiz 55(5):261–271

Rapisarda P, Rizzo M, Scuderi A (2015) Analysis of direct selling network of agrifood products. Ital J Food Sci 1:109–117

Rask M, Kragh H (2004) Motives for e-marketplace participation: differences and similarities between buyers and suppliers. Electron Mark 14(4):270–283

Scuderi A, Sturiale L (2014) Analysis of social network applications for organic agrifood products. Int J Agric Res Gov Ecol 10:176–189

Scuderi A, Sturiale L, Timpanaro G (2015) The importance of "origin" for on line agrifood products. Qual Access Success 16:260–266

Spiller N (2016) Le imprese e l'Everywhere shopper: un matrimonio da celebrare? Osservatorio Multicanalità 2016. Milan Polytechnic, Milan.

Stratigea A (2011) ICTs for rural development: potential applications and barriers involved. Netcom 25(3/4):179–204

Sturiale L (2000) Il commercio elettronico, vincoli ed opportunità con particolare riferimento al sistema agroalimentare. Economia Agro-Alimentare, Anno. 1:140–159

Sturiale L, Scuderi A (2001) Business to consumer e-commerce: problems and opportunities for some typical local products of the "mezzogiorno" of Italy. Proceedings of the 4th AIEA International Symposium, Perspectives of the Agrifood System in the New Millennium, Bologna, 5–8 Sep 2001, p. 153–174.

Sturiale L, Scuderi A (2011) Information and communication technology (ICT) and adjustment of the marketing in the agrifood system in Italy. In: Salampasis M., Matopoulos A. (eds), HAICTA 2011 5 International Conference on Information and Communication Technologies in Agriculture, Food and Environmental. p. 77-87, CEUR-Workshop.

Sturiale L, Scuderi A (2013) Evaluation of social media actions for the agrifood system. Procedia Tecnology 8:200–208

Sturiale L, Scuderi A (2015). Social commerce and marketing strategy for "made in Italy" food products. In: Andreopoulou Z, Bochtis D (eds) Proceedings of 7th International Conference on Information and Communication Technologies in Agriculture, Food and Environment, Kavala. p. 509-519.

Sturiale L, Scuderi A (2016) The digital economy: new e-business strategies for food Italian system. Int J Electron Market Retail 7(4):287–310

Sturiale L, Timpanaro G, La Via G (2017) The online sales models of fresh fruit and vegetables: opportunities and limits for typical italian products. Quality Access Success J 18(S2):444–451

UNATA (2017) The 2017 grocery e-commerce forecast: the time is now for eGrocery. www.unata.com

Verdouw CN, Vucic N, Sundmaeker H, Beulens AJM (2013) Future internet as a driver for virtualization, connectivity and intelligence of agrifood supply chain networks. Int J Food Syst Dyn 4(4):261–272

Wang C, Zhang P (2012) The evolution of social commerce: an examination from the people, business, technology, and information perspective. Commun Assoc Inf Syst 31(5):105–127

Warren MF (2002) Digital divide and adoption of information and communication technologies in the UK farm sector. Int J Inf Tecnol Mang 1(4):385–405

Evaluation of Farmer's e-Shops with an Emphasis on Services

Vladimír Očenášek, Jiří Vaněk, Pavel Šimek, Michal Stočes, Jan Jarolímek, and Jan Masner

Abstract The evaluation of websites at the Department of Information Technologies has so far been largely focused on the websites of towns, municipalities, or regional Internet portals. In the submitted article, we focused on evaluation of food and farmers' e-shops. As the new generation grows in the digital age and it is accustomed to search for and buy goods in this environment, we have decided to focus on this issue. To ensure that the user repeatedly returns, you need to recognize what services he requires. For the purposes of evaluation, we determined 25 parameters, in which 12 are new parameters focused on the service category and used them for evaluation of nine food and farmers' e-shops: 5 Czech, 2 British, and 2 Greek. It turned out that comparing Czech and foreign e-shops with the same parameters is not entirely appropriate. Some parameters can only be applied in the Czech environment (like cash on delivery), which was not even offered in evaluated foreign e-shops. Similarly, the option for free pickup at the designated place was also not available. If we would evaluate Czech and foreign e-shops separately, kosik.cz would be the winner among Czech e-shops and yolenis.com among foreign e-shops.

Keywords Digital governance in municipalities worldwide · Regional web portal · e-Shop · Consumer · Farmers' products · Minimal order price · Order tracking

V. Očenášek (✉) · J. Vaněk · P. Šimek · M. Stočes · J. Jarolímek · J. Masner
Department of Information Technologies, Czech University of Life Sciences, Prague, Czech Republic
e-mail: ocenasek@pef.czu.cz

A. Theodoridis et al. (eds.), *Innovative Approaches and Applications for Sustainable Rural Development*, Springer Earth System Sciences,
https://doi.org/10.1007/978-3-030-02312-6_15

1 Introduction

The Czech Republic has the highest number of e-shops per capita in Europe. Last year, the number of e-shops increased by 3900 to the total of 40,100. Since 2013, the number has stagnated. More and more brick and mortar stores are starting to sell online. In 2017, the income of e-shops increased by 18% to 115 billion Czech crowns.

The Czech Republic is a world leader in online sales; more than 43% of nonfood-related goods are sold via online stores. It is followed by the Netherlands (35%), Slovakia (34%), the United Kingdom (33%), and Germany (29%) according to data from GfK (Growth from Knowledge). The average of percentage of online sales in Europe is 24%. The average share of online sales in Europe is 24%. The most of the goods sold online in the Czech Republic fall within categories of information technology and cameras. The biggest year-on-year growth in online sales was in household appliances (24.5%) and consumer electronics (21.6%).

The interest in buying commodities online, which was so far only at a marginal interest of our consumers, is rising in Czech Republic. These include groceries and special farmers' products. Several specialized food e-shops operate in Czech Republic, and all of them report a significant increase in the number of customers. Ordered purchases are distributed mainly in cities and surrounding areas. Groceries can be delivered in 90 min, and shipping can be free on orders over certain price. These e-shops offer the same range of goods as supermarkets, but also specialized and gourmet goods – for example, products for vegetarians, gluten- or lactose-free food, etc. e-Shops also cooperates with farmers, which results in special offers of meat, sausages, cheese, bio-products, wine, etc.

In the literature, e-shopping has been around since the turn of the century. In 2001, Lee (2001) proposed a four-factor model of customer satisfaction, including after-sales service, customer service, commodity prices, and service quality. Srinivasan et al. (2002) identified from data from 1211 customers eight factors that potentially impact e-loyalty (customization, contact interactivity, care, community, convenience, cultivation, choice, and character). In 2002, Schubert and Dettling (2002) presented an evaluating tool EWAM (Extended Web Assessment Method) that included a set of possible criteria for assessing the quality and success of existing commercial websites. In 2003, van der Merwe and Bekker (2003) published an article in which they introduced the concept of evaluating commercial websites. They determined five categories for which they define criteria groups. For example, for the Content category, they defined these criteria: product-/service-related information, company and contact information, information quality, and interactivity. Gounaris et al. (2010) examined a sample of 1052 customers to determine the impact of service quality and customer satisfaction on behavior of online customers, namely, word of mouth, site revisit, and purchase intentions in the context of Internet shopping. In 2015 (Lin 2015), the index system of customer satisfaction was put together based on 19 indicators from five categories in e-shopping. In 2015 we evaluated regional Internet portals (Očenášek et al., 2015) and used as an inspiration for the choice of parameters the best known rating of web portals of capitals of the world's largest countries – digital governance in municipalities worldwide (Holzer et al. 2014).

2 Materials and Methods

Literature, especially recent (Lin 2015), shows that customer satisfaction is increasingly influenced, in addition to the quality of the commodity itself, by non-commodity elements. From the digital governance in municipalities Worldwide (Holzer et al. 2014), we have selected parameters for the security/privatization, usability, and content categories.

In the Service category, we have identified 12 most nonprofit parameters that, in our opinion, influence the e-shop customer satisfaction the most.

We newly included 12 parameters to the service category and checked whether they provide those information and services:

List of distribution points
Special offers via emails
Nutritional values of food
Minimal order price
Cash on delivery
Credit card payment on delivery
Free pickup (of the purchased food) at the designated place
e-Shop app
Optional registration
Order tracking
Social networks
FAQ

Individual parameters are evaluated on two scales. We score zero or one point in yes or no questions; for nine parameters up to three points were assigned. We evaluated five food e-shops offering special farmers' products from the Czech Republic and four foreign e-shops, two English e-shops and two Greek e-shops. The overall overview of parameters is shown in Tables 1, 2, 3, and 4.

3 Results

Whenever the consumer visits an e-shop, he expects a well-arranged product catalog with complete information and easy ordering, where it is clear what and why it has to be filled in. The consumer further expects a competitive price; standard payment options; fast, reliable, and cheap delivery; and constant control over the order.

The proposed evaluation methodology was verified on nine representatives of food and farmers' e-shops. From Czech representatives, we chose:

- nakupzfarmy.cz (motto: Direct sale and e-shop without middleman)
- svetbedynek.cz (motto: Farmers' products to your home)
- kosik.cz (motto: Bio fruits and Bio vegetables – taste it)
- sklizeno.cz (motto: Groceries at your doorstep)
- farmanemcova.cz (motto: Without chemicals and preservatives)

From foreign e-shops, we chose for comparison:

- jimmysfarm.com (motto: Free range produce fresh from the farm!) (Fig. 1)
- yolenis.com (motto: Carefully selected products from all over Greece) (Fig. 2)
- riverford.co.uk (motto: From one man and a wheelbarrow to an award-winning organic delivery company)
- trofos.com (motto: Greek Organic & Traditional Delicatessen e-Marketplace)

Parameters, score scale, and evaluation for Czech and foreign e-shops are shown in Tables 1, 2, 3, and 4.

Fig. 1 jimmysfarm.com – the butchery

Fig. 2 yolenis.com – product offerings

4 Security/Privacy

The Security/Privacy category deals with, e.g., if privacy policy determines which data are collected or whether the site allows access to nonpublic information. This category becomes significant with regard to the EU's General Data Protection Regulation.

5 Usability

The Usability category focuses on accessibility of e-shops for users with specific needs. For instance, in the category Usability of public administrations websites, the Czech Decree No. 64/2008 sb. defines the individual rules for the form of published information. The form of published information has to be adapted so that information is also available to the disabled users.

For the purposes of our evaluation, the following eight parameters were chosen:

6 Security/Privacy

1. Does the site have a privacy or security statement/policy?
2. Does the privacy policy identify the intended use of the data collected?
3. Does the privacy policy address the option to have personal information used to send unsolicited materials (e.g., mailing lists or another marketing that is targeted back to the consumer)?
4. Does the privacy policy address the use of digital signatures to authenticate users?

7 Usability

5. Does the site provide the date (month and year) of the last update of the website?
6. Does every page contain the link to the homepage?
7. Is it easy to navigate the content of the website?
8. Are language versions available?

The rating of this category is listed in Table 1.

Issues related to the use of digital signature for user authentication (question no. 4) are absolutely unnoticed not only on Czech but also on foreign e-shops. Similarly, the date of the last update of the website (question no. 5) is neglected. Website administrators have approached quite differently the problem of homepage link on every page of the website as well as seamless organization of information on the website (question nos. 6 and 7). More language variations are available only on Greek e-shop yolenis.com (English, German). The winners in those categories are Greek e-shop yolenis.com and Czech e-shop svetbedynek.cz.

8 Content

The Content category deals with contact information; online offering of documents, books and recipes, and calendar of additional events; and also the ability to view websites on different devices.

For the purposes of our evaluation, the following five parameters were chosen:

9. Does the site offer contact information (address, phone, email)?
10. Does the site offer documents, reports, books, or recipes online?
11. Does the site allow users to purchase or order documents, reports, books, or recipes?
12. Does the site offer a calendar of events?
13. Does the site have a responsive web design or is it accessed via smartphone (iOS, Android, Symbian)?

The rating of this category is shown in Table 2.

Table 1 Evaluation – security/privacy and usability

question	scale	nakupzfarmy. cz	svetbedynek. cz	sklizeno. cz	kosik. cz	farmanemcova. cz	jimmysfarm. com	yolenis. com	riverford.co. uk	trofos. com
1	0, 1	0	1	1	1	1	0	1	1	0
2	0, 1	0	1	1	1	0	0	1	0	0
3	0, 1	0	1	1	1	0	0	1	0	0
4	0, 1	0	0	0	0	0	0	0	0	0
5	0, 1	0	0	0	0	0	0	0	0	0
6	0, 1	1	1	1	1	1	1	1	1	1
7	0, 1	1	1	1	1	1	1	1	1	1
8	0, 1	0	1	0	0	0	0	1	0	0
Σ		**2**	**6**	**5**	**5**	**3**	**2**	**6**	**3**	**2**

Table 2 Evaluation – content

question	scale	nakupzfarmy. cz	svetbedynek. cz	sklizeno. cz	kosik. cz	farmanemcova. cz	jimmysfarm. com	yolenis. com	riverford.co. uk	trofos. com
9	0, 1	1	1	1	1	1	1	1	1	1
10	0, 3	0	3	0	0	0	0	0	3	0
11	0, 3	0	0	0	3	0	0	0	0	0
12	0, 3	0	0	0	0	0	3	0	0	0
13	0, 3	3	3	3	3	3	3	3	3	3
Σ		**4**	**7**	**4**	**7**	**4**	**7**	**4**	**7**	**4**

In this category all e-shops have met the condition of offering contact information (question no. 9). Question nos. 10 and 11 deal with the possibility of ordering documents, such as cooking books and recipes, or reading them online. This is possible online only on riverford.co.uk and svetbedynek.cz, while kosik.cz allows ordering documents, but not reading them online. Event calendar (question no. 12) is available only on jimmysfarm.com. All e-shops have websites that react on different screen sizes. However, only some offer specialized variation for smartphones with different operating systems (question no. 13).

Four e-shops have shared the victory in this category; all scored seven points – jimmysfarm.com, kosik.cz, riverford.co.uk, and svetbedynek.cz.

9 Service

We determined 12 new parameters from the services category:

14. List of distribution points
15. Special offers via emails
16. Nutritional values of food
17. Minimal order price
18. Cash on delivery
19. Credit card payment on delivery
20. Free pickup at the designated place
21. e-Shop app
22. Optional registration
23. Order tracking
24. Social networks
25. FAQ

Question no. 15 – sending special offers via email is available on every e-shop, with exception of trofos.com, and we can see similar results in question no. 17. Mobile app, question no. 21, is available only at kosik.cz as well as optional registration (question no. 22). Order tracking is possible only on yolenis.com (question no. 23). The link to social media (question no. 24) is missing solely on nakupzfarmy.cz. Every e-shop offers answers to frequently asked questions (question no. 25) (Table 3).

In this category, e-shop sklizeno.cz, which received 13 points, claimed the victory followed by e-shop kosik.cz with 11 points and e-shop farmanemcova.cz with 9 points.

Table 3 Evaluation – service

question	scale	nakupzfarmy. cz	svetbedynek. cz	sklizeno. cz	kosik. cz	farmanemcova. cz	jimmysfarm. com	yolenis. com	riverford.co. uk	trofos. com
14	0, 1	1	0	1	1	1	1	1	0	0
15	0, 1	1	1	1	1	1	1	1	1	0
16	0, 3	3	0	3	0	0	0	0	0	0
17	0, 1	1	1	1	1	1	1	1	1	0
18	0, 3	0	3	3	0	3	0	0	0	0
19	0, 3	0	0	0	3	0	0	0	0	0
20	0, 1	1	0	1	0	1	0	0	0	0
21	0, 3	0	0	0	3	0	0	0	0	0
22	0, 1	0	0	1	0	0	0	0	0	0
23	0, 3	0	0	0	0	0	0	3	0	0
24	0, 1	0	1	1	1	1	1	1	1	1
25	0, 1	1	1	1	1	1	1	1	1	1
Σ		**8**	**7**	**13**	**11**	**9**	**5**	**8**	**4**	**2**

10 Conclusion

A number of studies show that up to 30% of customers do not complete their order at an e-shop for various reasons. It is therefore very important to find out how e-shop works, whether it provides convenient purchase, easy payment for ordered goods, and convenient delivery of the purchase to the customer.

We focused on food and farmers' e-shops. We have evaluated nine e-shops, five of which were Czech and four foreign (two British and two Greek). The evaluation is based on the experience of evaluating the web portals of capitals of the world's largest countries – digital governance in municipalities worldwide (Holzer et al. 2014) – as well as evaluations of regional web portals. The original score is divided into four groups (privacy, usability, content, services). In our article, we focused on the service group most and created 12 new parameters due to the problematics of e-shops differing in many aspects from the original rating of cities or regional web portals. The original intention was to compare Czech and foreign e-shops. After evaluating e-shops, it can be said that this original intention cannot be fulfilled. The reason is simple. It turned out that the customs of Czech and foreign customers are somewhat different. For example, parameter 18 – cash on delivery – which is still expected in the Czech Republic, is not offered abroad. Just like parameter 20 – free pick up at the designated place – this option is also not offered abroad.

From evaluation of separate categories, we can state that category Security and privacy won the Greek e-shop yolenis.com and categories Usability won the Czech e-shop svetbedynek.cz.

In the Content category, the victory was shared between four e-shops with the same amount of points (Očenášek et al. 2015) – jimmysfarm.com, kosik.cz, riverford.co.uk, and svetbedynek.cz.

The Service category won e-shop Sklizeno.cz, which received 13 points, followed by e-shop kosik.cz with 11 points. The best foreign e-shop in this category was yolenis.com, which received eight points.

If we evaluate the Czech and foreign e-shops separately, we can announce the e-shop kosik.cz (23 points) as the winner of the Czech e-shops, followed closely by harvest.cz (22 points) and the svetbedynek.cz (20 points). Foreign e-shops won the Greek yolenis.com, which received 18 points.

The final results are in Table 4 and test winner in Fig. 3.

Table 4 Overall rating

Category	nakupzfarmy.cz	svetbedynek.cz	sklizeno.cz	kosik.cz	farmanemcova.cz	jimmysfarm.com	yolenis.com	riverford.co.uk	trofos.com
Privacy/usability	2	6	5	5	3	2	6	3	2
Content	4	7	4	7	4	7	4	7	4
Service	8	7	13	11	9	5	8	4	2
Total	**14**	**20**	**22**	**23**	**16**	**14**	**18**	**14**	**8**

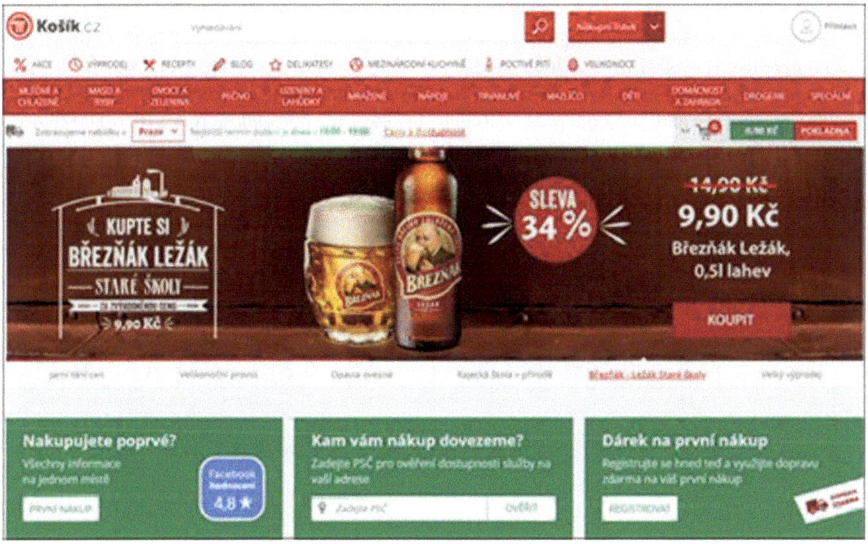

Fig. 3 kosik.cz – test winner

Acknowledgment The results and knowledge included herein have been obtained owing to support from the following institutional grants: Internal Grant Agency of the Faculty of Economics and Management, Czech University of Life Sciences, Prague, Grant No. 20181010.

References

Gounaris S, Dimitriadis S, Stathakopoulos V (2010) An examination of the effects of service equality and satisfaction on customers' behavioral intentions in e-shopping. J Serv Market 24/2:142–156. Emerald Group publishing limited [ISSN 0887-6045]. https://doi.org/10.1108/08876041011031118

Holzer M, Zheng Y, Manoharan A, Shark A (2014) Digital governance in municipalities worldwide – sixth global e-governance survey: a longitudinal assessment of municipal websites throughout the world the E-Governance Institute. The E-Governance Institute National Center for Public Performance School of Public Affairs and Administration Rutgers, the State University of New Jersey-Campusat Newark. Printed in the United States of America. ISBN: 978-0-942942-28-6

Lee MKO (2001) A trust model for consumer internet shopping, City University of Hong Kong, China, 2001. Int J Electron Comm 6(1):75–91. ISSN: 1086–4415

Lin Q (2015) Construction and evaluation of customer satisfaction index system in E-shopping. Fifth international conference on instrumentation and measurement, computer, communication and control. Qinhuangdao, China. ISBN: 978-1-4673-7723-2

Očenášek V, Ulman M, Vaněk J, Šilerová E (2015) Evaluation of regional web portals. In: Proceedings of the 7th international conference on information and communication technologies in agriculture, food and environment 17.09.2015, Kavala, Greece: CEUR-WS/Vol- 1498. pp 976–982

Schubert P, Dettling W (2002) Extended web assessment method (EWAM) – evaluation of e-commerce applications from the customer's view point. In: Proceedings of the 35th HICSS conference, Hawaii. ISBN: 0-7695-1435-9
Srinivasan SS, Anderson R, Ponnavolu K (2002) Customer loyalty in e-commerce an exploration of its antecedents and consequences. J Retail 78(1):41–50. https://doi.org/10.1016/S0022-4359(01)00065-3
van der Merwe R, Bekker J (2003) A framework and methodology for evaluating e-commerce Web sites. Internet Res 13(5):330–341. https://doi.org/10.1108/10662240310501612

ICT Tools by Farmers of Lucania Region in Italy

Gianluigi De Pascale, Piermichele La Sala, Nicola Faccilongo, and Claudio Zaza

Abstract This study aimed to investigate how farms in the Lucania region of Italy cluster according to the level of innovation adopted. It used a questionnaire to ask if farms adopted information and communications technology (ICT) tools and, if so, what types were involved in management and/or production processes. A cluster analysis was done on the collected data. The results showed that using a k-means clustering method, two clusters appeared: innovators and the remaining group. Using boxplot representation, there were three groups: innovators, early adopters, and laggards. These results will be used to identify good practices in terms of smart devices adopted, within the H2020 project titled Short Supply Chain Knowledge and Innovation Network (SKIN).

Keywords ICT · Clustering analysis · Lucania · Farms

1 Introduction

Agriculture is a field suffering from low efficiency of the processes undertaken to carry out related activities, for many reasons coming from quick changes of the surrounding scenario. Such changes have been fostered by new digital technologies. They appear in an integrated system named the Farm Management Information System (FMIS). Nowadays, the general directions to foster growth in Europe come from the European Commission (EC). In fact, being able to cope with daily problems means being able to engage synergies among tangible and intangible resources inspired by studies performed on such issues. They set out that smart growth can be put into practice by sharing knowledge and adopting innovations (Contò et al. 2015). Reducing distances among the available resources accelerates access to them. In addition, it has been stated that it is necessary to be in conformity

G. De Pascale (✉) · P. La Sala · N. Faccilongo · C. Zaza
Department of Economics, University of Foggia, Foggia, Italy
e-mail: gianluigi.depascale@unifg.it

A. Theodoridis et al. (eds.), *Innovative Approaches and Applications for Sustainable Rural Development*, Springer Earth System Sciences,
https://doi.org/10.1007/978-3-030-02312-6_16

with the needs of the ecosystem. To this end, information and communications technology (ICT) tools play an important role in achieving the aforementioned goals. Since the ecosystem is very dynamic and changes frequently occur, it is complex to manage all data emerging from daily activities and to take control of evolving scenarios. Such problems are much more evident in small-to-medium enterprises (SMEs), where information flows are often stressed by lack of capability to access ICT innovations (Contò et al. 2015). These tools allow farmers to reduce asymmetrical information, which is the main cause of moral hazard and adverse selection, mostly affecting firms operating in international markets. Information management, in turn, influences internal and external actions. Hence, the greatest concern has to be that of making accessible tools useful to reduce the information gap. The benefits will affect transactional costs. In fact, it is possible to find many farms that have obtained economic improvements by adopting such technologies. The demonstration of such claims comes from developing countries that show high levels of growth. To this end, the World Bank has included these data in their yearly reports since 2016. The reports show that firms in poor countries that adopt digital technologies get to be much more competitive but maintain a low profile in international markets because of lack of appropriate skills and infrastructure (World Bank 2016).

Researchers have devised complex management systems that bring together all elements that make up farms. The hardest challenge is to guarantee the right resource coordination and employment over a long period in order to attain the general goal of adding sustainable value for stakeholders. For these reasons, the study described in this chapter is aimed at investigating the degree of ICT adoption in a southern Italian region tackling important issues in terms of economic development. The study implies the necessity of investigating the features of agricultural enterprises that adopt ICT tools in undertaking their activities, from plantation to harvesting. Specifically, it puts forward a double analysis: firstly to group the sample according to ICT adoption; and secondly to provide insights into the determinants of income by highlighting the role of ICT.

Another perspective to lead the analysis for understanding of the determinants of the outline of a farm organization, according to the degree of the propensity to innovate and/or the reasons that increase that propensity, is explored through multivariate regression, assuming that income is a dependent variable, through use of a semi-log model to implement data method analysis.

This study has been undertaken as part of the research activities of the H2020 project titled Short Supply Chain Knowledge and Innovation Network (SKIN; grant agreement no. 728055), financed in 2016. The project aims to identify good practices adopted by organizations operating in a short food supply chain (SFSC). From the perspective of the project, good practices are those enabling economic players in the agro-food domain to reduce the distances along the supply chain, mainly between producers and consumers. In an SFSC, most operators are small farmers, constantly seeking to improve their efficiency. It has been shown that ICT adoption significantly improves efficiency, as well as connecting things and people, giving benefits even to those farmers who are struggling with tackling the pressures of market competition with large-scale operators embedded in longer food supply chains.

This chapter is structured into different sections, as follows: Sect. 2 describes the literature review; Sect. 3 gives an outline of the questionnaire used for collecting data from farms; and Sect. 4 explains the analytical method used for processing the data to extract information. Finally, the results are summarized and discussed, then the conclusions are presented.

2 Literature Review

Over the years, many changes have occurred with the advent of ICT, affecting, in particular, the efficiency of farms. Researchers have taken this opportunity to study the impacts of such extraordinary evolutions in depth. The first step to better understanding of the effects of ICT farming tools is related to the reasons that make it necessary to implement innovation of processes and knowledge uptake: they are based on the trend of costs level shown by farms so far. Nowadays, European and national policies direct organizations to realize cost reductions through the adoption of smart devices, in line with Industry 4.0 topics. To this extent, international organizations such as the World Bank have issued data analyses showing how costs decrease (World Bank 2016) with introduction of ICT tools for managing the growing complexity of activities due to the complexity of the competition, and vice versa. Have explained what type of problems prevent digital device adoption in developing countries. Obviously, many countries are divided into different areas; some demonstrate prominent growth, while others suffer lack of capability to stimulate growth. Many factors are involved. The significant ones relate to slowness in reforming business regulations and skills development systems. It is emphasized that building efficient information systems is the key to triggering sustainable growth for long periods. A signal of farms' (and, more generally, firms') efficiency is related to the impact of smart tools on the volume of production, which measure the total factor productivity growth. With regard to farming dynamics, Diederen et al. (2003) have distinguished between innovators, early adopters, and laggards. These three categories can represent the farmer profiles observed in European areas. Measurement of the degree of innovation within farms has been done using different assessment methods. One classification uses a matrix that differentiates between innovations as major, intermediate, or minor on the basis of their technological advancement. The matrix shows an innovation numerical index that represents the innovation level of each farm (Ariza et al. 2013). Other studies have explained agricultural income through multivariate regression analysis, considering income as a dependent variable (Ugwumba et al. 2010). On the other hand, their study issued that structural characteristics such as farm size, utilized agricultural area (UAA), and the farmer's age reflected the attitude and the willingness to choose to undertake and pursue innovation processes. According to these assumptions, they distinguished between innovators and the remaining group. The literature breakdown also includes a study using a more complex index. This index does not consider the adoption of a single technology; instead, it focuses on combined factors defining the innovations. The complexity is due to the variables that are taken into account, which do not focus only on tools and equipment (Chen et al. 2014; Esmeijer et al. 2015) for carrying out farm activities (e.g., ICT tools or tractors), but also on primary productive factors (e.g., seeds). Therefore, the index brings together different elements, combining their effects on the farming results and showing the benefits from emerging synergies. However, the latter is not the focus of this chapter.

3 Data Collection Methods

The questionnaire consists of 22 questions (Qs) in total. The survey is divided into two parts. The first part, dedicated to *General Information*, includes Q1 to Q7 and captures information on the general aspects of the farms involved in the survey. Depending on the response to Q7, regarding the use of ICT tools, the questionnaire either proceeds to the second part, dedicated to *Farms Using ICT* (from Q8 to Q22), or it ends in the case of a negative answer. In the second part there is a set of questions dedicated to analyzing what the most used ICT tools applied to farm management are, and the impact these technologies could have in decreasing agronomic input and manpower, and in increasing production.

The survey questions are as follows:

(Q1) Legal status. Possible answers: partnership; capital company; other.
(Q2) Time of constitution. Possible answers: less than 5 years; between 5 and 10 years; more than 10 years.
(Q3) Farmer's age. Possible answers: less than 35 years; between 35 and 50 years; more than 50 years.
(Q4) Utilized agricultural area (UAA). Possible answers: less than 10 hectares; between 10 and 50 hectares; more than 50 hectares.
(Q5) Crop type. Possible answers: tree crops, herbaceous crops, mixed crops.
(Q6) Income. Possible answers: between €0 and €50,000; between €50,001 and €120,000; between €120,001 and €250,000; between €250,001 and €500,000; between €500,001 and €1,000,000; more than €1,000,000.
(Q7) Do you use ICT tools? Possible answers: yes; no. (If the reply is positive, the farmer answers Q8 to Q22, otherwise the questionnaire ends here.)
(Q8) What type of management tools do you use? Possible answers: none; tools for the farm's notebook; tools for warehouse management; tools for management of the balance sheet; tools for management of invoicing; enterprise resource management; other. (Multiple answers are allowed.)
(Q9) What type of software for data management do you use? Possible answers: none; software for data storage; software for market analysis; decision support system software; software to analyze costs; other. (Multiple answers are allowed.)
(Q10) Do you use tools for precision agriculture? Possible answers: yes; no. (If the reply is positive, the farmer answers Q11, otherwise Q17.)

(Q11) Do you use environmental sensors? Possible answers: yes; no. (If the reply is positive, the farmer answers Q12, otherwise Q13.)

(Q12) Why do you use environmental sensors? Possible answers: fertilization; phytosanitary treatments; weeding; irrigation; sowing; soil management. (Multiple answers are allowed.)

(Q13) Do you use an unmanned aerial vehicle (UAV) or drone? Possible answers: yes; no. (If the reply is positive, the farmer answers Q14, otherwise Q15.)

(Q14) Why do you use a UAV? Possible answers: fertilization; phytosanitary treatments; weeding; irrigation; sowing; soil management. (Multiple answers are allowed.)

(Q15) Do you use satellite data? Possible answers: yes; no. (If the reply is positive, the farmer answers Q16, otherwise Q17.)

(Q16) Why do you use satellite data? Possible answers: fertilization; phytosanitary treatments; weeding; irrigation; sowing; soil management. (Multiple answers are allowed.)

(Q17) Do you use external data sources? Possible answers: yes; no. (If the reply is positive, the farmer answers Q18, otherwise Q19.)

(Q18) What types of data do you research? Possible answers: agrometeorological; market; legal aspects; phytosanitary bulletin; other. (Multiple answers are allowed.)

(Q19) What type of tool do you think is the most useful? Possible answers: external data sources; enterprise resource planning; software for data management; precision agricultural tools.

(Q20) Since you started to use ICT tools, have you detected a reduction in the use of agronomic inputs (pesticides, fertilizers, water, etc.)? To what extent? Possible answers: none; between 0% and 5%; between 6% and 10%; between 11% and 20%; more than 20%.

(Q21) Since you started to use ICT tools, have you detected a reduction in employed manpower? To what extent? Possible answers: none; between 0% and 5%; between 6% and 10%; between 11% and 20%; more than 20%.

(Q22) Since you started to use ICT tools, have you detected an increment in production? To what extent? Possible answers: none; between 0% and 5%; between 6% and 10%; between 11% and 20%; more than 20%.

For the cluster analysis presented in the next section, a subset of the total variables was taken into account (Table 1).

Further analysis of income was performed, as shown in Table 2.

Table 1 Variables used in the cluster analysis; the code is associated with a single answer in the analysis

Question number	Variable name and abbreviation	Answer	Code
Q1	Legal status	Partnership	1
		Capital company	2
Q3	Farmer's age (age)	Less than 35 years	1
		Between 35 and 50 years	2
		More than 50 years	3
Q4	Utilized agricultural area (UAA)	Less than 10 hectares	1
		Between 10 and 50 hectares	2
		More than 50 hectares	3
Q5	Crop type	Tree crops	1
		Herbaceous crops	2
		Mixed crops	3
Q6	Income	Between €0 and €50,000	1
		Between €50,001 and €120,000	2
		Between €120,001 and €250,000	3
		Between €250,001 and €500,000	4
		Between €500,001 and €1,000,000	5
Q7	Do you use ICT tools? (ICT)	Yes	1
		No	0

ICT information and communications technology

4 Data Analysis Methods

In this study, the collected data were analyzed using clustering analysis. For obtaining groups featured by homogeneous parameters, a k-means cluster method was used. The analysis returned acceptable results, setting two clusters. The choice of selecting two clusters was possible because:

- The k-means clustering method can be applied with both supervised and unsupervised methodology (Wagstaff et al. 2001).
- The use of three clusters did not return acceptable results.

In general, k-means is a method that is unsupervised. Therefore, the processing machine automatically calculates the smallest distances, respecting the set threshold between features (Zhang et al. 1997). The goal is to evaluate if the distances are such to consider the minimum sum of the squared error (SSE) within each group (Likas et al. 2003). The formula for the SSE is as follows:

Table 2 Variables used in the regression analysis; the code is associated with a single answer in the analysis

Question number	Variable name and abbreviation	Answer	Code
Q2	Time of constitution	Less than 5 years	1
		Between 5 and 10 years	2
		More than 10 years	3
Q3	Farmer's age (age)	Less than 35 years	1
		Between 35 and 50 years	0
		More than 50 years	0
Q4	Utilized agricultural area (UAA)	Less than 10 hectares	1
		Between 10 and 50 hectares	2
		More than 50 hectares	3
Q5	Crop type	Tree crops	1
		Herbaceous crops	2
		Mixed crops	3
Q6	Income	Between €0 and €50,000	25,000
		Between €50,001 and €120,000	85,000
		Between €120,001 and €250,000	185,000
		Between €250,001 and €500,000	375,000
		Between €500,001 and €1,000,000	750,000
Q7	Do you use ICT tools? (ICT)	Yes	1
		No	0

ICT information and communications technology

$$\mathrm{SSE} = \sum_{k=1}^{K} \sum_{\forall x_i \in C_k} \|x_i - \mu_k\|^2 \tag{1}$$

where C_k is the set of grouped data in cluster k, and μ_k is the vector mean of cluster k. Using an unsupervised method, it was found that there were two clusters. Nevertheless, to be in line with Diederen et al. (2003), the scope was to find three groups to be labeled as innovators, early adopters, and laggards. The test did not achieve this goal, so a supervised method was used to try and set three clusters. In turn, this test did not succeed, because clusters two and three presented identical features (this was the reason why the unsupervised method returned two clusters). Hence, it was chosen to apply a supervised method, selecting two clusters. At this stage, the test succeeded and the results were accepted to fulfill the two-groups theory (innovators and the remaining group) described by. The additional analysis, implemented with the multivariate regression model, was performed by considering the income variable as independent and the other variables, showed in Table 2, as dependent (Easterlin 2005). The multivariate model chosen was a log-linear one because of an assumption that income grows with a decreasing margin and the logarithmic trend aims to fulfill that purpose (Hastie 2017). On the other hand, the independent variables were supposed to have a linear evolution. The evaluation was conducted by first checking the R-squared value and the p value of the whole model and, after verification, the goodness of the method to consider at least one independent variable as fitting to explain the income, it was proceeded with checking of the distribution of the residuals, using a Breusch–Pagan test for a heteroscedasticity check. In addition, a further consideration upon the complexity of the phenomenon was taken into account. Indeed, the authors had a specific objective to check the influence of the aforementioned variables on the income, starting from the awareness that income is surely a result of more complex events than those represented by the regression hereby proposed, as will be shown hereafter (Press and Wilson 1978).The model, after transformation of the "income" variable into "ln income", is presented as follows (Wasserman 2004):

$$\ln \mathit{income} = \beta_0\ \mathit{time\ of\ constitution} + \beta_1\ \mathit{age} + \beta_3\ \mathit{uaa} + \beta_4\ \mathit{crop\ type} + \beta_5\ \mathit{ict} \tag{2}$$

This study used several simple methods for evaluating the results in terms of the innovation level. The first stage identified the groups through a clustering analysis, and such analysis revealed the existence of two main groups. The next step shed light on a multiple regression log-linear analysis. This examination highlighted that there were three groups, complying with the insights.

5 Results and Discussion

The questionnaire shows the answers from a sample within a producer organization (PO) in the Lucania region of Italy. There were 59 respondents. They represented an outline of the region in terms of the typology of farms and, in terms of scale, the composition of the farming population in the Lucania region.

In this section we present the results from the frequency distribution of Q20, Q21, and Q22, and the cluster analysis and boxplot. The section concludes with a discussion of the results, emphasizing the differences of the marks, based on insights from the literature mentioned in Sect. 2. The frequencies of the relevant selected variables and pertinent comments are also presented. The discussion provides insights to explain potential barriers that obstruct adoption of ICT tools, to be investigated in further studies. There are delivered considerations on what the farmers not ICTs skilled and, in consequence, not adopter, while oriented and inclined to adopting.

Figure 1 shows the frequency distributions related to Q20 (Fig. 1a), Q21 (Fig. 1b), and Q22 (Fig. 1c), regarding the 29 farms that used ICT tools. Q20 and Q21 focus on the reduction of agronomic inputs (such as fertilizers, pesticides, and water) and employed manpower, respectively, observed by the farmers since they started to adopt ICT tools. Analyzing the answers to Q20, 5 respondents (17.2%) did

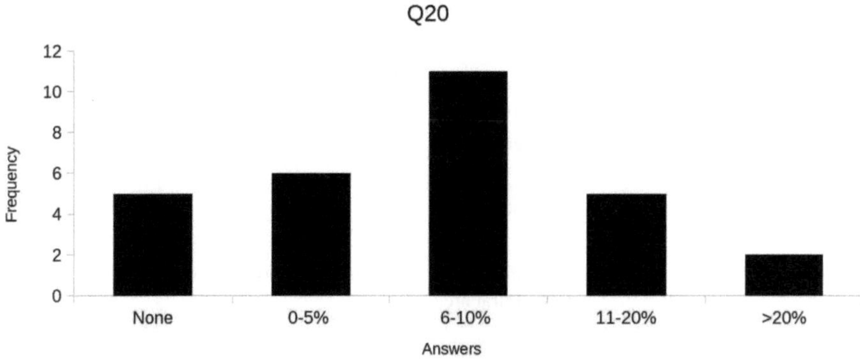

Fig. 1 Frequency distribution of answers related to (**a**) Q20 ("Since you started to use ICT tools, have you detected a reduction in the use of agronomic inputs (pesticides, fertilizers, water, etc.)? To what extent?"); (**b**) Q21 ("Since you started to use ICT tools, have you detected a reduction in employed manpower? To what extent?"); and (**c**) Q22 ("Since you started to use ICT tools, have you detected an increment in production? To what extent?"). None = none; 0–5% = between 0% and 5%; 6–10% = between 6% and 10%; 11–20% = between 11% and 20%; >20% = more than 20%. *ICT* information and communications technology; *Q20* question 20, *Q21* question 21, *Q22* question 22

not notice a difference in the application of agronomic inputs, while a small (0–5%) reduction was detected by 6 respondents (20.7%). The modal value, with 11 responses (37.9%) was the range of 6–10%, and the last two intervals (11–20% and >20%) accounted for 5 responses (17.2%) and 2 responses (6.9%), respectively. Taking into account the answers to Q21, 6 farmers (20.7%) did not notice any reduction in employed manpower related to the use of ICT tools, while a small decrease (0–5%) was perceived by 7 farmers (24.1%). Even in this case, the range 6–10% represented the modal value, with 10 answers representing 34.5% of the total, with the last two intervals (11–20% and >20%) each attracting 3 responses (10.3%). Finally, the modal value of the frequency distribution for Q22 was represented by the first class, with 10 farmers (34.5%) noticing no production increment associated with the use of ICT tools. The other ranges (0–5%, 6–10%, 11–20%, and >20%) accounted for 6 responses (20.7%), 7 responses (24.1%), 2 responses (6.9%), and 4 responses (13.8%), respectively.

As indicated in Sect. 3, the queries aimed to evaluate the number of farms that had adopted ICT tools and associated evidence on farm structure and the farmer's age. The cluster analysis showed two main clusters, characterized as shown in Table 3.

The resulting clusters presented different features concerning just three variables: crop type, income, and ICT. In general, both clusters appeared to consist of farms established as partnerships, with the median farmer age being 35–50 years and the UAA being between 10 and 50 hectares. The differences came from:

- *Crop type:* Cluster 1 featured herbaceous crops, while cluster 2 featured mixed crops (herbaceous and tree crops).

Table 3 Two emerging clusters after data processing; the results concern a selected number of variables from the survey

Variable	Cluster	
	1	2
Legal status	1	1
Crop type	2	3
Farmer's age	2	2
UAA	2	2
Incomes	2	4
ICT	0	1

ICT information and communications technology, *UAA* utilized agricultural area

Table 4 Number of cases in each cluster

Cluster number	Cases
1	38
2	21
Valid	59
Missing	0

- *Incomes:* The income in cluster 1 was between €50,000 and €120,000, whereas the income in cluster 2 was between €250,000 and €500,000.
- *ICT:* Cluster 1 was characterized by farms that did not adopt ICT tools, whereas cluster 2 featured farms that did adopt such tools.

Table 4 shows the frequencies within each cluster. The first cluster included 38 farms (Ariza et al. 2013) and the second cluster included 21 farms (Chen et al. 2014).

The majority of farms were concentrated in the first cluster. Although this evidence, the cluster analysis demonstrated—in accordance with what asserted—that there were two groups: innovators and remaining groups. Cluster 2, populated by innovators, registered much higher income (between €250,000 and €500,000) than cluster 1 (between €50,000 and €120,000). Furthermore, the farms in cluster 1 that adopted ICT were characterized by mixed crops. In this regard, crop diversification was associated with higher incomes (Di Falco and Zoupanidou 2017). Taking into consideration that the farmers in cluster 2 adopted ICT tools, the result in terms of incomes were significant. When farms adopt ICT, it seems to improve their performance. On the other hand, there is a dependency between those two variables, though the direction of it has not been defined; at this stage it is not clear what kinds of factors push farms to innovate with ICT. In fact, it can depend on the achievement of excessive dimension and, in view of the increasing complexity, farms need to improve their data collection and management phases; otherwise, it can depend on the need to improve revenue performance, and so the adoption of ICT tools results in an increase in income. To make this point even clearer, another analysis was done, assuming that the crop type variable was excluded because it was decided to conduct an analysis without considering such a qualitative agronomic variable. The focus remained on the economic aspects, exploring relations with economic parameters (Figs. 2, 3, 4, and 5).

Regression analysis was carried out, as shown in Table 5.

As shown in Table 5, the multiple regression assumed that the income variable, derived from Q6, was a dependent variable. For this analysis the variables included in the data set were transformed as described in Table 2. The farmer's age variable was discretized as young = 1 (less than 35 years) or old = 0 (between 35 and 50 years, or more than 50 years) to underline the difference between the young farmers and the others. Moreover, the income variable was transformed according to a semilogarithmic model (log-lin). As highlighted by the results of the model (R-squared = 0.6166), least one of the independent variables analyzed explain the

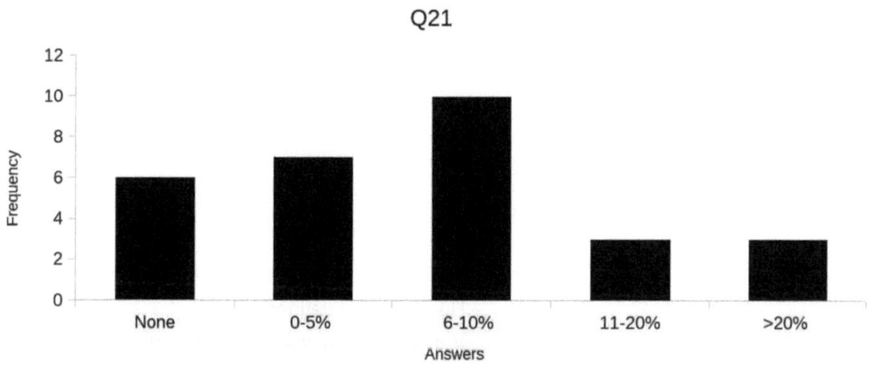

Fig. 2 Frequency of the perception of the reduction of manpower after adopting ICTs tools

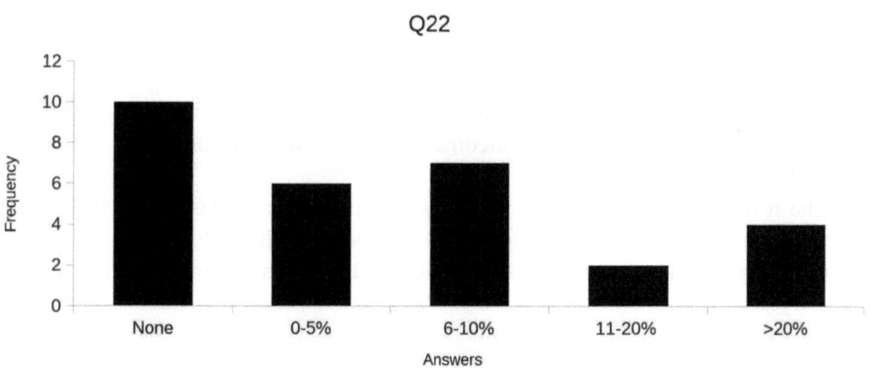

Fig. 3 Increment of production by adopting ICTs tools

Source	SS	df	MS
Model	31.1270881	5	6.22541762
Residual	19.358169	53	.365248472
Total	50.4852571	58	.870435467

Number of obs	=	59
F(5, 53)	=	17.04
Prob > F	=	0.0000
R- squared	=	0.6166
Adj R-squared	=	0.5804
Root MSE	=	.60436

lnincome	Coef.	Std. Err.	t	P>|t|	[95% Conf.	Interval]
time of constitution	-.0776958	.1102546	-0.70	0.484	-.2988384	.1434468
age	.131099	.2566027	0.51	0.612	-.3835809	.6457789
uaa	.7328532	.1284014	5.71	0.000	.4753127	.9903937
croptype	.181122	.0997929	1.81	0.075	-.0190372	.3812812
ict	.484792	.1710891	2.83	0.006	.1416308	.8279532
_cons	9.876301	.3299521	29.93	0.000	9.214501	10.5381

Fig. 4 Multiregression analysis

Breusch - Pagan / Cook-Weisberg test for heteroskedasticity
Ho : Constant variance
Variables : fitted values of lnincome

chi2 (1)	=	1.46
Prob > chi2	=	0.2265

Fig. 5 Heteroskedasticity test

dependent variable. In particular, the *p* values for UAA and ICT adoption were significant. This meant that the income variability was explained by these two variables.

The results excluding several variables (considering only UAA and ICT) were checked by use of a Breusch–Pagan test to check whether the distribution of the residuals was homogeneous or heterogeneous. The results are presented in Table 5.

Table 5 Coefficients in the multiple regression analysis

Source	**SS**	**df**	**MS**	**Number of obs = 59**		
				F(5, 53) = 17.04		
Model	31.1270881	3	6.22541762	Prob > F = 0.0000		
Residual	19.358169	53	.365248472	R-squared = 0.6166		
			.870435467	Adj R-squared = 0.5804		
Total	50.4852571	58	.870435467	Root MSE = .60436		
Income	**Coef.**	**Std. Err.**	*t*	*p* > *t*	**[95% Conf. Interval]**	
time of Constitution	−.0776958	.1102546	−0.70	0.484	−.2988384	.1434468
age	.131099	.2566027	0.51	0.612	−.383580[illegible]9	.6457789
uaa	.7328532	.1284014	5.71	0.000	.4753127	.9903937
croptype	.181122	.0997929	1.81	0.075	−.0190372	.3812812
ict	.484792	.1710891	2.83	0.006	.1416308	.8279532
_cons	9.876301	.3299521	29.93	0.000	9.214501	10.5381

6 Conclusions

The study described in this chapter mapped the profiles of farms in the Lucania region of Italy, highlighting good practices in terms of smart devices used for improving the efficacy and efficiency of the farmers' decisions and daily actions. Practitioners can improve their processes, which can be considered good practices when they reduce inputs and improve outputs, increasing revenue. In addition to other reasons that inspired this study, our survey also aimed to show good practices adopted by Lucania farmers. This goal fits with the objectives of the H2020 (EC 2015) project titled Short Supply Chain Knowledge and Innovation Network (SKIN), approved by the European Commission and started on November 1, 2016. The project consists of collection of information on good practices operating in a short food supply chain for involvement in building a European network to boost and facilitate knowledge transfer and real innovation uptake. The metrics indicators referenced within the project activities for collecting information on good practices show that farms also attain good practice if they adopt smart tools, such as information and communications technology (ICT) tools, to improve their economic, environmental, and/or social sustainability. Grouping of the collected data in clusters allows identification of the most significant features qualifying smart organizations. Innovators and early innovators are ready to get into the network to provide their experiences and gain from other farms' experiences. On the other hand, laggards can benefit after the network is built and synergies are engaged. They can align their profile to that of the smarter ones. Innovators in terms of ICT adoption are those who are able to promptly fit their farm's activities to the environmental complexity, and thus ICT tools play an important role in enabling farms to move into that category because they result in efficiency and efficacy (if they are implemented correctly). In contrast, despite growing environmental complexity (external factors that deflect implementation of the right activities if not correctly managed), laggards do not adopt solutions to simplify processes. However, innovations are needed in rural settings and in the agro-food transition to allow farms to become economically sustainable. Such a necessity is implied in the farm size, which is mainly small-to-medium and reduces competitiveness in finding profitable markets. Agricultural shocks are going to become increasingly frequent because of market uncertainty. Use of ICT tools facilitates information management and control of such shocks.

In addition, this chapter proposes another perspective of the proposed survey. Indeed, considering the income variable transformed logarithmically, it emerges that the significant variables in Lucania are represented by the utilized agricultural area (UAA) and use of ICT. This means that innovation through use of ICT should surely boost outputs.

In this way, the next step in this perspective of the study is aimed at finding out whether increases in income result from increasing throughput or from decreasing costs due to more efficient practices underlying ICT adoption. Additionally, crossing the results of both analyses, it emerges that ICT adoption plays a role in boosting income (on multiple regression analysis), and adoption of ICTs tools increases income significantly when the UAA is kept the same for both clusters.

Finally, further studies are recommended to investigate additional variables, such as the turnover index and the specific type of ICT tools adopted.

Acknowledgements The results presented in this chapter are part of the Short Supply Chain Knowledge and Innovation Network (SKIN) project (www.shortfoodchain.eu). This project has received funding from the European Union's Horizon 2020 Framework Program for Research and Innovation under grant agreement no. 728055.

References

Ariza C, Rugeles L, Saavedra D, Guatiero B (2013) Measuring innovation in agricultural firms: a methodological approach. Electr J Kn Manage 11:185–198 www.ejkm.com

Chen M, Mao S, Liu Y (2014) Big data: a survey. Mobile Netw Appl 19:171–209

Contò F, Faccilongo N, La Sala P (2015) The effects of cloud approach in short chain administration. Int J Agric Environ Inf Syst 6(1):19–31

Diederen P, van Meijel H, Wolters A, Bijak K (2003) Innovation adoption in agriculture: innovators, early adopters and laggards. Cah Econ Sociologie Rurales 67:30–50

Di Falco S, Zoupanidou E (2017) Soil fertility, crop biodiversity, and farmers' revenues: evidence from Italy. Ambio 46(2):162–172

Easterlin RA (2005) Diminishing marginal utility of income? Caveat emptor. Soc Indic Res 70 (3):243–255

Esmeijer J, Bakker T, Ooms M, Kotterink B (2015) Data-driven innovation in agriculture: case study for the OECD KBC2-programme. TNO Report 2015:R10154

European Commission (2015) Calls for proposals and related activities under the 2016–17 work programmes under Horizon 2020—the Framework Programme for Research and Innovation (2014–20) and under the Research and Training Programme of the European Atomic Energy Community (2014–18) complementing Horizon 2020. Off J Eur Union (Article number 2015/ C399/2)

Hastie TJ (2017) Generalized additive models. In: Statistical models in S. Routledge, Boca Raton, pp 249–307

Likas A, Vlassis N, Verbeek JJ (2003) The global k-means clustering algorithm. Pattern Recogn 36 (2):451–461

Press SJ, Wilson S (1978) Choosing between logistic regression and discriminant analysis. J Am Stat Assoc 73(364):699–705

Ugwumba COA, Okoh RN, Ike PC, Nnabuife ELC, Orji EC (2010) Integrated farming system and its effect on farm cash income in Awka south agricultural zone of Anambra state, Nigeria. Am Eur J Agric Environ Sci 8(1):1–6

Wagstaff K, Cardie C, Rogers S, Schrödl S (2001) Constrained k-means clustering with background knowledge. ICML 1:577–584

Wasserman L (2004) Log-linear models. In: All of statistics. Springer, New York, pp 291–301

World Bank (2016) World development report 2016: digital dividends. World Bank, Washington, DC

Zhang T, Ramakrishnan R, Livny M (1997) BIRCH: a new data clustering algorithm and its applications. Data Min Knowl Disc 1(2):141–182

Unlocking the Potential of the Internet of Things to Improve Resource Efficiency in Food Supply Chains

Sandeep Jagtap and Shahin Rahimifard

Abstract The food sector is under tremendous pressure to make its supply chains more resource efficient. In this context, the focus is on the reduction of the three nexus components which are food waste, energy and water. One of the key challenges identified in improving resource efficiency is the low availability of real-time data to all the actors of the food supply chain (FSC) which inhibits better decision-making capability. Having such an ability would increase the productivity of the supply chain. The Internet of things (IoT) concept, which has been around for a few years now, provides the possibility of monitoring, bringing in transparency and efficiency to supply chain activities. This paper investigates the practicability of utilising IoT concepts to improve the resource efficiency of FSCs. An IoT-based framework and methodology are proposed to integrate suitable data into supply chain decision-making processes for the reduction of nexus components.

Keywords Internet of things · IoT · Food supply chain · Resource efficiency · Energy · Water · Food waste

1 Introduction

The food supply chain (FSC) frequently faces the concerns such as resource scarcity, food wastage, non-consistent productivity and from time to time a lack of resilience (Parfitt et al. 2010). Consumer demand for fresh, high-quality and low-priced food products (Rahimifard et al. 2017) has increased the intricacies of global food supply chains (FSCs). On top of that, changes in consumption patterns and growing population, which is estimated to cross 9 billion by 2050, are adding pressure to

S. Jagtap (✉) · S. Rahimifard
Centre for Sustainable Manufacturing & Recycling Technologies (SMART), Loughborough University, Loughborough, UK
e-mail: S.Jagtap@lboro.ac.uk

A. Theodoridis et al. (eds.), *Innovative Approaches and Applications for Sustainable Rural Development*, Springer Earth System Sciences,
https://doi.org/10.1007/978-3-030-02312-6_17

the already strained supply chain (European Commission 2011). These issues are forcing FSCs to be more resource efficient.

One of the ways to overcome the resource constraint issue is by making the efficient use of resources and reducing the adverse environmental impact. Sustainable food production and consumption need consideration of all the stages of food production up until consumption. It is necessary to focus on sustainable lifestyles, sustainable diets, food losses and food waste management and recycling, sustainability standards and environmentally friendly actions and techniques to reduce resource consumption and adverse environmental impacts (FAO n.d.). Sustainable FSCs attempt to achieve high standards of environmental performance by reducing energy consumption and minimising resource inputs (DEFRA 2002). Some researchers have deduced that FSCs suffer due to unawareness of resource consumption and food losses and wastage could be avoided by using novel monitoring technologies (Jedermann et al. 2014). There are other benefits of implementing monitoring technologies such as financial savings, adhering to environmental regulations set by governments and fulfilling consumer demand for sustainable food products through sustainable production and consumption (Haight and Park 2015).

Access to real-time resource consumption data offers a new prospect of making the FSC more resource efficient. The Internet of things (IoT), which is a network of interconnected elements embedded with sensors, network connectivity, software and other electronics, has the capability of collecting real-time resource consumption data on behaviours and patterns. It can monitor, communicate and interact with various things almost 24/7 in real time. Currently, within the FSC, IoT applications play a crucial role in food safety and traceability solutions (Wang and Zhang 2014). They have been deployed to some extent in reducing food loss and waste (Jedermann et al. 2014), water (Kim et al. 2008) and energy (Haller et al. 2008) consumption (Combaneyre 2015).

This paper considers the benefits of adopting IoT to support and improve the resource efficiency of FSCs. It proposes an IoT-based framework and methodology for incorporating resource consumption data in the FSC decisions. The framework and methodology are expected to facilitate an improvement in supply chain practices by minimising water and energy use as well as a reduction in food wastage. Due to the significant range and type of activities of FSC actors, the scope of work reported in this paper is restricted to post-farm gate to retailer's shelf as depicted in Fig. 1 (highlighted in yellow).

2 Review of Resource Efficiency in FSCs

Prevailing FSC systems and methods are not sustainable in the face of continuously growing demand for important resources such as food ingredients, energy and water. Therefore, researchers and FSC practitioners are trying to develop a resource-efficient, sustainable FSC without harming overall supply chain productivity (Koh

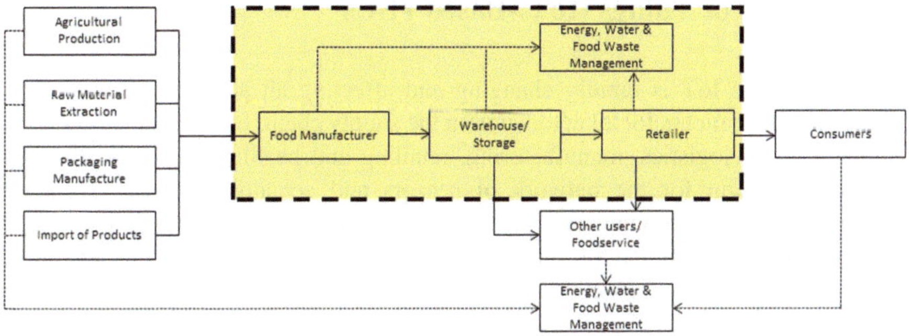

Fig. 1 Research scope (Jagtap and Rahimifard 2017)

2014). Managing resource consumption in FSCs can be a tough task because of the complexity involved arising from how the resources are consumed across various processes, with each process having individual resource consumption characteristics. Hence, at Nestle (like many other larger companies), an integrated and holistic approach is taken towards resource efficiency by considering energy, greenhouse gas emissions, water and materials (Brabeck-Letmathe 2016). There are significant opportunities for desirable, sustainable production through the improvement of communication between producers, retailers and consumers (Henningsson et al. 2004). But the absence of reliable and easily accessible data is reported as one of the major obstacles in achieving a high level of resource efficiency in FSCs (Lee et al. 2013). Another issue is that food actors are not fully aware of their resource consumption. For example, they are well informed of the total water consumed and water released in the form of effluent but usually oblivious of water consumed by the individual machine or process level (Webb 2016) which is also relevant to energy consumption (Thollander and Ottosson 2010).

A better understanding of nexus components can lead to a reduction in consumption as they are interlinked in such a way that changes to one component can impact one or both the components (Villamayor-Tomas et al. 2015). But then it is necessary to have access to meaningful, accurate and real-time data in order to reduce waste and inefficiencies in the FSC (Shahrokni et al. 2014) with regard to energy and water consumption and food waste produced by supply chain practices. Therefore, to make supply chains resource efficient, the first step is to be resource aware (Matopoulos et al. 2015), and for that real-time data is crucial as it optimises resource efficiency (Pitarch et al. 2017). The conventional methods of obtaining data using pen and paper are non-productive, slow and arduous. Hence, IoT-based applications can address those issues successfully as they can make the whole process transparent, fast and reliable. For example, IoT-based smart water metres are vital to water users as they can present real-time data on water usage, leakages and quality and, in some circumstances, could make water-efficient decisions by understanding their surrounding environment (Iotsens 2017).

3 Internet of Things (IoT)-Based FSCs

The rise of the IoT is rapidly changing and affecting all areas of businesses by increasing the benefits for all actors within the supply chain. It has been successfully adopted in the logistics, manufacturing, retailing and healthcare sectors. The IoT offers a platform for the network of sensors and actuators that can exchange information and communicate conveniently through an integrated framework. It can execute various functions such as ubiquitous sensing, data analytics and cloud computing to develop a seamless operation for enabling state-of-the-art applications (Gubbi et al. 2013). It offers the opportunity to continuously gather information and send it to cloud-based software tools to store, visualise and analyse data in real-time and enables better decision-making. The IoT usually relies on radio-frequency identification (RFID) and wireless sensor networks (WSN) technology to collect real-time data from various hotspots within the supply chain (Verdouw et al. 2016). It allows provision for continuous and massive data collection about machine availability, stock levels, traceability of products and resource consumption through multiple sensors and smart metres. RFID tags are widely used in logistics, pharmaceuticals, retailing and supply chain management for identifying, tracking and monitoring products and things (Amendola et al. 2014). WSN technology uses interconnected intelligent sensors to sense and track and finds wider applications in the area of environmental conditions, healthcare and industrial monitoring (Akkas 2016). Table 1 shows a list of companies adopting IoT technologies to improve resource efficiency in FSCs.

The application of the IoT concept for monitoring and tracking of resources in FSCs is at a relatively early stage when compared to the other manufacturing sectors (Verdouw et al. 2016). Few actors within FSCs from food producers to food retailers have adopted IoT to optimise a section or the entire process as shown in Fig. 2. Some of the pioneers have even implemented it for monitoring energy and water consumption and food waste management at equipment level as illustrated in Table 1.

The application of the IoT in FSCs has resulted in some benefits, namely, three areas which have been recognised as follows:

1. Operational efficiency – Some of the food companies have recognised the benefits of investing in the IoT and big data solutions. Utilising data analytics can help to ensure sustainable and profitable growth for the entire FSC. Data analytics provides insight into machine performance, predictive maintenance requirements and real-time inventory tracking (Satyavolu et al. 2014). It can help to identify inefficiencies and low productivity on the production shop floor and suggest real-time actionable solutions. For example, it supports the continuous improvement of FSCs' operational activities through decentralisation of the decision-making process by the generation of resource-oriented key performance indicators. Availability of 24/7 resource consumption patterns in real time enables stakeholders to plan and prioritise the efficient use of resources (Pang et al. 2012).

Table 1 IoT adoption by companies and its benefits (Jagtap and Rahimifard 2017)

Companies	Resource Monitoring	Parameters being checked for	Benefits of IoT adoption (resource efficiency related)	Practices enabled by IoT which lead to those benefits
Rova	Energy	Operation efficiency	Optimising truck routes and bin collection times	Provides vehicle location, traffic congestion and bin volumes
Siemens	Energy	Production and maintenance effectiveness	Increasing productivity	Provide alerts and schedules about machine maintenance to avoid future breakdowns.
Martec	Water	Cleaning process	Maintaining hygiene standards	Avoids over-cleaning and microbial contamination
2 Sisters	Water	Washing raw material	Water consumption, reduction in the effluent	Control system to deliver exact quantity needed
MyFresh	Water	Washing raw produce	Water consumption, reduced effluents	Reduced incoming water pressure, installation of efficient pumps
LeanPath	Food	Food waste	Reduction in food waste	Reasons for food waste, changing the production process
Winnow Solutions	Food	Food waste	Reduction in food waste	Reasons for food waste, recipe change, production planning

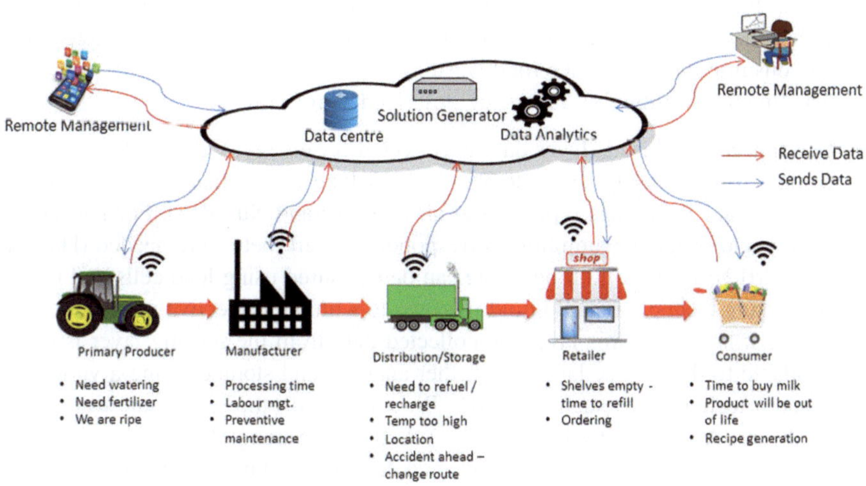

Fig. 2 IoT applications in FSC

2. Food safety – In 2017, a chicken safety scandal at a 2 Sisters factory in the UK prompted the need to focus on compliance, prevention and a proactive approach when it comes to food safety. Strengthening of food safety and food traceability throughout the FSC, i.e. from farm to fork, is necessary to reduce food-borne-related illnesses. The use of real-time tracking can support monitoring of food safety data points to enable compliance with local and global regulations. By implementing IoT-based automated hazard analysis and critical control point (HACCP) specifications during various food operation activities, supply chain actors have access to consistent, reliable, meaningful data that allow them to devise food safety solutions.
3. Transparency – The horsemeat scandal of 2013 severely damaged the consumer confidence in processed foods, and they now expect more openness from supply chain actors. Implementing full transparency and visibility along the complex global supply chain will aid the food sector to drive consumer trust back into food products. Combining blockchain with IoT technologies can secure the data so that it makes tough for anyone to forge the records (Jesus et al. 2018). Thus, bringing transparency to support FSC actors with added benefits such as greater trust, cost savings, better inventory management and better lead times.

4 IoT Architecture for Resource Efficiency in FSCs

The application of the IoT for resource efficiency consists of four layers: the sensing layer, the network layer, the service layer and the application layer (Ray 2016) as shown in Fig. 3. These layers form the IoT architecture and are designed in such a way that the system can meet the requirements of FSCs to minimise energy and water consumption and reduce food waste.

The functions of the four layers are described as follows:

- Sensing layer – In this layer data is collected for energy, water and food waste using various sensing technologies such as load cells, smart metres, sensors, cameras and RFID tags (Akyildiz et al. 2002; Li and Xu 2017). For measuring energy and water consumption, corresponding smart metres are needed (Hancke et al. 2012), while solid food waste can be measured using load cells and image processing technology and liquid food waste with the respective smart metre.
- Network layer – In this layer the collected data from the sensing layer is made available to the service layer for further analysis and storage, using a variety of modern technologies such as Wi-Fi, Bluetooth and other electronic devices or hardware (Arduino, Raspberry Pi, etc.) (Sethi and Sarangi 2017).
- Service layer – In this layer, the collected data is stored in the cloud or on a local server (Akyildiz et al. 2002). The collected data is analysed using data analytics platforms to extract meaningful information (Li and Xu 2017).
- Application layer – This layer provides user-friendly applications or services to stakeholders or users (Sethi and Sarangi 2017) with accurate data to manage projects on minimising resource consumption or waste.

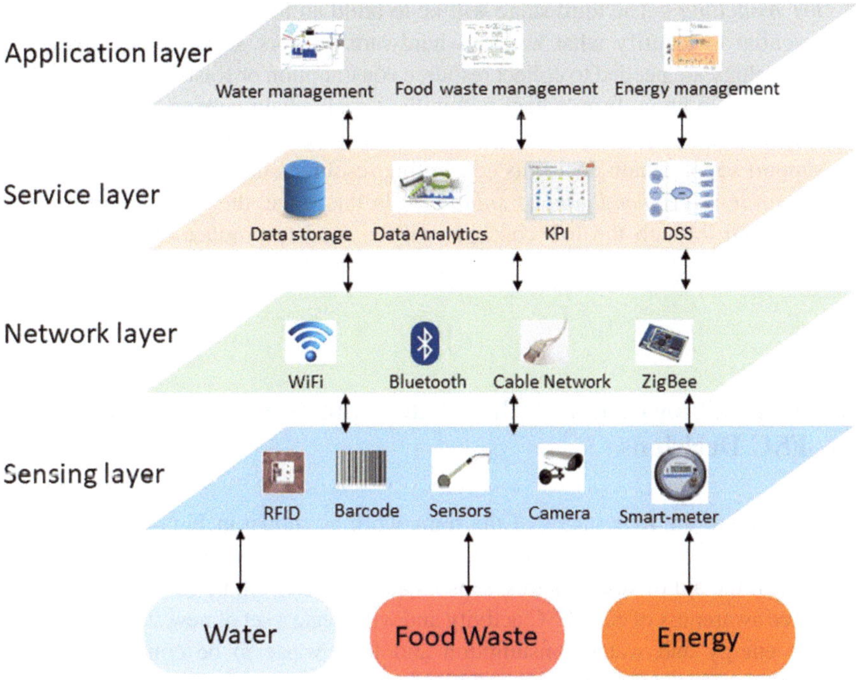

Fig. 3 IoT architecture for resource efficiency in FSCs (Jagtap and Rahimifard 2017)

5 IoT-Based Framework for Resource Efficiency in FSCs

Our literature review highlighted the urgent need for an IoT-based framework to be implemented in FSCs. Therefore, a four-stage IoT-based framework for resource efficiency in FSCs is developed and presented in Fig. 4. The four stages are described as follows:

1. *Establishing impactful resources* – In this stage, it is important to determine impactful resources. The literature review focused on three key nexus components to be addressed: food waste, energy and water. Other necessary criteria are to assess and interpret the resource-use impact on the environmental sustainability of FSCs and the strategies implemented by various actors in FSCs concerning resources.
2. *Supply chain process* – In this stage, it is essential to realise how resources flow within various actors of the supply chain. For example, if we consider resource flow at the factory level, it will be vital to learn the resource flow within different departments of the company, which may be further narrowed down to the machine level. These would help in interpreting the consumption behaviours and wastage of resources at various levels.

3. *IoT modelling* – The third stage will be to build an IoT model. In this stage, it is essential to identify what kinds of hardware, sensors, electronics, software or technology are needed to collect resource consumption or wastage data. Also, it is necessary to know from where within the supply chain network the data can be extracted and how this data will be filtered to get meaningful information to support supply chain decisions concerning resource efficiency.
4. *Generate recommendation or solution* – In this stage, the valuable information generated through the IoT concepts will be used to produce reports for better planning of resources in FSCs and the improvement of supply chain activities.

6 Methodology for IoT-Based Incorporation of Data into FSC Decisions

By integrating the four stages of the framework as shown in Fig. 4 into the FSC, resource efficiency can be improved. The real-time data produced using IoT concepts with the aid of smart metres, sensors and cameras will be used to increase the resource awareness of each FSC activity and will create a set of new standards. It will enable energy and water consumption and food waste to be considered during planning activities leading to the optimisation of resources, flexibility in production planning and control and better communication and decision-making at all actor levels. The three nexus components can be minimised through stepwise implementation of the IoT framework. Establishing what information needs to be collected regarding these resources is described below.

1. Water – Water used in food manufacturing can be categorised as production water and non-production water.

   - Production water – The water used directly by food production processes. Production water is further divided into two categories, process water and system water.

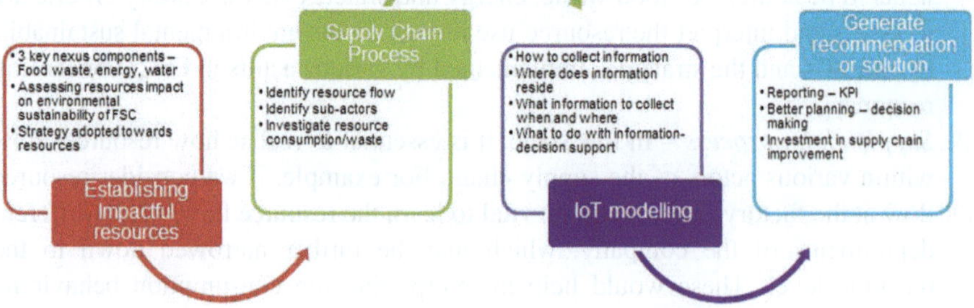

Fig. 4 IoT framework for resource efficiency in FSCs

Process water – The water required to convert raw ingredients into finished food products (e.g. washing fruits and vegetables, boiling, cooking)

System water – The water used to run production machines, tools and environment (e.g. cleaning of equipment, heat exchangers).

- Non-production water – The water used by facilities or infrastructures to support production activities such as heating and sanitation (Sachidananda 2013).

  Therefore, separate metres are required to record accurate data on water consumption.

2. Energy – Energy can be distinguished in two ways: direct and indirect energy.

   - Direct energy – The energy required by various processes within FSCs to make a finished food product available at the retailer's shelf (e.g. cleaning, washing, chopping, packing, chilling, transporting)
   - Indirect energy – The energy utilised by surroundings in which food production processes are carried out or food is stored and transported (e.g. lighting, ventilation, heating) (Seow 2011). It is crucial for the supply chain actors to install energy smart metres to track both types of energies in order to get a better understanding of energy consumption.

3. Food waste – Food waste is of three kinds: avoidable, unavoidable and possibly unavoidable waste.

   - Avoidable waste – Food and drink that is binned and which at some stage before its disposal was edible (e.g. bread loaves, meat, cheese, etc.)
   - Unavoidable waste – Waste generated from food and drink production that is non-edible under normal circumstances (e.g. bones, banana skins, egg shells, etc.)
   - Possibly avoidable waste – Food and drink that some people may consume and others cannot (e.g. bread crumbs, potato skins, etc.) (WRAP 2009)

   Measuring solid food waste is a complex process since it is a mixture of avoidable, unavoidable or possibly avoidable waste. Therefore, to overcome this issue, food waste smart metres to monitor and measure different types of solid food wastes with minimal human input can be deployed at various hotspots within FSC.

The data on the three nexus components collected by smart metres in real time is stored in the cloud or on a server and analysed to extract useful information as shown in Fig. 5. Data analytics techniques can be utilised to understand the consumption pattern and wastage behaviour of the three nexus components. In the next stage, the meaningful information created can be merged into FSC management systems and tools that support improvement in resource efficiencies, such as decision support systems (DSSs), key performance indicators (KPI) and real-time consumption

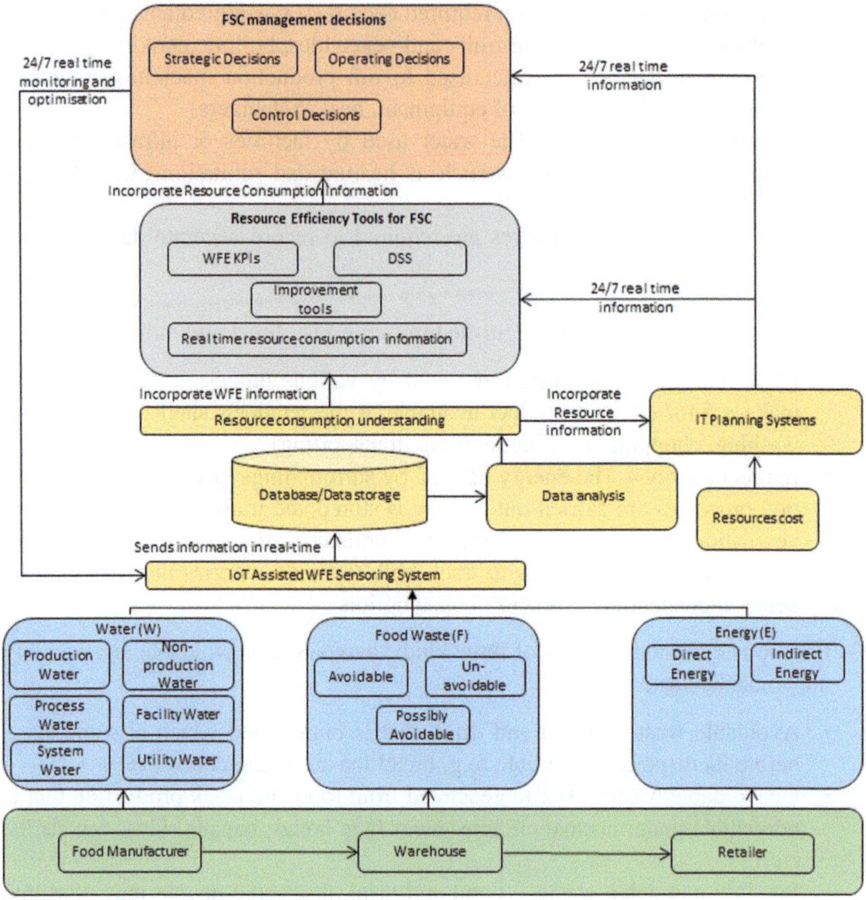

Fig. 5 Methodology for IoT-based incorporation of data into FSC decisions

dashboards. The useful information obtained from the data analytics would support high-level choices with regard to the strategic, operational and control decisions in the FSC's management systems.

In the future, the ability of IoT applications to provide on-time data will play a decisive role in creating production plans, updating production line status, tracking the current status of resource consumption, the wastage during each activity and monitoring production activities throughout the FSCs. The IoT can be integrated into IT planning systems to visualise real-time data on stock levels, stock movements, machine and labour availability, etc. enabling stakeholders to make better decisions (Satyavolu et al. 2014).

6.1 Data Storage and Analysis

To be resource aware with regard to resource consumption and waste, it is important to select the right sensors and smart metres and to install them at the right locations (e.g. production area or section, machine). The information collected via sensors and smart metres is stored and analysed at the actor level or stored in the cloud. The results through analysis of data are sent to the actors continuously in the real time or if there is an abnormal consumption or waste of resources. In short, data is summarised online, stored and audited for online or offline analysis.

6.2 Incorporating Data into IT Infrastructure

Resource consumption, as well as food waste data, should be incorporated into planning systems, to allow better decisions with regard to production planning, scheduling, storing, ordering and other resource management practices. Although the communication between sensors or smart metres and existing planning systems could be complicated, sensors can return results in different file formats such as XML, JSON and CSV files (Solaredge 2016). The planning system providers (enterprise resource planning (ERP) vendors) such as Epicor (Solarsoft), SAP and Sage X3 are upgrading their existing planning systems to accommodate resource management know-how (Compare Business Products n.d.). These solutions allow resources as well as waste data monitoring, a breakdown of resource usage and the generation of detailed reports. Hence, it can be seen that planning systems will play a significant role in resolving data incorporation problems.

6.3 Tools for Supporting Resource Efficiency

To be resource efficient, we need to be resource aware regarding what we measure and control. There are various tools to supplement resource-aware decision-making, such as water, food waste and energy KPIs (WFE KPIs), improvement tools, real-time resource consumption and decision support systems (DSSs). For instance, water, energy and food waste smart metres provide detailed information on water and energy consumption and the amount of food waste created, which can help in structuring a new set of WFE KPIs, to improve the performance of individual machines, sub-actors and actors.

The incorporation of improvement tools can play a significant role in enhancing resource efficiency. In many cases, resource usage and food waste are currently estimated, fixed or inaccurate. For example, food waste is hard to measure since it is hard to distinguish low-value food waste from high-value food waste when they are binned together. Traditional ways of tracking food waste such as pen and paper are not good enough as they are prone to errors and often estimated. But with the use of food waste smart metres, it is possible to identify different categories of waste, record their exact weights and establish a better high-value solution for the waste generated.

Real-time water and energy consumption and food waste data are essential since they enable stakeholders, operators and other management staff to understand consumption and wastage behaviour and patterns. Incorporating resource consumption and wastage data into FSC management decisions requires a decision support system (DSS) to support decision making with regard to resource usage and wastage. The function of a DSS is to equip stakeholders to support the efficient use of resources in FSCs and to provide solutions to operational needs, such as quicker response to external events (price changes, shortages, etc.).

6.4 Incorporation of Resource Data into FSC Management Decisions

At the top level, the framework supports FSC management decisions such as strategic, operating and control decisions that can be made more resource efficient when incorporated with water, energy and food waste data.

At the strategic level, decisions relating to products, processes and facilities are undertaken such as developing long-range production plans, process design, selecting and managing production technology, planning the arrangement of services and planning for the optimal distribution of scarce resources among product lines, sub-actors or actors and long-term production capacity. The improvement of resource efficiency can be achieved through techniques for efficient food processing and the redesigning of food manufacturing processes to reduce time.

The focus of operational management is on food production and meeting demand. It involves, inter alia, aggregate planning, production scheduling, planning and controlling finished goods inventories, planning materials and capacity requirements, short-range decisions about what to produce and when to produce and managing all facets of the material system. Knowledge of resources required by a particular food processing machine at various settings and configurations will enable the stakeholder to understand the most effective, efficient and economic setting for the machine. This information, when fed into the production scheduling systems, can reduce water and energy consumption as well as reduce food wastes.

In control decisions, there is more focus on planning and managing operations such as planning for the effective and efficient usage of human resources in operations, planning and controlling the quality of products and services, projects and maintaining the machines and facilities. Techniques such as line balancing, i.e. having the right number of people for a particular task and getting it right the first time, avoid recycling. The predictive maintenance of machines prevents unwanted breakdowns. All these techniques will ultimately lead to less consumption of water and energy and reduce food waste.

7 Conclusions and Further Work

In this paper, we have addressed the issue that stakeholders need to reflect upon while they are planning to improve resource efficiency in FSCs using IoT technologies. Although the IoT has brought a high level of visibility, transparency, awareness, flexibility and the ability to collect and process a large amount of data in the real-time, failure to properly implement it may result in financial disaster. Consequently, for this reason, it becomes essential to develop a strategy beforehand on how to integrate IoT resource monitoring system in the FSC initiatives to attain environmental sustainability. It focuses on the technology needed to gather and analyse IoT data and on tools that are required to support the system to achieve resource efficiency. It further illustrates an IoT-based framework and a methodology which provides a new perspective on incorporating water, energy and food waste data into FSC management and decision-making. Incorporating real-time data into supply chain planning systems such as SAP, APS (advanced planning system), MRP (material resource planning) and ERP will help stakeholders with better decision-making to optimise resource consumption and reduce wastage.

This paper has clearly added value that a well-implemented IoT-based resource monitoring system provides detailed insights into supply chain activities. And, the supply chain individuals do not have to spend enormous time gathering the essential data. It readily offers them with the available information needed to swiftly address resource-inefficient activities within FSC, as well as determining the root cause and fixing them resulting in cost savings. It provides the chance to monitor all existing supply chain activities and the data related to three nexus components so that analysis can be carried out in the context of the whole supply chain. Detailed information leads to better decision making – decisions that result in resource-efficient supply chains. More research is needed to explore and develop other IoT concepts for improving the resource efficiency of FSCs and embedding them in supply chain planning and control to support decision-making processes. However, the methodology presented in this paper will be utilised in future prototype development to address the issues outlined and recommendations suggested in this work.

Acknowledgements This work was supported by the Engineering and Physical Sciences Research Council (EPSRC) Centre for Innovative Manufacturing in Food [Reference: EP/K030957/1].

References

Akkas M (2016) Using wireless underground sensor networks for mine and miner safety. Wirel Netw 23:1–10

Akyildiz I, Su W, Sankarasubramaniam Y, Cayirci E (2002) A survey on sensor networks. IEEE Commun Mag 40(8):102–114

Amendola S et al (2014) RFID technology for IoT-based personal healthcare in smart spaces. IEEE Internet Things J 1(2):144–152

Brabeck-Letmathe P (2016) Climate change, resource efficiency and sustainability. In: Biswas A, Tortajada C (eds) Water security, climate change and sustainable development. Springer, Singapore, pp 7–26

Combaneyre F (2015) Understanding data streams in IoT. Cary: SAS, New York

Compare Business Products (n.d.) Enterpriseinnovation.net. [Online] Available at: http://www.enterpriseinnovation.net/files/whitepapers/top_10_erp_vendors.pdf. Accessed 05 Jan 2017

DEFRA (2002) The strategy for sustainable farming and food. Facing the Future. [Online] Available at: http://www.randd.defra.gov.uk/Document.aspx?Document=DO107_9829_FRP.pdf. Accessed 27 Sept 2016

European Commission (2011) Sustainable food consumption and production in a resource-constrained world. European Commission, Brussels

FAO (n.d.) Sustainable food consumption and production. [Online] Available at: http://www.fao.org/ag/ags/sustainable-food-consumption-and-production/en/. Accessed 26 Sept 2016

Gubbi J, Buyya R, Marusic S, Palaniswami M (2013) Internet of things (IoT): a vision, architectural elements, and future directions. Futur Gener Comput Syst 29(7):1645–1660

Haight J, Park H (2015) IoT analytics in practice. Blue Hill Research, Boston

Haller S, Karnouskos S, Schroth C (2008) The internet of things in an enterprise context. Springer, Berlin-Heidelberg, pp 14–28

Hancke G, Silva B, Hancke G Jr (2012) The role of advanced sensing in smart cities. Sensors 13 (1):393–425

Henningsson S, Hyde K, Smith A, Campbell M (2004) The value of resource efficiency in the food industry: a waste minimisation project in East Anglia, UK. J Clean Prod 12(5):505–512

Iotsens (2017) Smart water metering. [Online] Available at: http://www.iotsens.com/smart-water-metering/. Accessed 13 Mar 2017

Jagtap S, Rahimifard S (2017) Utilization of internet of things to improve resource efficiency of food supply chains. Greece., CEUR-WS.org, Chania, pp 8–19

Jedermann R, Nicometo M, Uysal I, Lang W (2014) Reducing food losses by intelligent food logistics. Philos Trans R Soc A Math Phys Eng Sci 372(2017):20130302

Jesus E, Chicarino V, de Albuquerque C, Rocha A (2018) A survey of how to use Blockchain to secure internet of things and the stalker attack. Secur Commun Netw 2018:1–27

Kim Y, Evans R, Iversen W (2008) Remote sensing and control of an irrigation system using a distributed wireless sensor network. IEEE Trans Instrum Meas 57(7):1379–1387

Koh L (2014) Advanced resource efficiency centre (AREC) – creating the supply chain of the future. [Online] Available at: https://www.sheffield.ac.uk/polopoly_fs/1.445286!/file/ARECBrochure2015.pdf. Accessed 27 Sept 2016

Lee P et al (2013) The role of Lean thinking in increasing resource efficiency in the UK food and drink supply chain.. [Online] Available at: http://www.oakdenehollins.co.uk/media/270/11668_20131114RevisedLeanresearchreport-DEFR01270FO0425_from_defra_website.pdf. Accessed 21 Sept 2016

Li S, Xu L (2017) Securing the internet of things. Syngress, Cambridge

Matopoulos A, Barros A, van der Vorst J (2015) Resource-efficient supply chains: a research framework, literature review and research agenda. Supply Chain Manag Int J 20(2):218–236

Pang Z, Chen Q, Han W, Zheng L (2012) Value-centric design of the internet-of-things solution for food supply chain: value creation, sensor portfolio and information fusion. Inf Syst Front 17 (2):289–319

Parfitt J, Barthel M, Macnaughton S (2010) Food waste within food supply chains: quantification and potential for change to 2050. Philos Trans R Soc B Biol Sci 365(1554):3065–3081

Pitarch J et al (2017) Optimisation of the resource efficiency in an industrial evaporation system. J Process Control 56:1–12

Rahimifard S et al (2017) Forging new frontiers in sustainable food manufacturing. In: Campana G, Howlett R, Setchi R, Cimatt B (eds) Sustainable design and manufacturing 2017. Springer, Bologna, pp 13–24

Ray P (2016) A survey on internet of things architectures. J King Saud Univer Comput Inf Sci 30:291–319

Sachidananda M (2013) A framework for modelling and reduction of water usage in the manufacturing industry. Loughborough University, Loughborough

Satyavolu P, Setlur B, Thomas P, Iyer G (2014) Designing for manufacturing's 'internet of things'. Cognizant, Teaneck

Seow Y (2011) A framework for modelling embodied product energy to support energy efficient manufacturing. Loughborough University, Loughborough

Sethi P, Sarangi S (2017) Internet of things: architectures, protocols, and applications. J Electr Comput Eng 2017:1–25

Shahrokni H, van der Heijde B, Lazarevic D, Brandt N (2014) Big data GIS analytics towards efficient waste management in Stockholm, 2nd edn. International Conference on ICT for Sustainability, Stockholm

Solaredge (2016) Solar edge monitoring server API. [Online] Available at: http://www.solaredge.com/sites/default/files/se_monitoring_api.pdf. Accessed 27 Oct 2016

Thollander P, Ottosson M (2010) Energy management practices in Swedish energy-intensive industries. J Clean Prod 18(12):1125–1133

Verdouw C, Wolfert J, Beulens A, Rialland A (2016) Virtualization of food supply chains with the internet of things. J Food Eng 176(2016):128–136

Villamayor-Tomas S et al (2015) The water-energy-food security nexus through the lenses of the value chain and the institutional analysis and development frameworks. Water Alternatives 8 (1):735–755

Wang A, Zhang Q (2014) Application of the internet of things technology in food safety monitoring and controlling system. Adv Mater Res 1006–1007:534–537

Webb P (2016) Modelling, monitoring and management of water consumption in food manufacture. Manufacturing Food Futures Conference, Loughborough

WRAP (2009) Household food and drink waste in the UK. [Online] Available at: http://www.wrap.org.uk/sites/files/wrap/Household_food_and_drink_waste_in_the_UK_-_report.pdf. Accessed 3 Aug 2017

Part IV
New Methods for High Quality Agricultural Products and Environmental Protection

Molecular Detection of Resistance to Biotic Stress Conditions in Spring Bread Wheat Cultivars

Anatolii Karelov, Natalia Kozub, Igor Sozinov, Oksana Sozinova, Athanasios G. Mavromatis, and Ioannis N. Xynias

Abstract Biotic stress conditions are the most serious obstacle in bread wheat cultivation resulting in yield reduction and safety problems (toxin production) for consumers. For this reason, identification of resistant cultivars and their respective genes is a main prerequisite in most breeding programs. Into this study and in order to exploit the benefits of molecular technology, 9 Hellenic and 93 Ukrainian bread wheat cultivars were analyzed to determine the responsible alleles for some important diseases in wheat. Among the Hellenic cultivars, carriers of resistance-associated alleles of the *Lr34/Yr18/Pm38/Sr57/Bdv1* (a rare gene among the European bread wheat cultivars), *Tsn1*, *Tsc2*, and *TDF_076_2D* genes were identified. Four of the studied Hellenic cultivars carry the 1BL.1RS wheat-rye chromosome translocation of the 'Kavkaz' type proved to be resistant to the stem rust and especially to the dangerous Sicilian race TTTTF due mainly to the presence of the *Sr31* gene. Among the Ukrainian cultivars, the toxin A insensitivity allele of the *Tsn1* gene is predominant. Of the nine Ukrainian spring cultivars containing wheat-rye translocations, seven were found to possess the 1BL.1RS and two the 'Amigo'-type 1AL.1RS translocation. The previous two cultivars carry the gene $Sr1RS^{Amigo}$ conferring resistance to dangerous stem rust Ug99 biotypes. One cultivar ('Vyshyvanka') was found to carry a 1BL.1RS translocation distinct from

A. Karelov · N. Kozub
Institute of Plant Protection, National Academy of Agrarian Sciences of Ukraine, Kyiv, Ukraine

Institute of Food Biotechnology and Genomics, National Academy of Sciences of Ukraine, Kyiv, Ukraine

I. Sozinov · O. Sozinova
Institute of Plant Protection, National Academy of Agrarian Sciences of Ukraine, Kyiv, Ukraine

A. G. Mavromatis
Laboratory of Genetics and Plant Breeding, School of Agriculture, Aristotle University of Thessaloniki, Thessaloniki, Greece

I. N. Xynias (✉)
School of Agricultural Technology & Food Technology and Nutrition, Western Macedonia University of Applied Sciences, Florina, Greece
e-mail: ixynias@teiwm.gr

A. Theodoridis et al. (eds.), *Innovative Approaches and Applications for Sustainable Rural Development*, Springer Earth System Sciences,
https://doi.org/10.1007/978-3-030-02312-6_18

1BL.1RS of the 'Kavkaz' type and may carry novel rye disease resistance genes. The results of the examined germplasm regarding disease resistance gene alleles could be helpful for breeders who are looking for sources of disease resistance genes.

Keywords Fungal diseases · Bread wheat · Resistance genes · Wheat-rye translocations

List of Abbreviations

ATP	Adenosine triphosphate
AFLP	Amplified fragment length polymorphism
ABC	ATP-binding cassette
APR	Adult plant resistance
KASP	Competitive allele-specific PCR
MAS	Marker-assisted selection
NAAS	National Academy of Agrarian Sciences
PCR	Polymerase chain reaction
PDR	Pleiotropic drug resistance
RFLP	Restriction fragment length polymorphism
RAPD	Random amplified polymorphic DNA
QTL	Quantitative trait locus
SCAR	Sequence characterized amplified region
SSR	Short sequence repeat

1 Introduction

Methods of studying the genetics of bread wheat (*Triticum aestivum* L.) have undergone significant changes in the past few decades (Sears 1969; Langridge et al. 2001; Kiszonas and Morris 2018). The first attempts to understand population genetics of wheat involved the use of electrophoresis of storage proteins and isozymes (Nevo and Payne 1987). The scientists were afterwards moved on to DNA analysis through RLFP markers to proceed with the polymerase chain reaction (PCR) methods using RAPD, ALFP, SSR, or SCAR markers (Tanksley et al. 1989; Devos and Gale 1992; Tian et al. 2015). Nowadays powerful techniques as next-generation sequencing (Jia et al. 2018), SNP genotyping (Ganal and Röder 2007), and KASP markers (Rasheed et al. 2016) based on whole-genome sequencing (Gill et al. 2004; Muthamilarasan and Prasad 2014) provide a huge amount of data for statistical analysis for understanding basic principles of population genetics as well as tools for marker-assisted selection (MAS) of wheat (Gupta et al. 2010). Breeders use markers localized in target genes or closely linked to genes controlling important traits, to accumulate ("pyramid") them in particular lines and cultivars in a comparatively small period of time with considerable effectiveness (Aktar-Uz-Zaman et al. 2017). The majority of genes used for MAS confer resistance in biotic stress factors (Yadav et al. 2015).

The biotic stress factors of the plants are represented by a wide range of pathogens, including viruses, nematodes, fungi, pests and even other plants (Duveiller et al. 2007; Figueroa et al. 2017). Among them, fungal pathogens are the most widespread and dangerous: they can move across significant distances and survive through unfavorable weather conditions as spores, cause significant yield losses, and even result in poisoning humans by their toxins (Goyal and Prasad 2010). According to the type of nutrition, all plant pathogens are commonly divided into two general groups, namely, biotrophs and necrotrophs (Glazebrook 2005). Biotrophs are obligate parasites of plants which have developed the ability to penetrate host cells and obtain their nutrition without killing them (Tang et al. 2018). Necrotrophic pathogens, on the other hand, cause in one or another way the death of a target cell to obtain nutrition (Glazebrook 2005; Zhang et al. 2013). Plants therefore developed different mechanisms to protect or avoid different types of pathogens (Glazebrook 2005).

Resistance to biotrophic pathogens is provided by several signal cascades, which include the production of reactive oxygen, programmed death of infected cells, cell wall decomposition, and synthesis of proteins related to pathogenesis (Martin et al. 2003; Glazebrook 2005; Bent and Mackey 2009). The so-called R proteins play a key role in the detection of plant pathogens and induction of resistance to biotrophs in plants (Bent and Mackey 2009; Glowacki et al. 2010). The resistance conferred by this gene family is referred to as gene-for-gene interaction and belongs to the "strong" type, which is easily recognized in the field (Ellis et al. 2014). Among the most widespread biotrophic pathogens of wheat are the leaf rust pathogen *Puccinia recondita* Dietel & Holw), *Puccinia graminis* Pers. f. sp. *tritici* causing stem rust, *Puccinia striiformis* var. *tritici* Westend. causing stripe rust, and the powdery mildew pathogen *Blumeria graminis* (DC.) Speer (Figueroa et al. 2017). Nowadays the list of identified disease resistance genes (R-genes) in wheat includes more than 71 leaf rust resistance (*Lr*) genes, 78 stripe rust (*Yr*) resistance genes, 77 stem rust (*Sr*) resistance genes, and 58 powdery mildew *(Pm)* resistance genes, many of which were introgressed from wild and cultivated relatives (McIntosh 2013; McIntosh et al. 2017).

However, the effectiveness of R-genes is commonly lost through the time, due to the appearance of new virulent races of pathogens (Glazebrook 2005). Another group of resistance genes includes race-nonspecific resistance genes providing moderate but more stable and durable resistance. Their effects are associated with the prolonged latent period for development of the disease. In particular, race-nonspecific rust resistance genes of wheat provide partial resistance and show slow rusting: the longer the latent period, the smaller the number and sizes of uredinia in the first 2 weeks after the infection in comparison with susceptible plants (Kolmer 1996). Such genes are referred to as adult plant resistance (APR) genes and commonly express resistance phenotypes only in adult plants (Kolmer 1996; Lagudah

2011; Ellis et al. 2014). Another benefit of APR genes is the fact that some of them provide multiple resistance against different pathogens (Lagudah 2011). Because of their low level of expression and inability to withstand the artificially created infectious background, APR genes are perfect targets for MAS (Yadav et al. 2015). The most well-known APR gene is *Lr34* conferring moderate leaf rust resistance (Dyck 1977; Krattinger et al. 2009). In addition, *Lr34* was found to be pleiotropic to the APR gene *Yr18* providing moderate resistance to stripe rust (McIntosh 1992; Singh 1992). It also co-segregates with the *Pm38* gene providing tolerance to powdery mildew (Dyck 1993). The locus is also associated with tolerance to barley yellow dwarf virus as *Bdv1* (Singh 1993). It was also found to confer race-nonspecific moderate resistance to stem rust as the *Sr57* gene (Dyck 1987).

Mechanisms of wheat resistance to necrotrophic fungal pathogens might be divided into two groups. The first one is based on the ability to avoid the effect of toxins to cause changes in physiological processes eventually to provoke forced cell death (Glazebrook 2005; Lemmens et al. 2005; Mesterhazy et al. 2008). The second group avoids the forced cell death which is stimulated by necrotrophs at the key stages of triggering by additional regulation, cell wall decomposition, and other mechanisms somehow similar to APR against biotrophic pathogens (Diethelm et al. 2012, 2014; Glazebrook 2005; Han et al. 2005; Golkari et al. 2009; Schweiger et al. 2013). Among widespread necrotrophic pathogens of wheat are *Fusarium* species (*F. graminearum*, *F. sambucinum*, *F. culmorum*, *F. avenaceum*, *F. sporotrichiella*, and *F. moniliforme*) (Summerell et al. 2010), *Pyrenophora tritici-repentis* (Died.) Drech., and *Parastagonospora nodorum* (Berk.) Quaedvlieg, Verkley & Crous (Figueroa et al. 2017).

In addition to yield losses due to *Fusarium* head blight, fungi of this genus are able to produce mycotoxins zearalenone, fumonisins, moniliformin, and trichothecenes, which are harmful for humans and animals (Nesic et al. 2014). Factors of resistance against *Fusarium* fungi are considered to show additive effects providing the measured level of resistance in comparison to plants lacking such factors (Kollers et al. 2013). It was discovered that certain alleles of the homoeologous *NPR1*-like genes on wheat chromosomes 2D and 2A (*TDF_076_2D* та *TDF_076_2A*) conferred type II resistance (resistance against symptom spread in the head) against *F. graminearum* та *F. culmorum* on a level of 14.2% and 3%, respectively (Diethelm et al. 2014).

P. tritici-repentis causing tan spot may be responsible for yield losses of about 5–10% and more than 50% under favorable conditions among susceptible cultivars and significantly affects yield quality (Shabeer and Bockus 1988; Fernandez et al. 1998). There are three main host-selective toxins produced by *P. tritici-repentis*, namely, PtrToxA (causes necrosis in plants of the wheat cultivar 'Glenlea'), PtrToxB (causes chlorosis in plants of the wheat line '6B662'), and PtrToxC (causes chlorosis in plants of the wheat line '6B365') (Lamari et al. 2003).

P. nodorum fungi produce their own toxin A (SnToxA), which is functionally identical to PtrToxA and lends half or less of *P. nodorum* virulence (Faris and

Friesen 2009). PtrToxA and SnToxA sensitivity in wheat plants is governed by a single dominant gene (*Tsn1*) mapped on the long arm of wheat chromosome 5B (Faris et al. 1996, 2010). Another single gene localized on the short arm of wheat chromosome 2B named *Tsc2* is responsible for the sensitivity to PtrToxB (Friesen and Faris 2004).

It was previously shown that resistance-associated alleles of the *Lr34/Yr18/Pm38/Sr57/Bdv1*, *Tsn1*, *Tsc2*, *TDF_076_2D* genes, as well as the wheat-rye translocations 1BL.1RS and 1AL.1RS with respective resistance genes, are common in Ukrainian winter bread wheat cultivars (Karelov et al. 2011; Kozub et al. 2017). Spring wheat cultivars originating from Ukraine and Hellas have not yet been studied for the resistance gene markers. Thus, the purpose of this research was to study Ukrainian as well as Hellenic spring bread wheat germplasm using the same set of molecular genetic markers of the genes conferring resistance against different biotic stress factors.

2 Material and Methods

2.1 Plant Material

The Hellenic commercial spring bread wheat cultivars produced at the Cereal Institute of Thessaloniki, 'Yecora E', 'Elissavet', 'Xenia', 'Acheron', 'Strymonas', 'Louros', and 'Lydia', the experimental cultivar 'Chios' which was developed at the Aristotle University of Thessaloniki, and the Russian-derived cultivar 'KVZ/Cgn' were used as the genetic material for this study. Spring bread wheat cultivars from Ukrainian breeding program (93 in total), developed in different climatic zones for different periods of time, were also studied with the same molecular genetic approach. The cultivar 'Chinese Spring' was used as the control for the *tr* allele of the marker *Xfcp623* for the *Tsn1* gene (associated with ToxA insensitivity) (Faris et al. 2010), the *tsr* allele of the marker *XBE444541* for the *Tsc2* gene (associated with PtrToxB insensitivity) (Abeysekara et al. 2010), and the allele *R* of the *Lr34* gene (the presence of resistance) (Lagudah et al. 2009). The cultivar 'Katepwa' was used as the control for the *Ts* allele of the marker *Xfcp623* (associated with the ToxA toxin sensitivity), the *Tss* allele of the marker *XBE444541* (associated with PtrToxB toxin sensitivity), and the allele *S* of the *Lr34* gene (the absence of resistance). For the *TDF_076_2D* gene, the cultivar 'Mironovskaya 808' was used as the control for the allele *R* (associated with moderate resistance to *Fusarium* head blight) and the cultivar 'Chinese Spring' – as the control for the sensitivity-associated allele *S* (Diethelm et al. 2014). The collection of Ukrainian cultivars and the cultivars used as controls was kindly provided by the National Center for Plant Genetic Resources of Ukraine of NAAS (Kharkiv).

2.2 Molecular Methods

DNA was extracted from a sample of 25–30 mg obtained from grinding 5–7 seeds with a Diatom™ DNA Prep100 DNA isolation kit (the sales representative in Ukraine is Neogene® Company) following the standard protocol. PCR was performed using GenPak® PCR Core Kits (the sales representative in Ukraine is Neogene® Company) according to the manufacturer's recommendations. PCR was performed in an amplifier 2720 GeneAMP System.

The resistance genes, markers, and respective primer sequences are given in Table 1.

The marker *Xfcp623* for the *Tsn1* gene has two alleles: the amplified fragment of 379 bp in the length is associated with the allele *Ts* for the toxin A sensitivity, and the absence of amplification was associated with the allele *tr* also the toxin A insensitivity (Faris et al. 2010) (Fig. 1).

Table 1 Details for the markers used for the research

Gene	Marker name	Primer sequence	Reference
Tsn1	*Xfcp623*	5'ctattcgtaatcgtgccttccg3'	Faris et al. (2010)
		5'ccttctctctcaccgctatctcatc3'	
Tsc2	*XBE444541-STS*	5'tggaccagtatgaga3'	Abeysekara et al. (2010)
		5'ttctggaggatgttgagcac3'	
Lr34/Yr18/Pm38/Sr57/Bdv1	*caISBP1*	5'catatcgagcttgccaaacg3'	Dakouri et al. (2010)
		5'tcagccacacaatgttccat3'	
		5'cgtgagcacagagaaaacca3'	
	caSNP12	5'tccccagtttaaccatcctg3'	
		5'cattcagtcacctcgcagc3'	
TDF_076_2D	*INDEL1*	5'tcatgcagtgttgcttgatct3'	Diethelm et al. (2014)
		5'ccattcacttgagcaacttcc3'	

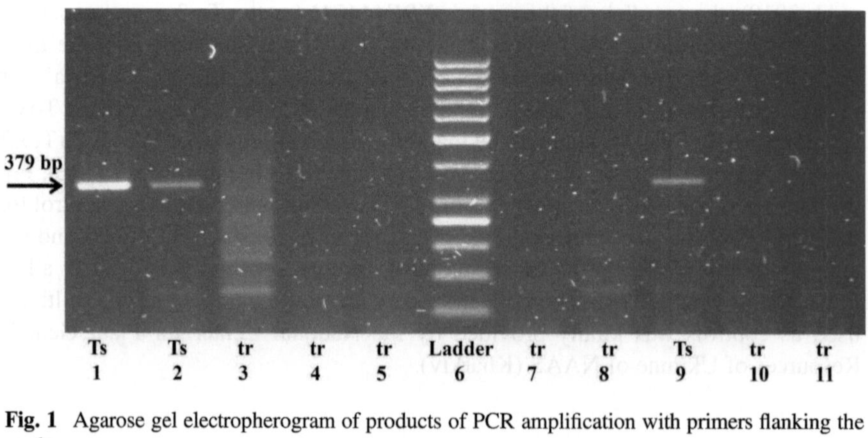

Fig. 1 Agarose gel electropherogram of products of PCR amplification with primers flanking the marker

The marker *XBE444541-STS* for the *Tsc2* gene (Table 1) has two alleles: the toxin B sensitivity-associated allele (*Tss*) producing the amplified fragment of 340 bp and the insensitivity-associated allele (*tsr*) producing the amplified fragment of 509 bp masked on agarose gel by nonspecific PCR products (Abeysekara et al. 2010).

To determine the allelic state of the *Lr34* gene, a combination of the gene-localized marker *caSNP12* and the closely linked marker *caISBP1* was used (Table 1) (Dakouri et al. 2010). The amplified fragments of 509 and 234 bp are associated with the resistance allele *R* and the fragments of 391 bp – with the susceptibility allele *S* (Fig. 2).

For the *TDF_076_2D* gene responsible for moderate *Fusarium* head blight resistance, the intron-localized marker *INDEL1* was used (Table 1) (Diethelm et al. 2014). In the case of the resistance-associated allele *R*, the amplified fragment of 212 bp was obtained and in case of the susceptibility-associated allele *S* – fragments of 212 bp and 221 bp.

The annealing temperature was lowered to 42 °C for the primer pair flanking the marker *XBE444541*. For the combination of the markers *caSNP12* and *caISBP1*, the conditions were the following: dissociation/activation of the hot start polymerase at 95 °C for 7 min and then 32 cycles with denaturation phase at 94 °C for 30 s, annealing phase at 62.5 °C for 40 s, elongation at 72 °C for 40 s, and final elongation – for 5 min. Besides this, PCR was performed according to the conditions described in the literature (Diethelm et al. 2014; Abeysekara et al. 2010; Faris et al. 2010).

PCR results were visualized by electrophoresis in 2–2.5% agarose gel in 0.5 x TBE buffer with subsequent staining with ethidium bromide or (in the case of the *INDEL1* marker) in 8% polyacrylamide gel with subsequent staining with $AgNO_3$ and the use of the gel visualization system Vision Gel.

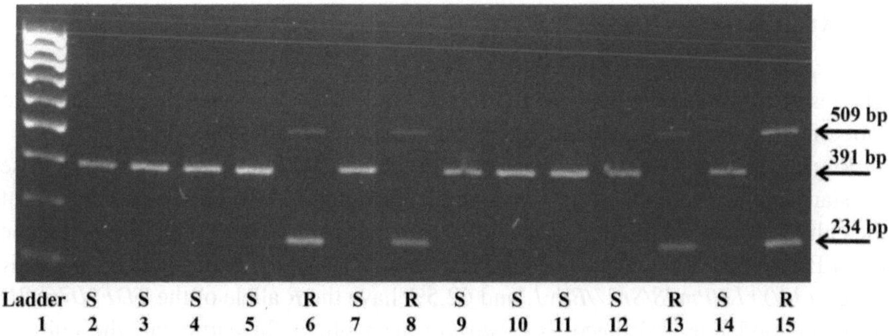

Fig. 2 Agarose gel electropherogram of products of PCR amplification with primers flanking the markers *caISBP1* and *caSNP12* of the gene *Lr34*: R is the resistance allele, S is the susceptibility allele.(1) O'GeneRuler™ 100 bp DNA Ladder; (2) 'Hordynia'; (3) 'Etiud'; (4) 'Myroslavna'; (5) 'Torchynska'; (6) 'Yubileynaya yarovaya'; (7) 'Stavyska'; (8) 'Mironovskaya 5'; (9) 'Bukovynka'; (10) 'Elehiya'; (11) 'Kvorum'; (12) 'Luhanskaya 4'; (13) 'Kuibyshevska 2'; (14) 'Kievskaya 77'; (15) 'Khersonska 183'

2.3 Electrophoresis of Storage Proteins and Identification of Wheat-Rye Translocations

Acid-polyacrylamide gel electrophoresis of gliadin of 7–20 single seeds from each variety was carried out by the procedure described by Kozub et al. (2009). The presence of a specific block of components on the electropherogram of alcohol-soluble proteins designated as *Gli-B1l* (GLD 1B3) is a marker for the 1BL.1RS translocation of the 'Kavkaz' type (Sozinov and Poperelya 1980; Metakovsky 1991). The presence of a specific block of components on the electropherogram of alcohol-soluble proteins, designated as *Gli-A1w* (Gld 1A17) is a marker for 1AL.1RS translocation of the Amigo type (Sobko and Poperelya 1986; Kozub et al. 2009). Finally, SDS-electrophoresis of total seed proteins was performed according to Laemmli (1970).

2.4 Statistical Analysis

Calculation of the numeric data was performed with software under the Linux operating system. Associations between alleles of different loci were estimated with the use of the phi coefficient (φ) and the Pearson's χ^2 test (Yates 1934; http://vassarstats.net/tab2x2.html).

Pedigrees of cultivars were analyzed using the GRIS database (http://wheatpedigree.net/).

3 Results and Discussion

The results of molecular analysis for genes and their alleles conferring resistance to biotic factors for Hellenic bread wheat cultivars are presented in Table 2.

Regarding the Hellenic cultivars studied, combinations of two or more disease resistance genes were mainly revealed at the loci analyzed (Table 2). About 44% of the cultivars carry the toxin A insensitivity allele *tr* of the *Tsn1* gene, 56% show the toxin B insensitivity allele *tsr* of the *Tsc2* gene, 33% carry the *R* allele of the APR gene *Lr34/Yr18/Pm38/Sr57/Bdv1*, and 62.5% have the *R* allele of the *TDF_076_2D* gene for moderate resistance to *Fusarium* head blight. In particular, the cultivar 'Elissavet' carries genes conferring resistance to tan spot due to insensitivity to toxins A and B of *P. tritici-repentis* and resistance to rusts and powdery mildew due to the presence of the wheat-rye 1BL.1RS translocation of the 'Kavkaz' type (Xynias et al. 2006) and the gene *Lr34/Yr18/Pm38/Sr57/Bdv1*, which confers moderate race-nonspecific resistance to a number of biotrophic pathogens, including yellow dwarf virus. The 'Yecora E' cultivar has the gene for resistance to tan spot (insensitivity to toxin B) and the gene for moderate race-nonspecific resistance to rusts and other

Table 2 Alleles of genes conferring resistance to biotic stressing factors in Hellenic spring bread wheat cultivars

Cultivar	*Tsn1*	*Tsc2*	*TDF_076_2D*	*Lr34*	Presence of the wheat-rye translocation[a]
'Yecora E'	*Ts*	*tsr*	*S*	*R*	
'Elissavet'	*tr*	*tsr*	*S*	*R*	1BL.1RS of the 'Kavkaz' type[a]
'Xenia'	*Ts*	*tsr*	*R*	*S*	–
'Acheron'	*Ts*	*tsr*	*R*	*S*	1BL.1RS of the 'Kavkaz' type
'Strymonas'	*tr*	*tsr*	*R*	*S*	–
'Louros'	*tr*	*Tss*	*S*	*S*	–
'Lydia'	*Ts*	*Tss*	*R*	*S*	–
'Chios'	*tr*	*Tss*	*R*	*S*	1BL.1RS of the 'Kavkaz' type /–
'KVZ/Cgn'	*Ts*	*Tss*	*S*	*R*	1BL.1RS of the 'Kavkaz' type

[a]With *Pm8*, *Sr31*, *Lr26*, *Yr9*

pathogens *Lr34/Yr18/Pm38/Sr57/Bdv1*. The cultivar 'Acheron', which also carries the 1BL.1RS translocation (Xynias et al. 2006), has the respective resistance genes, as well as the gene for insensitivity to *P. tritici-repentis* (tan spot) toxin B and moderate resistance to *Fusarium* head blight. The cultivar 'Strymonas' is characterized by three genes conferring resistance to necrotrophic pathogens (tan spot, *Septoria nodorum* blotch, and *Fusarium* head blight). At least two important disease resistance genes were detected in the cultivars 'Xenia', 'Chios', and 'KVZ'.

As mentioned before (see introduction), among the resistance genes described in the wheat gene pool (McIntosh 2013), a large proportion is comprised by genes introgressed from different wild and cultivated relatives. However, only some of them have been widely used in breeding practice. The most successful example is the 1BL.1RS translocation from the rye *Secale cereale* L. 'Petkus', which carries a number of disease resistance genes (Rabinovich 1998; Schlegel 2016): the *Pm8* gene for resistance to powdery mildew, the *Sr31* gene for resistance to stem rust, the *Lr26* gene for resistance to leaf rust, and the *Yr9* gene for resistance to stripe rust. In virtually all such commercial varieties, 1RS is derived from one source – the line 'Riebesel 47-51', developed by G. Riebesel in the 1930s (Rabinovich 1998). Whereas the *Lr26*, *Yr9*, and *Pm8* genes have lost their effectiveness shortly after the large-scale deployment of cultivars with the 1BL.1RS translocation (McIntosh et al. 1995; Limpert et al. 1987; Parks et al. 2008), the *Sr31* gene remains effective against all known stem rust races, except for the Ug99 race (TTKSK), which was first identified in 1999 in Uganda (Pretorius 2000). Since then Ug99 biotypes virulent to *Sr31* (TTKST, TTKSK, PTKSK, or PTKST) were revealed in Kenya, Ethiopia, Sudan, Yemen, Tanzania, Eritrea, Rwanda, Egypt, the South African Republic, and Iran (Singh et al. 2015). However, the *Sr31* gene is still important, in particular for European countries. This was recently demonstrated by the outbreak of a new, highly virulent variant of stem rust race TTTTF in Sicily (Bhattacharya 2017), which may present a threat to wheat production in the Mediterranean region. It was detected that race TTTTF has complex virulence but is avirulent on gene *Sr31* (Anonymous 2017). Thus, it is highly probable that the Hellenic cultivars

'Elissavet', 'Acheron', 'Chios', and 'KVZ' possess resistance to the Sicilian race TTTTF of stem rust, which is currently potentially dangerous for the Mediterranean region, as they carry the *Sr31* gene on the 1BL.1RS translocation. In addition, the positive effect of the presence of the 1BL.1RS translocation on grain yield and development of the root system was demonstrated in many studies (Villareal et al. 1991; Moreno-Sevilla et al. 1995; Ehdale et al. 2003; Howell et al. 2014).

The data for genes and alleles related to resistance in biotic stressing factors for the Ukrainian spring bread wheat cultivars are presented in Table 3.

The proportion of Ukrainian spring bread wheat cultivars carrying alleles for partial resistance to necrotrophic fungal pathogens is comparatively high: the *tr* allele of the *Tsn1* gene was found in 74% of the sample, 48% possess the *tsr* allele of the *Tsc2* gene, and 46% carry the *R* allele of the *TDF_076_2D* gene. The *R* allele of the *Lr34/Yr18/Pm38/Sr57/Bdv1* gene is less frequent among the Ukrainian spring cultivars studied, and it was found in only 11% of the collection. The cultivars with any wheat-rye translocation comprise about 10% of the sample studied. Among them, seven cultivars possess the 1BL.1RS translocation, and two carry 1AL.1RS. The translocation 1AL.1RS originating from rye cultivar 'Insave' is another wheat-rye translocation that has been actively exploited in wheat commercial varieties (Rabinovich 1998; Schlegel 2016). The first commercial bread wheat variety with 1AL.1RS was cultivar 'Amigo' released in the USA in 1976 (Sebesta and Wood 1978; Sebesta et al. 1995). This translocation carries the powdery mildew resistance gene *Pm17* and the stem rust resistance gene $Sr1RS^{Amigo}$. Whereas *Pm17* was overcome in many areas (Parks et al. 2008), the gene $Sr1RS^{Amigo}$ is still of importance as it is effective against the known variants (biotypes) of the race Ug99 (Singh et al. 2015). In addition to the disease resistance genes, 1AL.1RS of the 'Amigo' type also carries pest resistance genes: the *Gb2* gene for resistance to greenbug *Schizaphis graminum* (Rondani) biotypes B and C as well as the wheat curl mite *Aceria tosicheilla* (Keifer) resistance gene *Cm3* (McIntosh 2013), which provide additional protection against biotic stresses. Thus, the cultivars 'Etiud' and 'Struna myronivska' possess the $Sr1RS^{Amigo}$ gene conferring resistance to stem rust race Ug99 biotypes.

Six cultivars ('Kardynal', 'Klara', 'Melnykivska', 'Ostynka', 'Provintsialka', and 'Rubin') carry the widespread translocation 1BL.1RS from the rye 'Petkus' of the 'Kavkaz' type (Table 3). However, it has been revealed that in the spring cultivar 'Vyshyvanka', the 1BL.1RS translocation is distinct from 1BL.1RS of the 'Kavkaz' type, as evidenced by differences in the secalin pattern both on APAG and SDS electropherograms (Fig. 3). It should be noted that the most prominent difference between their secalin patterns is the presence of two secalin components below the high-molecular-weight glutenin subunits in the cultivar 'Vyshyvanka' on the SDS electropherogram (Fig. 3b), in contrast to the patterns of carriers of the 1BL.1RS translocation of the 'Kavkaz' type. Most likely these secalin components visible on the SDS electropherogram are encoded by the secalin locus *Sec-N*, which is located distally from the *Sec-1* locus on the arm 1RS (Kozub et al. 2014). Because of this, it is highly probable that the translocation in 'Vyshyvanka' carries novel rye disease resistance genes.

Table 3 Alleles of genes conferring resistance to biotic stressing factors in Ukrainian spring bread wheat cultivars

Cultivar	*Tsn1*	*Tsc2*	*Lr34*	*TDF_076_2D*	The presence of the wheat-rye translocation
'Anshlag'	*tr*	*tsr*	*S*	*R*	–
'Artemovka'	*tr*	*Tss*	*S*	*S*	–
'Azhurnaya'	*Ts*	*tsr*	*S*	*S*	–
'Bezostaya 4'	*tr*	*Tss*	*S*	*S*	–
'Bukovynka'	*Ts*	*Tss*	*S*	*S*	–
'Cheremshyna'	*tr*	*Tss*	*R/S*	*R*	–
'Dneprianka'	*tr*	*Tss*	*R/S*	*S*	–
'Dublianka 4'	*tr*	*Tss*	*S*	*S*	–
'Elehiya myronivska'	*Ts*	*tsr*	*S*	*R*	–
'Etiud'	*tr*	*tsr*	*S*	*R*	1AL.1RS of the 'Amigo' type[a]
'Fontan'	*Ts*	*tsr*	*S*	*S*	–
'Heroyinia'	*tr*	*tsr*	*S*	*R*	–
'Hordynia'	*tr*	*Tss*	*S*	*R*	–
'Ivanovskaya 61'	*Ts*	*tsr*	*S*	*S*	–
'Kardynal'	*tr*	*tsr*	*S*	*R*	1BL.1RS of the 'Kavkaz' type[b]
'Kharkivska 10'	*tr*	*Tss*	*S*	*S*	–
'Kharkivska 12'	*tr*	*Tss*	*S*	*S*	–
'Kharkivska 14'	*tr*	*Tss*	*S*	*S*	–
'Kharkivska 16'	*tr*	*Tss*	*S*	*S*	–
'Kharkivska 18'	*tr*	*Tss*	*S*	*S*	–
'Kharkivska 2'	*tr*	*Tss*	*S*	*S*	–
'Kharkivska 22'	*tr*	*Tss*	*S*	*S*	–
'Kharkivska 24'	*tr*	*Tss*	*S*	*R*	–
'Kharkivska 26'	*tr*	*Tss*	*S*	*R*	–
'Kharkivska 28'	*tr*	*Tss*	*S*	*R*	–
'Kharkivska 30'	*tr*	*Tss*	*S*	*R*	–
'Kharkivska 4'	*tr*	*Tss*	*R/S*	*S*	–
'Kharkivska 6'	*tr*	*Tss*	*S*	*S*	–
'Kharkivska 75 yarova'	*Ts*	*tsr*	*S*	*S*	–
'Kharkivska 93'	*tr*	*Tss*	*S*	*S*	–
'Khersonska 183'	*tr*	*tsr*	*R*	*S*	–
'Kievskaya 77'	*Ts*	*Tss*	*S*	*S*	–
'Klara'	*tr*	*tsr*	*S*	*R*	1BL.1RS of the 'Kavkaz' type
'Kolektyvna 1'	*tr*	*Tss*	*S*	*R*	–
'Kolektyvna 2'	*tr*	*tsr*	*S*	*S*	–
'Kolektyvna 5'	*tr*	*Tss*	*S*	*R*	–
'Kollektyvnaya'	*tr*	*Tss*	*S*	*S*	–
'Komsomolskaya 29'	*Ts*	*tsr*	*S*	*S*	–
'Krasa Polissia'	*Ts*	*tsr*	*S*	*R*	–
'Kuibyshevska 2'	*Ts*	*Tss*	*R*	*S*	–
'Kvorum'	*tr*	*tsr*	*S*	*R*	–

(continued)

Table 3 (continued)

Cultivar	*Tsn1*	*Tsc2*	*Lr34*	*TDF_076_2D*	The presence of the wheat-rye translocation
'Luhanskaya 3'	*Ts*	*tsr*	*S*	*S*	–
'Luhanskaya 4'	*tr*	*tsr*	*S*	*S*	–
'Luhanskaya 5'	*tr*	*tsr*	*R*	*S*	–
'Luhanskaya 6'	*tr*	*tsr*	*S*	*S*	–
'Lutescens 28'	*Ts*	*tsr*	*R*	*S*	–
'Lutescens 491'	*tr*	*Tss*	*S*	*S*	–
'Mazhor'	*tr*	*tsr*	*S*	*R*	–
'Melnykivska'	*tr*	*tsr*	*S*	*R*	1BL.1RS of the 'Kavkaz' type
'Milturum 162'	*tr*	*Tss*	*S*	*S*	–
'Mironovskaya 3'	*Ts*	*tsr*	*S*	*S*	–
'Mironovskaya 4'	*tr*	*tsr*	*R*	*S*	–
'Mironovskaya 5	*tr*	*tsr*	*S*	*S*	–
'Mironovskaya krupnozernistaya'	*tr*	*Tss*	*R*	*S*	–
'Mironovskaya ranniaya'	*tr*	*Tss*	*S*	*S*	–
'Mironovskaya yarovaya'	*tr*	*Tss*	*S*	*S*	–
'Myroslavna'	*tr*	*tsr*	*S*	*R*	–
'Nedra'	*tr*	*Tss*	*S*	*R*	–
'Odesskaya 13'	*Ts*	*tsr*	*S*	*S*	–
'Osoblyva'	*tr*	*Tss*	*S*	*R*	–
'Ostynka'	*Ts*	*tsr*	*S*	*R*	1BL.1RS of the 'Kavkaz' type
'Otechestvennaya'	*tr*	*Tss*	*S*	*S*	–
'Otrada'	*Ts*	*tsr*	*S*	*S*	–
'Panianka'	*tr*	*tsr*	*S*	*R*	–
'Pecherianka'	*tr*	*tsr*	*S*	*R*	–
'Podarunok'	*tr*	*Tss*	*S*	*R*	–
'Provintsialka'	*tr*	*Tss*	*R/S*	*R*	1BL.1RS of the 'Kavkaz' type/–
'Rannia 93'	*tr*	*Tss*	*S*	*S*	–
'Ranniaya 73'	*tr*	*tsr*	*S*	*R*	–
'Rovenskaya 60'	*tr*	*Tss*	*S*	*S*	–
'Rubin'	*tr*	*tsr*	*S*	*R*	1BL.1RS of the 'Kavkaz' type
'Simkoda myronivska'	*tr*	*tsr*	*S*	*R*	–
'Siuita'	*tr*	*tsr*	*S*	*R*	–
'Skorospilka 82'	*tr*	*tsr*	*S*	*S*	–
'Skorospilka 95'	*tr*	*tsr*	*S*	*S*	–
'Skorospilka 98'	*tr*	*tsr*	*R/S*	*R*	–
'Skorospilka 99'	*Ts*	*Tss*	*R/S*	*R*	–
'Sperantsa'	*Ts*	*tsr*	*S*	*R*	–
'Sriblianka'	*Ts*	*tsr*	*S*	*R*	–

(continued)

Table 3 (continued)

Cultivar	*Tsn1*	*Tsc2*	*Lr34*	*TDF_076_2D*	The presence of the wheat-rye translocation
'Stavyska'	*tr*	*Tss*	*S*	*R*	–
'Struna myronivska'	*tr*	*tsr*	*S*	*R*	1AL.1RS of the 'Amigo' type
'Svitanok'	*Ts*	*tsr*	*S*	*S*	–
'Taina'	*tr*	*Tss*	*S*	*R*	–
'Torchynska'	*Ts*	*tsr*	*S*	*R*	–
'Uliublena'	*tr*	*Tss*	*S*	*R*	–
'Veselopodolianskaya 12'	*tr*	*Tss*	*S*	*S*	–
'Vitka'	*tr*	*Tss*	*S*	*R*	–
'Volynska Yara'	*Ts*	*tsr*	*S*	*R*	–
'Vyshyvanka'	*tr*	*Tss*	*S*	*R*	1BL.1RSnew[c]
'Yevdokiya'	*tr*	*Tss*	*S*	*R*	–
'Yubileynaya'	*tr*	*tsr*	*S*	*S*	–
'Yubileynaya yarovaya'	*Ts*	*Tss*	*R*	*S*	–
'Zhuravka'	*Ts*	*Tss*	*S*	*S*	–

[a]With $Sr1RS^{Amigo}$, *Pm17*, *Gb2*, *Cm3*
[b]With *Pm8*, *Sr31*, *Lr26*, *Yr9*
[c]A new translocation

Thus, genotyping the samples of Hellenic and Ukrainian spring bread wheat cultivars revealed the presence of resistance (or toxin insensitivity)-associated alleles. An interesting finding is the presence of the resistance allele of the *Lr34/Yr18/Pm38/Sr57/Bdv1* gene with moderate frequencies, 33% and 11%, in the Hellenic and Ukrainian samples, respectively. This allele, as it was reported by previous investigations, was very rare among European cultivars (Kolmer et al. 2008). It is characteristic that among the representative collection of Western European bread wheat cultivars, the resistance allele of *Lr34* was found only in cv. Kavkaz. However, the resistance allele of the *Lr34/Yr18/Pm38/Sr57/Bdv1* gene proved to be widespread among American, Canadian, and Australian cultivars (Kolmer et al. 2008), as well as among Ukrainian winter bread wheat cultivars (Karelov et al. 2011). The *Lr34/Yr18/Pm38/Sr57/Bdv1* gene is derived from the old Italian wheat cultivars 'Ardito' and 'Mentana', developed in the 1900s, and retains its effectiveness for about a century (Kolmer et al. 2008). Krattinger et al. (2009) performed physical and genetic mapping of the *Lr34* locus and proposed its functional role in the wheat genome. According to their data, this gene encodes a pleiotropic drug resistance-like (PDR-like) ATP-binding cassette (ABC) transporter, which determines the presence of the corresponding type of resistance in wheat plants.

The probable origin of the resistance-associated allele of the *Lr34* gene for Ukrainian spring bread wheat cultivars might be diverse. For example, according to pedigrees in GRIS database (http://wheatpedigree.net/), the cultivar 'Kuibyshevska 2' has obtained the *R* allele from 'Penjamo T62'. 'Luhanskaya 5'

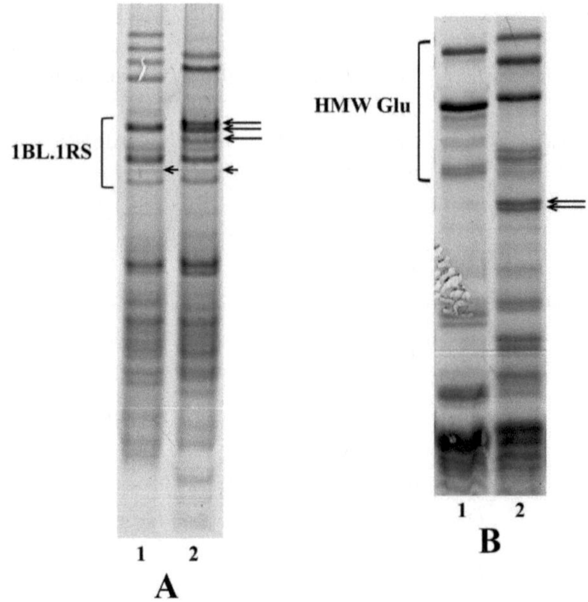

Fig. 3 Electrophoretic patterns of storage proteins of the bread wheat line B-16, a carrier of the 1BL.1RS translocation of the 'Kavkaz' type (1), and the cultivar 'Vyshyvanka' with a novel 1BL.1RS translocation (2). (**a**) APAG electropherogram of ethanol-soluble proteins, the zone of secalins encoded by the *Sec-1* (syn. *Gli-R1*) locus is marked by a bracket (1BL.1RS); the arrows indicate bands by which there are differences in this zone. (**b**) SDS electropherogram of total seed proteins, the zone of high-molecular-weight glutenin (HMW Glu) subunits is marked by a bracket; the arrows indicate the possible products of the *Sec-N* locus (Kozub et al. 2014) in 'Vyshyvanka'

has 'Bezostaya 1' (with the *R* allele of the *Lr34* gene) as its direct parent. 'Kharkivska 4' might have obtained it from the Russian wheat gene pool. As for 'Mironovskaya 3', the *R* allele of the *Lr34* gene has been possibly transferred from 'Siete cerros 66'.

The other genes of interest into this research, *Tsn1* and *Tsc2* conferring (in) sensitivity to the toxins of *P. tritici-repentis* and *P. nodorum*, have not been studied as precisely as *Lr34/Yr18/Pm38/Sr57/Bdv1*. The reason might be that the virulence factors of *P. tritici-repentis* and *P. nodorum* are not only PtrTox A/SnTox A and PtrTox B (Lamari et al. 2003). However, the key role of the toxin insensitivity for the resistance against both of pathogens was shown (Friesen and Faris 2004; Faris and Friesen 2009). It has been revealed that the toxin Ptr ToxA has conservative regions rich in arginine, glycine, and aspartic acid, causes necrotic reaction, and is able to take part in the interaction with the transmembrane receptors that are similar to the animal integrins (Ruoslahti 1996; Meinhardt et al. 2002; Lamari et al 2003; Sarma et al. 2005). This toxin was found to have dual functions, one of which is competing against host chitinases (Liu et al. 2016). It also has been suggested that the functional product of the *Tsn1* gene is essential for necrosis triggering (Manning et al. 2007; Tai et al. 2007). The *Tsn1* gene was sequenced and found to encode a protein similar to that encoded by genes conferring race-specific resistance having serine-protein kinase domains, nucleotide binding sites, and leucine-rich repeats (Faris et al. 2010; Glazebrook 2005).

The allele *tr* of the *Tsn1* gene, in most of the Ukrainian spring wheat cultivars, is probably derived from the local gene pool. For example, according to their pedigrees, 'Artemovka' has probably obtained it from 'Kremenchug'; 'Bezostaya 4, from 'Lutescens 17' or 'Skorospelka 2'; 'Veselopodolianskaya 12', from the cultivar 'Veselii Podol'; 'Dublianka 4', from the line 'LV-WEST-UKR'; 'Miltrum 162', from the line 'LV-UKR'; and 'Kollektyvnaya', from the cultivar 'Artemovka' or 'Otechestvennaya'. Further 'Kollektyvnaya' could have been the source of the allele *tr* for the cultivars 'Dneprianka', 'Ranniaya 73', 'Rannia 93', 'Panianka', 'Skorospilka 95', and 'Skorospilka 98'. For some of the cultivars studied, the possible source of the allele *tsr* of the *Tsc2* gene might have been the local gene pool, as well. That might be said about 'Odesskaya 13', Khersonska 183', and 'Svitanok'.

The allele *R* of the gene *TDF_076_2D* conferring moderate resistance to *Fusarium* head blight was found to almost half of the cultivars studied in this work. Among them are cvrs. 'Artemovka', 'Bezostaya 4', 'Veselopodolianskaya 12', and 'Kollektyvnaya', which have been mentioned in the pedigree of the majority of spring and many winter wheat cultivars from Ukraine. Such result might testify to the local origin of the allele. However, for most of the 'Kharkivska' cultivars, 'Elehiya myronivska', 'Mironovskaya 4', 'Yubileynaya', 'Mazhor', 'Kolektyvna 1', 'Kolektyvna 2', 'Kolektyvna 5', and 'Etiud', possible sources of the *R* allele might have been foreign cultivars such as 'Selkirk', 'pPG 56', 'Turaco', 'TAM 200', 'Siete Cerros', etc.

In the cultivar 'Struna myronivska', the 1AL.1RS translocation is derived from the winter wheat cultivar 'Ekspromt' with 'Amigo' in its pedigree. 'Etiud' is derived from the 'TAM 200' line, and it has 'Amigo' in its pedigree as well. For all the Ukrainian spring cultivars with the 1BL.1RS translocation of the 'Kavkaz' type and 'Vyshyvanka' with a new type of the 1BL.1RS translocation, the pedigree has not been reported, so the possible sources of the translocations are unknown. The Hellenic spring cultivar 'Elissavet' might have inherited the 1BL.1RS translocation together with the *R* allele of the *Lr34* gene from 'Kavkaz' through 'Kavko'. It is interesting that for the sample of Ukrainian spring cultivars, a medium-effect association between the allelic state of the markers for the *Tsn1*and *Tsc2* genes was revealed ($\chi^2 = 9.17$; $\varphi = +0.31$, $p = 0.002$). It turned out that 45.2% of the Ukrainian spring cultivars carried the combination of the *tr* allele of the *Tsn1* gene and the *Tss* allele of the *Tsc2* gene. The amount of cultivars carrying the insensitivity alleles of both genes made up only 29.0%; 19.4% of the cultivars carried the *Ts* allele of the *Tsn1* gene and the *trr* allele of the *Tsc2* gene, whereas carriers of two sensitivity alleles comprised less than 6.5%. The obtained data might denote that some selection pressure was present and led to this disequilibrium.

4 Conclusions

Hellenic and Ukrainian spring bread wheat cultivars were studied with the use of molecular markers for biotic stress resistance genes. The frequency of the resistance-associated allele of the *Lr34/Yr18/Pm38/Sr57/Bdv1* gene among the Hellenic and Ukrainian spring wheat germplasm was moderate but higher than the respective previously described value for the wheat cultivars of the European collection (Kolmer et al. 2008). Both groups of cultivars had high frequencies of alleles of the genes *Tsn1*, *Tsc2*, and *TDF_076_2D* conferring partial resistance to the necrotrophic pathogens. Cultivars with combinations of disease resistance/toxin insensitivity alleles were identified. The Hellenic cultivars 'Elissavet', 'Acheron', 'Chios', and 'KVZ' possess resistance to the Sicilian race TTTTF of stem rust, which is currently potentially dangerous for the Mediterranean region, as they carry the *Sr31* gene on the 1BL.1RS translocation. Two Ukrainian cultivars with 1AL.1RS translocation of the 'Amigo' type possess the gene $Sr1RS^{Amigo}$ conferring resistance against stem rust Ug99 biotypes. Cultivar 'Vyshyvanka' was found to carry a 1BL.1RS translocation distinct from 1BL.1RS of the 'Kavkaz' type and may carry novel rye disease resistance genes. The results obtained from the analysis of the Hellenic and Ukrainian wheat germplasm regarding disease resistance gene alleles could be helpful for breeders who are looking for new sources for breeding disease-resistant cultivars.

Literature

Abeysekara NS, Friesen TL, Liu Z, McClean PE, Faris JD (2010) Marker development and saturation mapping of the tan spot Ptr ToxB sensitivity locus *Tsc2* in hexaploid wheat. Plant Genome 3:179–189

Aktar-Uz-Zaman M, Tuhina-Khatun M, Hanafi MM, Sahebi M (2017) Genetic analysis of rust resistance genes in global wheat cultivars: an overview. Biotechnol Biotechnol Equip 31 (3):431–445

Anonymous (2017) Caution: risk of wheat stem rust in Mediterranean basin in the forthcoming 2017 crop season following outbreaks on Sicily in 2016. http://wheatrust.org/news-and-events/news-item/artikel/caution-risk-of-stem-rust-in-eastern-mediterranean-basin-in-the-forthcoming-2017-season-due-to-ou/

Bent AF, Mackey D (2009) Elicitors, effectors, and R genes: the new paradigm and a lifetime supply of questions. Annu Rev Plant Biol 60:379–406

Bhattacharya S (2017) Deadly new wheat disease threatens Europe's crops. Nature 542:145–146. https://doi.org/10.1038/nature.2017.21424

Dakouri A, McCallum BD, Walichnowski AZ, Cloutier S (2010) Fine-mapping of the leaf rust *Lr34* locus in *Triticum aestivum* (L.) and characterization of large germplasm collections support the ABC transporter as essential for gene function. Theor Appl Genet 121:373–384

Devos KM, Gale MD (1992) The use of random amplified polymorphic DNA markers in wheat. Theor Appl Genet 84:567–572. https://doi.org/10.1007/BF00224153

Diethelm M, Rhiel M, Wagner C, Mikolajewski S, Groth J, Hartl L, Friedt W, Schweizer G (2012) Tissues of European winter wheat after challenge with *G. zeae*. Euphytica 186(1):103–114

Diethelm M, Schmolke M, Groth J, Friedt W, Schweizer G, Hartl L (2014) Association of allelic variation in two *NPR1*-like genes with *Fusarium* head blight resistance in wheat. Mol Breed 34:31–43

Duveiller E, Nicol JM, Singh RS (2007) The challenges of maintaining wheat productivity: pests, diseases, and potential epidemics. Euphytica 157:417–430

Dyck PL (1977) Genetics of leaf rust reaction in three introductions of bread wheat. Can J Genet Cytol 19:711–716

Dyck PL (1987) The association of a gene for leaf rust resistance with the chromosome 7D suppressor of stem rust resistance in bread wheat. Genome 29:467–469

Dyck PL (1993) Inheritance of leaf rust and stem rust resistance in 'Roblin' wheat. Genome 36:289–293

Ehdale B, Whitkus RW, Waines JG (2003) Root biomass, water-use efficiency, and performance of wheat rye translocations of chromosomes 1 and 2 in spring bread wheat 'Pavon'. Crop Sci 43:710–717

Ellis JG, Lagudah ES, Spielmeyer W, Dodds PN (2014) The past, present and future of breeding rust resistant wheat. Front Plant Sci 5:641: 1–13. https://doi.org/10.3389/fpls.2014.00641

Faris DJ, Anderson JA, Francl LJ, Jordah JG (1996) Chromosomal location of a gene conditioning insensitivity in wheat to a necrosis-inducing culture filtrate from *Pyrenophora tritici-repentis*. Phytopathology 86:459–463

Faris JD, Friesen TL (2009) Reevaluation of a tetraploid wheat population indicates that the *Tsn1*–ToxA interaction is the only factor governing *Stagonospora nodorum* blotch susceptibility. Phytopathology 99:906–912

Faris JD, Zhang Z, Lu H, Lu S, Reddy L, Cloutier S, Fellers JP, Meinhardt SW, Rasmussen JB, Xu SS, Oliver RP, Simons KJ, Friesen TL (2010) A unique wheat disease resistance-like gene governs effector-triggered susceptibility to necrotrophic pathogens. Proc Natl Acad Sci USA 107:13544–13549

Fernandez MR, DePauw RM, Clarke JM, Fox SL (1998) Discoloration of wheat kernels by *Pyrenophora tritici-repentis*. Can J Plant Pathol 20:380–383

Figueroa M, Hammond-Kosack KE, Solomon PS (2017) A review of wheat diseases—a field perspective. Mol Plant Pathol 19(6). https://doi.org/10.1111/mpp.12618

Friesen TL, Faris JD (2004) Molecular mapping of resistance to *Pyrenophora tritici-repentis* race 5 and sensitivity to PtrToxB in wheat. Theor Appl Genet 109:464–471

Ganal MW, Röder MS (2007) Microsatellite and SNP markers in wheat breeding. In: Varshney RK, Tuberosa R (eds) Genomics-assisted crop improvement. Springer, Dordrecht, pp 1–24

Gill BS, Appels R, Botha-Oberholster A-M, Buell CR, Bennetzen JL, Chalhoub B, Chumley F, Dvořák J, Iwanaga M, Keller B, Li W, McCombie WR, Ogihara Y, Quetier F, Sasaki T (2004) A workshop report on wheat genome sequencing: international genome research on wheat consortium. Genetics 168(2):1087–1096. https://doi.org/10.1534/genetics.104.034769

Glazebrook J (2005) Contrasting mechanisms of defense against biotrophic and necrotrophic pathogens. Annu Rev Phytopathol 43:205–227

Glowacki S, Macioszek VK, Kononowicz AK (2010) R proteins as fundamentals of plant innate immunity. Cell Mol Biol Lett 16(1):1–24. https://doi.org/10.2478/s11658-010-0024-2

Golkari S, Gilbert J, Ban T, Procunier JD (2009) QTL-specific microarray gene expression analysis of wheat resistance to *Fusarium* head blight in Sumai-3 and two susceptible NILs. Genome 52 (5):409–418

Goyal A, Prasad R (2010) Some important fungal diseases and their impact on wheat production. In: Araya A, Edith Perelló A (eds) Management of fungal plant pathogens CABI, pp 362–374

Gupta P, Langridge P, Mir R (2010) Marker-assisted wheat breeding: present status and future possibilities. Mol Breed 26:145–161. https://doi.org/10.1007/s11032-009-9359-7

Han FP, Fedak G, Ouellet T, Dan H, Somers DJ (2005) Mapping of genes expressed in *Fusarium graminearum*-infected heads of wheat cultivar 'Frontana'. Genome 48(1):88–96

Howell T, Hale I, Jankuloski L, Bonafede M, Gilbert M, Dubcovsky J (2014) Mapping a region within the 1RS.1BL translocation in bread wheat affecting grain yield and canopy water status. Theor Appl Genet 127:2695–2709

Jia M, Guan J, Zhai Z, Geng S, Zhang X, Mao L, Li A (2018) Wheat functional genomics in the era of next generation sequencing: an update. Crop J 6(1):7–14

Karelov AV, Pirko YV, Kozub NA, Sozinov IA, Pirko NN, Litvinenko NA, Lyfenko SF, Koliuchii VT, Blume YB, Sozinov AA (2011) Identification of the allelic state of the *Lr34* leaf rust resistance gene in soft winter wheat cultivars developed in Ukraine. Cytol Genet 45(5):271–276. https://doi.org/10.3103/S0095452711050069

Kiszonas AM, Morris CF (2018) Wheat breeding for quality: a historical review. Cereal Chem 95:17–34. https://doi.org/10.1094/CCHEM-05-17-0103-FI

Kollers S, Rodemann B, Ling J, Korzun V, Ebmeyer E, Argillier O, Hinze M, Plieske J, Kulosa D, Ganal MW, Röder MS (2013) Whole genome association mapping of *Fusarium* head blight resistance in European winter wheat (*Triticum aestivum L.*). PLoS ONE 8(2):e57500. https://doi.org/10.1371/journal.pone.0057500

Kolmer JA (1996) Genetics of resistance to wheat leaf rust. Annu Rev Phytopathol 34:435–455

Kolmer JA, Singh RP, Garvin DF, Viccars L, William HM, Huerta-Espino J, Ogbonnaya FC, Raman H, Orford S, Bariana HS, Lagudah ES (2008) Analysis of the *Lr34/Yr18* rust resistance region in wheat germplasm. Crop Sci 48:1841–1852

Kozub NA, Motsnyi II, Sozinov IA, Blume YB, Sozinov AA (2014) Mapping a new secalin locus on the rye 1RS arm. Cytol Genet 48(4):203–207

Kozub NA, Sozinov IA, Sobko TA, Kolyuchii VT, Kuptsov SV, Sozinov AA (2009) Variation at storage protein loci in winter bread wheat cultivars of the Central Forest-Steppe of Ukraine. Cytol Genet 43:55–62

Kozub NA, Sozinov IA, Karelov AV, Blume YB, Sozinov AA (2017) Diversity of Ukrainian winter bread wheat varieties with respect to storage protein loci and molecular markers for disease resistance genes. Cytol Genet 51(2):117–129

Krattinger SG, Lagudah ES, Spielmeyer W, Singh RP, Huerta-Espino J, McFadden H, Bossolini E, Selter LL, Keller B (2009) A putative ABC transporter confers durable resistance to multiple fungal pathogens in wheat. Science 323:1360–1363

Laemmli UK (1970) Cleavage of structural proteins during the assembly of the head of bacteriophage T4. Nature 227(5259):680–685

Lagudah ES (2011) Molecular genetics of race non-specific rust resistance in wheat. Euphytica 179:81–91

Lagudah ES, Krattinger SG, Herrera-Foessel S, Singh RP, Huerta-Espino J, Spielmeyer W, Brown-Guedira G, Selter LL, Keller B (2009) Gene-specific markers for the wheat gene *Lr34/Yr18/Pm38* which confers resistance to multiple fungal pathogens. Theor Appl Genet 119:889–898

Langridge P, Lagudah ES, Holton TA, Appels RS, Sharp PJ, Chalmers KJ (2001) Trends in genetic and genome analyses in wheat: a review. Crop Pasture Sci 52:1043–1077. https://doi.org/10.1071/AR01082

Lamari L, Strelkov SE, Yahyaoui A, Orabi J, Smith RB (2003) The identification of two new races of *Pyrenophora tritici-repentis* from the host center of diversity confirms a one-to-one relationship in tan spot of wheat. Phytopathology 93:391–396

Lemmens M, Scholz U, Berthiller F, Dall'Asta C, Koutnik A, Schuhmacher R, Adam G, Buerstmayr H, Mesterházy A, Krska R, Ruckenbauer P (2005) The ability to detoxify the mycotoxin deoxynivalenol colocalizes with a major quantitative trait locus for *Fusarium* head blight resistance in wheat. Mol Plant Microbe Interact 18:1318–1324

Limpert E, Felsenstein FG, Andrivon D (1987) Analysis of virulence in populations of wheat Mildew in Europe. J Phytopathol 120:1–8

Liu Z, Gao Y, Kim YM, Faris JD, Shelver WL, de Wit PJ, Xu SS, Friesen TL (2016) SnTox1, a *Parastagonospora nodorum* necrotrophic effector, is a dual-function protein that facilitates infection while protecting from wheat-produced chitinases. New Phytol 211(3):1052–1064. https://doi.org/10.1111/nph.13959

Manning VA, Hardison LK, Ciuffetti LM (2007) Ptr ToxA interacts with a chloroplast-localized protein. Mol Plant Microbe Interact 20(2):168–177

Martin BG, Bogdanove AJ, Sessa G (2003) Understanding the functions of plant disease resistance proteins. Annu Rev Plant Biol 54:23–61

McIntosh RA (1992) Close genetic linkage of genes conferring adult-plant resistance to leaf rust and stripe rust in wheat. Plant Pathol 41:523–527

McIntosh RA (2013) Catalogue of gene symbols. Gene Catalogue. http://www.shigen.nig.ac.jp/wheat/komugi/genes/download.jsp. MacGene

McIntosh RA, Dubcovsky J, Rogers WJ. Morris C, Appels R, Xia XC (2017) Catalogue of gene symbols for wheat: 2017 Supplement. URL:: https://shigen.nig.ac.jp/wheat/komugi/genes/macgene/supplement2017.pdf

McIntosh RA, Wellings CR, Park RF (1995) Wheat rusts: an atlas of resistance genes. Plant Breeding Institute, The University of Sydney CSIRO Australia

Meinhardt SW, Cheng W, Kwon CY, Donohue CM, Rasmussen JB (2002) Role of the Arginyl-Glycyl-Aspartic motif in the action of Ptr ToxA produced by *Pyrenophora tritici-repentis*. Plant Physiol 130:1545–1551

Metakovsky EV (1991) Gliadin allele identification in bread wheat II Catalogue of gliadin alleles in bread wheat. J Genet Breed 45:325–344

Mesterhazy A, Bartok T, Mirocha CG, Komoroczy R (2008) Nature of wheat resistance to *Fusarium* head blight and the role of deoxynivalenol for breeding. Plant Breed 118:97–110

Moreno-Sevilla B, Baenzinger PS, Peterson CJ, Graybosch RA, McVey DV (1995) The 1BL/1RS translocation: agronomic performance of F3-derived line from a winter wheat cross. Crop Sci 35 (4):1051–1055

Muthamilarasan M, Prasad MJ (2014) An overview of wheat genome sequencing and its implications for crop improvement. J Genet 93:619–622. https://doi.org/10.1007/s12041-014-0455-z

Nevo E, Payne PI (1987) Wheat storage proteins: diversity of HMW glutenin subunits in wild emmer from Israel. Theor Appl Genet 74:827–836. https://doi.org/10.1007/BF00247564

Nesic K, Ivanovic S, Nesic V (2014) Fusarial toxins: secondary metabolites of *Fusarium* fungi. Rev Environ Contam Toxicol 228:101–120

Parks R, Carbone I, Murphy JP, Marshall D, Cowger C (2008) Virulence structure of the Eastern US wheat powdery mildew population. Plant Dis 92(7):1074–1082

Pretorius ZA (2000) Detection of virulence to wheat stem rust resistance gene *Sr31* in *Puccinia graminis* f sp *tritici* in Uganda. Plant Dis 84(2):203

Rabinovich SV (1998) Importance of wheat-rye translocations for breeding modern cultivars of *Triticum aestivum* L. Euphytica 100:323–340

Rasheed A, Wen W, Gao F, Zhai S, Jin H, Liu J, Guo Q, Zhang Y, Dreisigacker S, Xia X, He Z (2016) Development and validation of KASP assays for genes underpinning key economic traits in bread wheat. Theor Appl Genet 129:1843. https://doi.org/10.1007/s00122-016-2743-x

Ruoslahti E (1996) RGD and other recognition sequences for integrins. Annu Rev Cell Dev Biol 12:697–715

Sarma GN, Manning VA, Ciuffetti LM, Karplus PA (2005) Structure of PtrToxA: an RGD-Containing host-selective toxin from *Pyrenophora tritici-repentis*. Plant Cell 17:3190–3202

Sears ER (1969) Wheat cytogenetics. Annu Rev Genet 3:451–468

Sebesta EE, Wood EA (1978) Transfer of greenbug resistance from rye to wheat with X-rays Agron Abstr. American Society of Agronomy, Madison, WI, pp 61–62

Sebesta EE, Wood EA, Porter DR, Webster JA, Smith EL (1995) Registration of amigo wheat germplasm resistant to greenbug. Crop Sci 35:293. https://doi.org/10.2135/cropsci1995001l183X003500010074x

Schlegel R (2016) Current list of wheats with rye and alien introgression. V05–16: 1–18. http://wwwrye-gene-map.de/rye-introgression

Schweiger W, Steiner B, Ametz C, Siegwart G, Wiesenberger G, Berthiller F, Lemmens M, Jia H, Adam G, Muehlbauer GJ, Kreil DP, Buerstmayr H (2013) Transcriptomic characterization of two major *Fusarium* resistance quantitative loci (QTLs), *Fhb1* and *Qfhs.ifa-5A*, identifies novel candidate genes. Mol Plant Pathol 14(8):772–785

Shabeer A, Bockus WW (1988) Tan spot effects on yield and yield components relative to growth stage in winter wheat. Plant Dis 72:599–602

Singh RP (1992) Genetic association of leaf rust resistance gene *Lr34* with adult plant resistance to stripe rust in bread wheat. Phytopathology 82:835–838

Singh RP (1993) Genetic association of gene *Bdv1* for tolerance to barley yellow dwarf virus with genes *Lr34* and *Yr18* for adult plant resistance to rusts in bread wheat. Plant Dis 77:1103–1106

Singh RP, Hodson DP, Jin Y, Lagudah ES, Ayliffe MA, Bhavani S, Rouse MN, Pretorius ZA, Szabo LJ, Huerta-Espino J, Basnet BR, Lan C, Hovmøller MS (2015) Emergence and spread of new races of wheat stem rust fungus: continued threat to food security and prospects of genetic control. Phytopathology 105:872–884

Sobko TA, Poperelya FA (1986) The frequency of alleles of gliadin-coding loci in different cultivars of winter bread wheat. Visnik Silskogospodarskoi Nauki 5:84–87 (in Ukrainian)

Sozinov AA, Poperelya FA (1980) Genetic classification of prolamines and its use for plant breeding. Annales de Technologie Agricole 29:229–245

Tai Y-S, Bragg J, Meinhardt SW (2007) Functional characterization of ToxA and molecular identification of its intracellular targeting protein in wheat. Am J Plant Physiol 2:76–89

Tang C, Xu Q, Zhao M, Wang X, Kang Z (2018) Understanding the lifestyles and pathogenicity mechanisms of obligate biotrophic fungi in wheat: the emerging genomics era. Crop J 6 (1):60–67

Tanksley SD, Young ND, Paterson AH, Bonierbale MW (1989) RFLP mapping in plant breeding: new tools for an old science. Bio/Technology 7:257–264. https://doi.org/10.1038/nbt0389-257

Tian J, Chen J, Chen G, Wu P, Zhang H, Zhao Y (2015) Genetic analyses of wheat and molecular marker. Assisted Breeding Volume 2. Springer Press, The Netherlands 1st ed, XXVII, 321 p

Villareal RL, Rajaram S, MuJeeb-Kazi A, Del-Toro E (1991) The effect of chromosome 1B/1R translocation on the yield potential of certain spring wheats (*Triticum aestivum* L). Plant Breed 106:77–81

Xynias I?, Kozub NO, Sozinov IA (2006) Seed storage protein composition of Hellenic bread wheat cultivars. Plant Breed 125:408–410

Yadav PS, Mishra VK, Arun B, Chand R, Vishwakarma MK, Vasistha NK, Mishra AN, Kalappanavar IK, Joshi AK (2015) Enhanced resistance in wheat against stem rust achieved by marker assisted backcrossing involving three independent *Sr* genes. Curr Plant Biol 2:25–33

Yates F (1934) Contingency table involving small numbers and the χ^2 test. Suppl J R Stat Soc 1 (2):217–235

Zhang Y, Lubberstedt T, Xu M (2013) The genetic and molecular basis of plant resistance to pathogens. J Genet Genomics 40:23–35

Proposed OLEA Management System with Farming Monitoring Processes for Virgin Olive Oil Production Traceability and Assessment

Dimitra Papaefthimiou, Sotirios Kontogiannis, Angelos Chatzimparmpas, George Kokkonis, and Stavros Valsamidis

Abstract This paper proposes a cloud application architecture called OLEA, for monitoring the olive oil production chain. OLEA system deployment follows a divide-and-conquer management logic, which maintains olive tree clusters. On each cluster, NFC technology is used for monitoring plant protection practices and fertilization. Apart from on-site monitoring services, the system is also equipped with virgin oil management services. It uses an OLEA system controller that interconnects with sensors on oil mills, for the procurement of quantitative and qualitative olive oil characteristics, during the industrial extraction process. OLEA system services and management algorithms are controlled by a cloud application server, where collected data uploads and notifications are sent to the end users using a mobile phone application. This paper presents the OLEA system technical characteristics as well as the structure of OLEA communication protocols. Furthermore, a case study of the OLEA system data mining capabilities is presented examining the application of such efforts to the improvement of systematic cultivation, branding, and product exports.

D. Papaefthimiou
Department of Biological Applications and Technology, University of Ioannina, Ioannina, Greece

S. Kontogiannis (✉)
Department of Mathematics, Laboratory of Distributed Micro-Computers, University of Ioannina, Ioannina, Greece
e-mail: skontog@cc.uoi.gr

A. Chatzimparmpas
Department of Computer Science and Media Technology, Linnaeus University, Växjö, Sweden

G. Kokkonis
Department of Business Administration, TEI of Western Macedonia, Grevena, Greece

S. Valsamidis
Department of Accounting and Finance, TEI of East Macedonia and Thrace, Kavala, Greece

A. Theodoridis et al. (eds.), *Innovative Approaches and Applications for Sustainable Rural Development*, Springer Earth System Sciences,
https://doi.org/10.1007/978-3-030-02312-6_19

Keywords *Olea europaea* · Olive tree cultivation processes · Virgin oil characteristics · Olive tree management system · Virgin olive oil monitoring · Olive tree traceability · NFC technology · IoT applications · Data mining algorithms

Abbreviations and Acronyms

ACK	Acknowledgement
AS	Application server
EPROM	Erasable programmable read-only memory
CoAP	Constrained Application Protocol
EU CAP	European Union Common Agricultural Policy
EVOO	Extra-virgin olive oil
HTTP	HyperText Transfer Protocol
GIS	Geographic information system
GPIO	General-purpose input/output
I2C	Inter-integrated circuit (I-squared-C)
IOC	International Olive Council
IoT	Internet of things
MCU	Microcomputer unit
Md	Median of defect
Mf	Median of fruity
MQTT	Message Queuing Telemetry Transport
NDEF	NFC data exchange format
NFC	Near-field communication
OPEKEPE	Greek Payment Authority of Common Agricultural Policy Aid
PWM	Pulse width modulation
ReST	Representational State Transfer
RFID	Radio-frequency identification
SPI	Serial Peripheral Interface
UDP	User Datagram Protocol
URI	Uniform Resource Identifier

1 Introduction

Olive oil is a standard food source of the Mediterranean diet which has been correlated with a low frequency of cardiovascular diseases, neurological disorders, breast and colon cancers, as well as with antioxidant properties. It is consumed worldwide due to its highly nutritious composition, including its fatty acid structure and other functional food components, such as polyphenols (Vekiari et al. 2010).

Focusing on the Greek virgin oil industry, more than 80% of the yearly production of olive oil in Greece is extra-virgin oil, and 90% of the Greek olive oil is exported to European countries (Bettini and Sloop 2014; Mylonas 2015). Olive farms in Greece extent mainly to the areas of Peloponnese, Crete, and the Aegean and the Ionian Islands, where up to 80% of Greece's olive oil is produced. Nevertheless, while an overall 84% worldwide production of olive oil is achieved in Europe, Spain remains the largest producer in the world, followed by Italy and then Greece.

According to the 2015 World's reports for the ten best olive oil brands (WBOOR 2015), Greek Messenia's branded "Lia" oil is placed at the 9th position, after Italian, Spanish, and US brands. While the USA produces only 2% of virgin olive oil worldwide, it has managed to preserve the highest number of individual olive oil brands in the global market after Spain, due to its production monitoring and marketing processes and infrastructure.

To better understand the conditions that make Greek olive oil less competitive in the global market, we should review the practices used throughout production. More than 400 olive oil pressing companies, coalitions, and partnerships are located in Greece. Regarding virgin oil extraction, most olive decanter mills in Greece are of small capacity and not as technologically advanced as those in Spain. Traditional grindstone removing processes are still used and preferred from total olive oil pressing techniques. This of course leads to manual operations and higher milling costs, while modern decanter systems decrease oil leakage from the traditional milling process up to 15–20%. A few more modernized olive mills located in the area of Peloponnese use the two-phase decanter centrifugation technique for oil extraction. These systems are much more efficient extraction machines for producing olive oil of higher volume and comparable quality to that of the old-fashioned grindstone milling process (Niaounakis and Halvadakis 2006). Other methods such as the three-phase decanter, Sinolea method, or cold extraction methods are not used (Niaounakis and Halvadakis 2006).

In Italy despite the small milling capacities, the production process is normalized and monitored with the use of information and communication technology monitoring systems, well-established distribution and marketing processes. Management and monitoring methods used in Italy helped its virgin oil industry bypass obstacles like the disseminated structure of olive oil corporations and to achieve esteemed quality olive oil branded products that match exportation requirements. However, the overall practices used by the Greek virgin oil industry, such as the fragmented and technologically outdated coalitions' mechanisms regarding farming, standardization and product quality control further increases the risks of inability to introduce to the market already premium quality olive oil (Mylonas 2015).

Farming normalization and pesticides control on an olive tree area are the two major farming factors of success. However, such control needs to cope with crop rotation methods, olive compost or green manure use, and biological pest control. Such techniques are essential for the well-known olive oil organic cultivation. Organic cultivation prohibits or rigorously restricts the use of manufactured fertilizers, pesticides (which include herbicides, insecticides, and fungicides), plant growth regulators such as hormones, and genetic modifications (Camarsa et al. 2010; Ehaliotis et al. 2011). Nowadays, there are a small number of farmers in Greece that are shifting cultivation practices to organic production following EU directives. However, cultivation policy remains unsupervised due to the lack of modern farming techniques, cultivation monitoring systems, and farmers' educational training.

Despite the immersive difficulties of monitoring and marketing olive oil products in Greece, the future of the Greek olive oil economy looks encouraging due to the globally increasing needs for virgin oil products. These demands are expected to lead to increased Greek exports by 2020 up to 40–50% more than in 2015 (210,000 tons – 2015) (International Olive council World 2012; European Commission Agricultural Market Briefs 2012). Even if Greek reports from authorities are pessimistic toward the rise of Greek olive oil production, due to progressive reduction of EU CAP grants and the distributed small-scale production chain, these are obstacles that can be confronted with the use of production monitoring and traceability technologies. Furthermore, following the Italian industrial model and incorporating new technologies shall improve merchandise exportation which is already of high-quality standards (European Commission Agricultural Market Briefs 2012).

The organization of the remaining paper is as follows: Sect. 2 presents Greek virgin olive oil attributes and classification. Section 3 presents the related work of existing olive oil quality monitoring systems. Section 4 presents the proposed OLEA system high-level architecture; Sect. 1 focuses on knowledge mining capabilities of the OLEA system, presents a new clustering algorithm, and includes a proof of concept study for virgin oil merging and branding. Finally, Sect. 6 concludes this paper.

2 Virgin Olive Oil Processes, Guidelines, Classification, and EU Legislation

Farming, cultivation, and harvesting of olive trees are the key steps that will eventually affect the final yield of olives and olive oil and their quality characteristics such as the fatty acids profile, as well as color and flavor (Aparicio and Harwood 2013; Vossen 2013). The standards of the International Olive Council (IOC) (COI/T.15/NC No 3/Rev.12 and COI/OH/DOC.1) provide a guideline for farmers and laboratories involved in the processes of farming, cultivating, and harvesting of olives designated for oil production (oil olives). This way, standards are set for taking all possible measures to prevent any type of contamination through the farming and production processes. The next steps in olive oil production concern extraction and processing leading to the final oil product, ready for consumption. According to COI/T.15/NC No 3/Rev.12, virgin olive oils are defined by their direct mechanical or other physical extraction from the fruit of the olive tree (*Olea europaea* L.). Therefore, the only acceptable processing steps include washing, decantation, centrifugation and filtration. The harvested olives are first washed and then crushed to break their tissues and release oil droplets from their containing mesocarp cell vacuoles into the olive paste (Aparicio and Harwood 2013). This process is important for the establishment of the eventual phenolic profile of the particular virgin oil. Before further processing, malaxation (mixing) of the olive paste is used to enable higher oil yields. Extraction of oil from other paste constituents is performed by either continuous pressing in stone mills, percolation with steel blades, or a centrifugation system employing two- and three-phase decanters. A final step of filtration may also be included before virgin olive oil is ready for storage and bottling.

The quality of the resulting virgin olive oil will depend on both the quality of the olives as well as all the processing steps described above. Virgin olive oils are further distinguished as the top-grade "extra-virgin" and the "ordinary virgin" oils, according to specific quality characteristics. In addition, since olive oil quality is prompt to deterioration during all stages of processing and storage, product monitoring is of major importance. Quality of olive oils is therefore determined by sensory evaluation (organoleptic tests) and the measurement of certain physico-chemical characteristics established by international legislations. Acceptable quality levels are mandated by the FAO/WHO Codex Alimentarius Commission (CODEX STAN 33-1981 revisions and amendments) and other international committee legislations and are implemented by the International Olive Council (IOC), the intergovernmental organization that administers the provisions of the UN's International Agreement on Olive Oil and Table Olives (TD/OLIVE OIL.11/5) to 98% of the world's olive and olive oil producers. In Greece, the legislations for olive oil compliant with international legislations are available at the national level by the Food and Drink Code (Chap. 8) issued by the Greek State General Chemical Laboratory.

Efficient separation of the different olive oil types is necessary in order to ensure their authenticity and quality. In this respect, standard values have been established for both physical and chemical characteristics, as well as other control parameters specific for each oil category including the organoleptic characteristics of virgin oils (Commission Delegated Regulation (EU) 2015/1830, IOC's COI/T.15/NC No 3/Rev. 12).

Organoleptic classification of extra-virgin and virgin olive oils is based on the median of any defect (Md) and median of fruity attributes (Mf) and is according to the following:

- Extra-virgin olive oil, $Md = 0$ and $Mf > 0$
- Virgin olive oil, $0 < Md < 3.5$ and $Mf > 0$

In addition, good-quality extra-virgin and virgin oils should have yellow to greenish color and no viscosity at 20 °C.

The most important physical and chemical characteristics distinguishing extra-virgin and virgin oils according to the above international legislations are listed in Table 1.

According to the Greek association of industries and processors of olive oil (Sevitel), Greece comes third in global olive production, accounting for over

Table 1 Physicochemical characteristics for extra-virgin and virgin olive oils

Attribute	Analysis (concentration)	Standards	
		Extra-virgin olive oil	*Virgin olive oil*
Oleic acid acidity	Oleic acid (g/100 gr)	≤0.8	<2.0
Peroxide index	Peroxide oxygen (mEq O_2/kg)	<20	<20
Moisture and volatile matter	Content (% m/m)	<0.2	<0.2
UV absorbance (K 1%)	270 nm (cyclohexane)/268 nm (isooctane)	<0.22	<0.25
	ΔK (K268-[(K262+K274)/2])	<0.01	<0.01
	232 nm	<2.50	<2,60
Insoluble impurities in light petroleum	Total content (m/m)	<0.1	<0.1
Trace metals	Iron (mg/kg)	<3.0	<3.0
	Copper (mg/kg)	≤0.1	≤0.1
Fatty acid esters	Fatty acid methyl esters FAMEs)	∑(FAME+FAEE) <75 mg/kg (2012/2013 crop year)	–
	Fatty acid ethyl esters (FAEEs)	≤35 mg/kg	–
Biophenol content	Determined according to COI/T.20/Doc. No 29		
Waxes	Total content (mg/kg)	C42 + C44 + C46 ≤ 150	C42+C44+C46 ≤ 150
Stigmastadienes	Total content (mg/kg)	≤0.05	≤0.05
Total sterols	Total content (mg/kg)	≥1 000	≥1 000
Total trans-oleic isomers	Total content (%)	≤0.05	≤0.05

170 million trees and producing an annual average of more than 300,000 tons olive oil, with over 80% of it being extra-virgin. Greece is also the largest global exporter of extra-virgin olive oil to destinations mostly covering European Union countries, where Italy imports about three-quarters of total Greek olive oil exports. In general, 30 Greek branded olive oils have officially been recognized and regulated under EU regulations, including 19 with protected designation of origin (PDO) and 11 with protected geographical indication (PGI) designations. The basic olive species cultivated for virgin oil production in Greece (Niaounakis and Halvadakis 2006; European Commission Agricultural Market Briefs 2012) are the following: (a) Koroneiki (*Olea europaea* var. *microcarpa alba*), though it's a small fruit variety, represents 60% of Greece's production. Its seed weight is 0.3–1 g and height 12–15 mm. Olive oil is of high quality and has a clear greenish-yellow color, fruity taste, and excellent fruit aroma; (b) Athinolia (*Olea europaea* var. *Mamillaris sub.*

minima) is an olive variety that produces low-acidity oil and has medium size, 2.2–2.9 g seed weight, and height of 7.5–25 mm. It matures slowly and has intense aromatic and fruity characteristics as well as really high phenolic concentration; (c) Tsounati (Mastolia, Mouratolia) (*Olea europaea* var. *Mamillaris*) covers almost 15% of Greece's production variety, with seed weight of 1.2 g and height of 10–16 mm. It gives high oil quantity with fine and amber color and full bitter and spicy taste; (d) Kalamon-Chalkidikis (ChondroElia) has large fruit, smooth and meaty in flavor, covers about 1–2% of Greece's production, and is used mainly as table or paste oils; and (e) Manaki (Lianolia, Koutsourelia) has small- to medium-sized fruit, which matures slowly and produces fine olive oil with a rich sweet and fruity taste and smooth texture. It constitutes 10% of Greece's production and is considered as a high-altitude variety (acidity and oil quantity).

3 Olive Oil Monitoring Systems

Currently, the most common approach for olive oil qualitative characteristics monitoring is still chemical analysis of collected samples. Such offline sampling requires expensive infrastructure, such as proper electrical balances, automatic particle counters, gas and liquid chromatographs, and highly educated human personnel. In addition, significant limitations of such processes are the delay or even prevention of timely failure diagnosis and application of countermeasures.

Cannazza et al. (2011) suggested the use of a customized system for quality control based on dielectric spectroscopy analysis. The ultimate purpose of this study was to determine a methodology for an alternative analysis practice that would be able to differentiate virgin oils, to certify oil quality and origin, and to detect any presence of adulterants. They proposed a system with low-frequency and high-frequency measurements performed through an LCR meter and time-domain UV refractometer, providing an examination table of the behavioral range of edible oils. The adoption of a single, specifically-designed probe (used for all the measurements), along with the possibility of using portable instrumentation, makes the proposed system an appealing candidate for the realization of a real-time, on-site system for quality monitoring of virgin oils production chain. Toward that direction, Taouil et al. (2008) proposed a video monitoring system for quality control in the olive oil production chain. The video monitoring system can be used for achieving real-time color constancy sampling.

Mailer and Beckingham (2006) recommended a sensory (organoleptic) testing system, suggesting that the most crucial test to guarantee that the oil is tolerable for consumption is sensory quality. Aroma, flavor, pungency, and bitterness are the organic parts that the proposed olive oil methodology evaluates using evaluation matrices and can be performed either manually or with real-time sensors.

However, according to the International Olive Council (IOC) standards, extra-virgin olive oil, for example, should have some fruitiness and should not have any form of deficiencies. Therefore, to achieve proper classification, more attributes and metrics should be included in the sensory quality control proposed by Mailer and Beckingham (2006). Furthermore, an intermediate control system should be developed that can easily accomplish different types of sensors interconnection. Additionally, past knowledge of regional virgin olive oil varieties attributes and qualitative response should also be available to a production chain management system. Such system may take into account the accumulated knowledge and have some features like the ability to identify oil origin, farming discrepancies, provide suggestions to farmers, detect merchandise production chain flaws, and offer statistical trends. At a more advanced level, such system could be capable of virgin oil classification to serve particular market needs and the capability to evaluate the combination of oils per farmer's contribution that would accurately meet consumers' specifications.

Gao et al. (2004) suggested a multiple sensors system with centralized control. Their target was to analyze and improve the automation of commodity operations. As a result, a single-sensor system was estimated to be a worse approach compared to multiple-sensor real-time controlled system for the process of virgin oil characterization. Reliability of multiple sensors systems could be explained by the sensitivity and granularity ability of each sensor to distinguish flaws and perform measurements. Therefore, virgin oil characteristics that are being collected by the system can be observed and calculated online by single-function sensors. The proposed oil measurement system could be divided into three main parts: (i) multiple sensors for temperature, pressure, moisture, and viscosity; (ii) signal acquisition channels with signal amplification, a diverter switch for digital-analog signals and an A/D converter; and (iii) finally the analysis and processing (control unit) made up of an industrial processing unit.

Regarding farming monitoring processes, Kiama et al. (2014) proposed a low-cost RFID-based palm monitoring system. In their work, a passive RFID tag identified every palm oil tree in the plantation area, and their fresh fruit bunch (FFB) production was monitored by scanning the passive RFID tags using RFID handheld scanners. The technology behind that proposed method enables to control harvesting information and statistical harvesting reports per area. The overall system architecture can be split into three main components which are the RFID scanner-controller, the application server, and the GIS.

Finally, an integrated and advanced monitoring method was suggested by Kontogiannis et al. (2016) for the sheep industry, implementing a new system architecture and test-bed application. The Sheep Manager system uses cheap NFC (near-field communication) technology and custom NFC data exchange format records for the identification and attribute recording of sheep inside a flock. It also utilizes sensors for real-time measurements and recordings of raw milk extraction per ewe. All recorded information are then stored into an application server. Following their previous proposal for the sheep industry, the authors modified their previous approach to better suit the virgin oil industry. That is, the use of novel system architecture called OLEA is presented at Sect. 3.

4 OLEA System High-Level Architecture

The authors propose an olive oil monitoring and quality assurance methodology called OLEA framework, supported by a test-bed system called OLEA system. OLEA system methodology includes the OLEA clustering algorithm and the steps that the OLEA system implements, while OLEA system architecture is presented in Fig. 1. In the following subsections, the OLEA system service functionality is outlined, and technical implementation details are presented.

4.1 OLEA Services and Service Functionalities

The proposed OLEA system includes the following capabilities as denoted by system services:

1. **Traceability service**: OLEA traceability service divides cultivation regions following two distinctive layer processes:

   Layer 1-Microclimate unification areas: Each olive field is a separate system entity described by the spatial geographical location GIS system maintained by OPEKEPE (Greece EU CAP funding organization) (OPEKEPE 2017). Then each

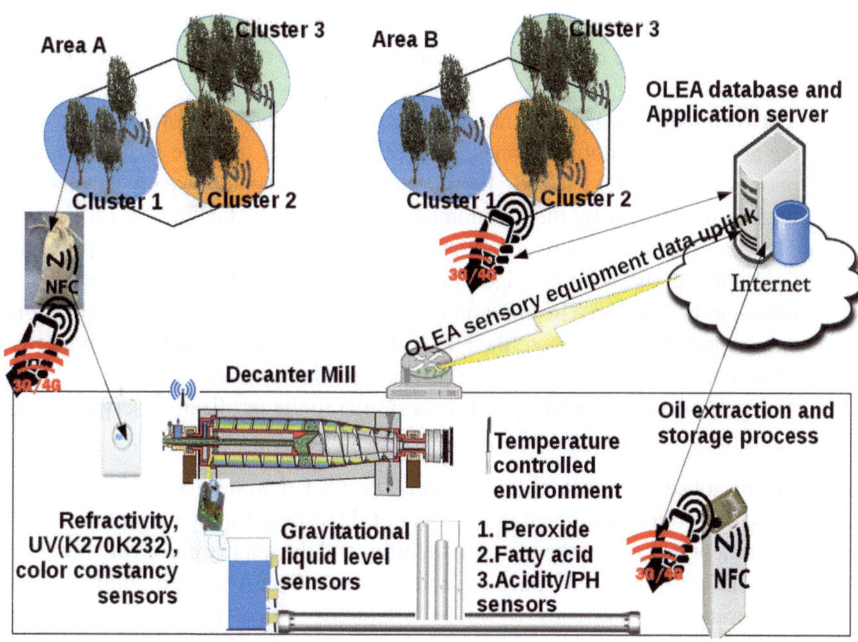

Fig. 1 OLEA high-level system architecture

field is grouped by the OLEA system into areas (see Fig. 1) with each area pertaining similar microclimate characteristics (Theodosiou et al. 2012).

Layer 2-Tree clustered sections: The process covers olive tree traceability functionality, presented in Fig. 1 as rounded areas of tree clusters. Such form of traceability is achieved with the use of passive NFC tags placed each tree cluster (each farmer claims for OPEKEPE funding). Apart from ID traces, these tags are also capable of recording yearly fertilization and pesticide activities (date of appliance, active ingredient, agronomist approval for organic, integrated, or conventional cultivation) following an NDEF message formulation similar to the one presented by Kontogiannis et al. (2016). From 2015 and on, the use of pesticides requires agronomist prescription and farmer certification; thus the recording of olive trees needs to be performed in a more controlled way (Ministry of Agriculture 2015).

2. **Olive tree monitoring services:** The OLEA system olive tree monitoring services include three basic services: The cluster identification service, the cluster soil fertilization service, and the pest control service. Information related to these services are exchanged and updated per tree cluster with the use of a single NFC tag mounted into one tree of the tree cluster accordingly.

   - **The identification service** includes recorded olive tree field location characteristics defined and recorded by the OLEA traceability service at the NFC tags into an NDEF identification record. The NDEF record fields include location characteristics acquired by the protected designation of origin (PDO), protected geographical indication (PGI) or protected origin production names or protected origin names of highest quality (OPAP), olive field microclimate area characteristics, and agronomist reports of cultivation type are recorded.
   - **The fertilization service** includes NFC message exchange information between the agronomists' mobile phones, the tree cluster NFC tag, the farmers' mobile phones, and the OLEA AS. Fertilization service includes yearly fertilization records initially set by the agronomist into each cluster tag and the OLEA AS. These records are then read by the farmer's OLEA mobile phone application, field applied and acknowledged in each cluster tag (NDEF fertilization ACK message) per cluster. The agronomist then with the use of OLEA mobile phone application reads once more the NFC information and when Internet connectivity is available updates the AS fertilization service monitoring manager.
   - **The pest control service** operation is similar to the fertilization service, and it monitors pesticides' use per tree cluster updating the OLEA AS.

3. **Olive oil sensory monitoring service:** The OLEA system includes sensory-attached controller equipment (shown in Fig. 1) installed at the oil presses for the measurement process of the olive oil quantity and the quality clustering based on metrics. Due to its special purpose and function, OLEA system controller and controller-sensor interfaces could be nationally patented as well. With the OLEA oil sensory monitoring service and real-time sensory system, olive oil quantity and quality characteristics of mill-extracted olive oil per cluster are kept at the OLEA AS.

In order to record the extracted olive oil quantity and chemical consistency (quality attributes) per tree cluster, the proposed OLEA framework includes a multisensory equipment placed similarly to the ones proposed by Scarafia (Scarafia 2011), in the olive fruit decanter equipment (see Fig. 1). This multisensory equipment includes sensors such as (1) gravity liquid level sensors for monitoring produced oil quantities, (2) refractivity UV sensors for K232 and K270 measurements of UV absorbance, (3) color sensors for monitoring color constancy, and (4) peroxide and acidity sensors (see Fig. 1). These measurements are then uploaded in real time to the OLEA application service. The capture attributes from this automatic sensory process are the following: Oxide rancidity protection, additives, and refinement process metrics are not taken into account by the authors, since refinement, if it occurs, is a process that follows the oil chemical characterization. Oil refinement is done by using chemicals that are harmful to humans. This means that the oil is treated with acid, or purified with an alkali, or bleached. It can also be neutralized, filtered, or deodorized. This paper focuses only on virgin oils characterization where no intermediate refinement or additive processes are allowed.

4. **Olive oil management smart service engine and modules**

   Logging and reporting: The OLEA database shall record olive oil production quantity and quality attributes connected to cultivation characteristics of tree cluster areas. The database shall also provide information for tree clusters regarding quantities produced, their quality, and market prices. Such information attained nationwide could be proven very helpful for the development of the oil industry and therefore stimulation of the national economy.

   Data analysis algorithms: The OLEA management service is capable of statistical information processing and visualization, regarding production characteristics for each farmer. The OLEA system database collects recorded data from each farmer regarding visualization of statistical trends as well as data analysis and knowledge extraction. An example of how the OLEA system can be used for product management and branding of uniform products is the olive oil clustering algorithm proposed by the authors and presented at Sect. 3.3. The implementation of data analysis and more specifically data mining algorithms in the OLEA system in the form of scavenging agents formulates the systems' future capabilities of suggestive services, quotation reports, and prediction services.

4.2 OLEA System NFC Traceability Fertilization and Pest Control Services

The proposed OLEA system uses the NFC technology for the purpose of tree field identification and assessment. In depth, it uses appropriate NFC NDEF identification messages for each tree cluster monitored by the system (OLEA application server – OLEA AS). It also makes use of appropriate NDEF messages placed into tree clusters' NFC tags for monitoring fertilization and pest control.

To achieve the abovementioned services, appropriate NFC protocol and NDEF message records have been designed and used (NFC Forum NDEF 2006). The authors propose NFC because the technology is accessible in the majority of mobile phones, PDAs, PCs, and tablets (Kontogiannis et al. 2016). NFC tags used are passively activated and have 4Kb of storage EPROM. Each tag storage area is divided into sectors (40), where each sector can be configured with different access rights. Each sector includes blocks. The first 32 sectors include 4 blocks each, and each block contains 64 bytes of data. The last 8 sectors include 16 blocks/sector (256 bytes per sector). OLEA system uses type 3 (FeliCa) and type 4 tags (SONY company 2014). Tag sectors store one NDEF message, and the NDEF message includes multiple NDEF records (Kontogiannis et al. 2016). OLEA NFC tags include a unique read-only identification number at block 0. FeliCa Lite are the NFC tags used by the OLEA system, which are 5 cm in diameter and circular (SONY company 2014). In general NFC tag size extends from 3 cm in diameter down to 5 mm. Each OLEA tag is mounted up to the central tree of every tree cluster and includes cluster information as well as fertilization and pest control NDEF records. In the following subsections, the OLEA services of identification, fertilization, and pest control that use the near-field communication (NFC) technology and OLEA NFC data exchange format (NDEF) message structures are presented.

OLEA NFC Tag-Based Services

NFC is a specialized subset of the family of RFID (radio-frequency identification) technology. Specifically, NFC is a branch of high-frequency (HF) RFID, and both operate at the frequency of 13.56 MHz. AES (advanced encryption standard) for NFC is mainly used to encode data for this technology and implemented to the OLEA system.

The OLEA NFC tag type categories are the following: (a) Tag 1 type that is based on the ISO14443A standard and has a communication speed of 106 kbit/s. These tags have read and rewrite capabilities, while the users can modify the tag to be read-only. They include 96 bytes available memory, which is more than enough to save a website's URL or few other data. Nevertheless, memory capacity is expandable up to 2 Kbyte. (b) Tag 2 type is based on the same standard as the previous category and with the same communication speed. These NFC tags also have the same capabilities with tag 1 type and a memory size of only 48 bytes, although this can also be extended to 2 Kbyte. (c) Tag 3 type is based on the Sony FeliCa system and has an integral 2 Kbyte memory capacity with data communication speed of 212 kbit/s. (d) Tag 4 type is compatible with ISO14443 A and B standards. These NFC tags are pre-configured at the production level to either be read/read-write or read-only. Furthermore, their memory capacity can be up to 32 Kbytes, and the communication speeds vary from 106 kbit/s to 424 kbit/s.

NFC data exchange format (NDEF) is a binary structured message format, which may include several types of records and can be used to exchange specific application-determined payloads (in form of records) between NFC-capable mobile phone devices and NFC tags. Every NDEF message includes per-record header that holds the metadata about the encapsulated NDEF records, such as the type, length, and payload. NDEF messages start with a type name format (TNF) that shows the NDEF record structure types. A well-known NFC record-type name format (TNF = 0x01 - NFC Forum RTD (2006)) is used which includes several pre-defined record-type formats, such as URI and smart poster, and finally the text record-type format, which the OLEA system uses to store its record types to the NFC tag.

Each OLEA NDEF record is of record-type text ("T"-0x54), with no ID field set (record IL = 0). Moreover each NDEF record used by the OLEA system has the short record bit set (SR = 1). This means that the payload length field is 1 byte long, limiting the per-record maximum size to 255 bytes. The OLEA NDEF record structure is presented in Fig. 2.

In Fig. 2, the ME (message end) flag indicates if this is the last record in the message, and the MB (message begin) flag indicates if this is the start of the NDEF message. The IL flag points out if the ID length field (IL field) is present or not. In the case of OLEA, this is set to 0 since the IL field is omitted. The SR (short 1 byte record) flag is set to 1, and the OLEA record payload length field is 1 byte (payload less than 255 bytes long). This permits for more compressed records. If the SR field

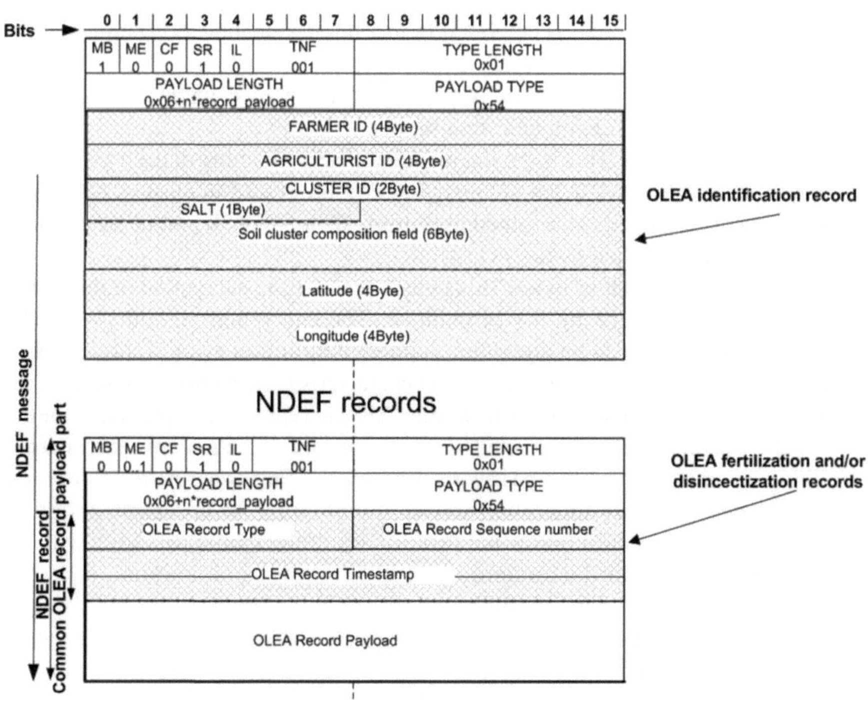

Fig. 2 OLEA NDEF records structures for identification, fertilization and pest control

is zero, then the payload length value would be 4 bytes. The CF (chunk flag) is also set to 0 in OLEA records, since record chunks are not used.

OLEA NDEF Tree Field Identification Message and Functionality

The OLEA system NFC traceability mechanism functions as follows:

Initially, the agronomist responsible for each olive tree cluster places the NFC tags (one into each clustered field) and initializes them by passing the NDEF identification message values, using the OLEA mobile phone application. The OLEA mobile phone application connects to the OLEA application server (OLEA AS), and the agronomist can download the identification messages for each cluster prior to per cluster NFC tree placement and NFC identification message write operation.

The identification record is the first record of the NFC tag NDEF message, set by the agronomist, and the NFC tag is placed in the central tree of each cluster in a visible spot on the tree trunk. The farmer cannot modify these attributes of the NDEF message due to the read-only permission set for the recorded data. The identification record message is a text-type NDEF message that contains the following text-encoded attributes, illustrated in Fig. 2:

- Farmer ID (4 bytes). This is the 32-bit farmer unique identification number set by the OLEA application service.
- Agronomist ID (4 bytes). This is the 32-bit agronomist unique identification number set by the OLEA application service.
- Cluster ID (2 bytes). This is the unique identification number of each tree cluster and is of 16 bits long. It can uniquely identify up to 65,536 clusters per farmer ID. If a cluster of trees is assumed as *K*-trees, then 65536*k unique trees can be included and identified by their clusters.
- Secret passphrase-salt (1 byte). This byte is used for the encryption of the cluster identification message by the agronomist. The encryption algorithm used for payload is an AES 128-bit algorithm, supported by almost every Android mobile phone. With this approach a cluster's ID characteristics can only be written by the agronomist and be read-only to the farmer. The encryption-decryption key for this 3DES operation is constructed by the binary addition of the Farmer ID/Agronomist ID and the password used by the farmer/agronomist to access the manager and user application service, followed by the concatenation of the XOR operation on each one of the first 3 bytes of the farmer password with the passphrase byte padded with zeros.
- Field soil chemical composition based on the latest chemical analysis can be found encoded in the identification tag. That is, the following fields are kept at the OLEA application server database for each cluster: (1) soil pH that has a range of values from 6 to 9 (4-bit encoded granularity of 0.2), (2) ground drainage that can be measured from 0 (bad) to 4 (excellent) (2 bits), (3) salinity value that starts from 0 (bad) to 4 (very high) (2 bits), (4) B value greater than 0.3 ppm with 6 ppm

as the maximum limit (8-bit value), (5) P value from 15 to 30 ppm (8-bit value), (6) K value from 0.5 to 1 meg/100 g (8-bit value), (7) Ca value from 6 to 12 meg/100 g (8-bit value), and (8) Mn value from 1 to 3 meg/100 g (8-bit value). The stored data length for total soil composition of the field is 6 byte.
- Latitude and longitude information of the cluster position expressed as integer fields (8 byte).

The minimum data size for an OLEA NDEF message is 29 bytes (only identification record present), and the size of an NDEF message with one fertilization and pest control records is around 64 bytes. The fertilization and pest control records have similar payload fields such as the record type (0x01 for fertilization and 0x02 for pest control), a sequence number field (starting from value 1), and a 4-byte timestamp field as illustrated in Fig. 2. A detailed description of these records follows.

OLEA NDEF Tree Field Fertilization and Pest Control Service Message and Functionality

The OLEA system NFC fertilization and pest control monitoring functionality is as follows: (a) First OLEA AS records are updated by the agronomist responsible for each cultivation area cluster, with the use of the OLEA mobile phone application. (b) Then the OLEA notification service notifies the farmer. For that purpose, same-type NDEF records are used: one for fertilization and one for plant protection processes. At this stage, it is important to point out that the agronomist can write additional NFC tag comments and prescription notes to the farmer in each cluster, if the agronomist thinks that additional information is required. (c) The farmer receives the notification and on the cluster field acknowledges the task by setting the ACK record bit field to 1, from the OLEA application on mobile phone to the corresponding NDEF record at the NFC tag (ACK bit NDEF payload). (d) Finally, the agronomist would later monitor the tags to check which tasks have been performed-acknowledged by the farmer.

The pest control record extended payload includes the following fields (Fig. 2):

- Prescription ID (2 bytes field).
- Pesticide dosage (g/cluster): This is a 2-byte field indicating the amount of fertilizer dose for each cluster.
- Chemical sprayed ID: This is used to measure and analyze the ground and is 2 bytes long.
- Starting date of pesticide appliance (4 bytes).
- Closing date of appliance (4 bytes). Beyond this date, a new prescription is required by the agronomist.

Comments or notes: 200 bytes can be used for comments or messages if the agronomist wants to inform the farmer. For the fertilization treatment of modern olive tree fields, thorough studies have been performed for establishing the need of specific chemical compounds and elements per average olive tree. Olive trees respond spectacularly to nitrogen fertilization, especially in soils of moderate fertility. Boron is an important element in the slightly sandy, acidic, and/or calcareous soils. In general, the average nutrient needs per tree are nitrogen (N) 1–1.5 kg, phosphorus (P_2O_5) 0.2–0.4 kg, potassium (K_2O) 1–1.5 kg, and boron (B) 40–45 g. In a basic application of fertilizer, half the nitrogen, phosphorus, potassium, and boron are supplied. The rest of the nitrogen is given as surface fertilization in two doses, just before the fruiting and during the sowing of the endocarp (Sakellariou (2017), GmbH tech. report).

The NDEF message for the fertilization extended payload includes the following fields (Fig. 2):

- Fertilization work ID (2 bytes)
- Fertilization element or elements (1 byte), with a 5-bit code where each bit corresponds to the element (N, P, K_2O, B, Mn) that will be included in the fertilization compound
- Fertilization N quantity, if set, that starts from 500 to 1000 gr/tree/year (2 bytes)
- Fertilization P quantity, if set, that ranges from 300 to 500 gr/tree/year (2 bytes)
- Fertilization K_2O quantity, if set, that ranges from 600 to 1200 gr/tree/year if required (2 bytes)
- Fertilization B quantity, if set, that ranges from 0 to 450 gr/tree/year (2 bytes)
- Fertilization Mn quantity, if set, that ranges from 0 to 300 gr/tree/year, if required (2 bytes)

Usually in the NFC tag, at least 1 to 2 messages are placed for fertilizing and four to eight for yearly spraying. This means that the tag can hold messages in a depth of 3–4 years prior to record cleanup. As a conclusion, the services supported by the NFC tags are (i) notification service, (ii) fertilization monitoring service, (iii) pest control monitoring service, and (iv) identification management service.

4.3 OLEA System Oil Extraction Sensory Service

For recording the extracted olive oil quantity and chemical consistency (quality attributes), the proposed OLEA framework includes a multisensory equipment. This sensory equipment used by the OLEA sensory service is illustrated on Fig. 1 and includes the following sensors:

Sensor 1 Waterproof temperature sensor: For the needs of our system, we need a waterproof temperature sensor to monitor temperature in extracting the quality of the produced olive oil. The DS18B20 digital temperature probe (Maxim Integrated Products, Inc.), which is applicable in thermostatic controls, industrial systems, consumer products, thermometers, and any thermally sensitive system, was considered as the best fit for our system requirements. It operates in a temperature that covers the range between 10 °C and 40 °C, with an accuracy of ±0.5 °C that can be improved up to 0.1 °C using a multiple DS18B20 sensor configuration.

Sensor 2 Gravity liquid level sensors are used for recording the produced oil quantities: The capacity change liquid level sensors are placed in the oil extraction container, where the virgin olive oil is concentrated before bottling. The number of sensors used depends on the container's size. The minimum measurement granularity designed for a container is one sensor per lt virgin oil extracted. Capacity change liquid level sensors include a digital output connected to the OLEA controller either directly (maximum of 18 input pins) or using SPI 8-bit digital input serializers (more than 18 input pins).

Sensor 3 Acidity/stiffness sensor: Regarding acidity, since pH measurements correspond to water-soluble solutions, the correct way to monitor the acidity is with the use of an acidometer probe that determines oil acidity on-site (TuryanYa et al. 1996; Grossi et al. 2014). Usually analog spear tips are employed, dipped inside a virgin oil sample test tube, and then cleaned with the use of a dry cleaning cloth. For the process of acidity measurements, a robotic servo-controlled arm is used, by a PWM digital RPi3 and an 8-bit analog to SPI converter for processing the acidometer probe analog values. Such sensor probe operates between 20 and 25 °C (controlled by the temperature sensor) with an accuracy of $\pm$ 0.05 pH and the use of computational logic to estimate content acidity. It should be mentioned that acidometer probes require the use of an additional component to convert the output signal from analog to 8-bit SPI digital.

Sensor 4 I2C color constancy sensor uses an open tube with reflective surfaces that an RGB transponder has been placed. Through the open tube, part of the extracted filtered virgin oil passes. The RGB receiver used includes records in real-time virgin oil color constancy.

Sensor 5 Refractivity, ΔK UV measurements for K268/K232: The UV absorbance ΔK (K232-K268) measurements are manually performed at the time being, with the use of a refractometer that operates in the UV between 220 and 360 nm, since such equipment with MCU external connection interface (SPI, i2C, Wi-Fi, Bluetooth) is not yet available in the market for frequencies less than 380 nm.

Sensors 6 and 7 Peroxide and mean fatty acid concentration sensory measurements require specialized reagents and treatment and are performed at the laboratory, passed to the OLEA AS at a later time with the use of the OLEA AS web interface.

The MCU controller of the oil sensory service is used to interconnect all sensory equipment sensors (1–5) mentioned above. The MCU controller is an RPi3 model B with quad core ARM-64 1.2 GHz board CPU that includes 4 USB ports and HDMI display interface and DSI touchscreen display interface, a CSI camera port interface, Wi-Fi and Bluetooth (BLE) transponders, and a GPIO interface with 1 I2C (pins 3,5), 1 SPI interface (pins 19–26), 3 PWM digital outputs (pins 12,33,35), and 16 generic digital outputs (2 used for UART interfaces).

4.4 OLEA Sensory Service Communication Protocol

CoAP has been chosen to be used by the OLEA management system protocol rather than its main competitor MQTT (Thangavel et al. 2014). MQTT is mostly a many-to-many communication protocol, while CoAP is best suited for one-to-one communication. Moreover, the CoAP provides support for observing resources and is best suited for a state transfer model rather than an event-based model, where the MQTT applies.

OLEA protocol over CoAP (Shelby et al. 2014) is used for the sensory data transmission via the MCU controller from the decanter mill to the cloud application server (AS). CoAP is a specialized web transfer protocol for use with constrained nodes and constrained (e.g., low power, lossy) networks. It is a transport protocol designed for machine-to-machine (M2M) communication for the IoT industry. The communication protocol functional states follow the ReST model, where the AS make resources available via URIs and clients access these resources using the HTTP protocol methods over UDP.

CoAP is a document transfer protocol and can carry most of the existing data model formats as XML, JSON, or CBOR. It is designed to be used by IoT, and it asynchronously carries out small packet sizes, in respect to the HTTP TCP synchronous flows. Since OLEA protocol runs over CoAP, it uses connectionless datagrams over an existing IP network infrastructure. Since UDP is an unreliable protocol, OLEA protocol reliability is enforced at the application layer using retransmits and flow control. There are two types of messages used by the OLEA protocol: the "confirmable" and the "nonconfirmable." Confirmable messages are acknowledged by the receiver with an ACK packet. If the ACK packet is not received by the sender in a pre-defined period of time called "retransmission timeout," packet retransmission is performed for at least 6 times. Since OLEA protocol runs over CoAP, for security, it enforces the CoAP Datagram Transport Layer Security (DTLS) with RSA authentication and AES 128-bit encryption.

5 OLEA Proposed Clustering Algorithm and Proof of Concept Scenario

OLEA system implementation includes the capability of modular smart agents and algorithm utilization, upload, and interface with the OLEA application server engine. These modules may include a front-end web user and/or configuration interface written PHP and a backend database and logic written either in R or Python. Similar to content management systems, an administrative role may upload and install an OLEA module.

Smart agents and services are the next step of the deep learning evolution and will rig the OLEA platform with novel capabilities. However, many of these agents need to be designed and implemented, and many others have to be invented. As a first step, the OLEA system provides the engine means (easy to use upload and install package manager at the AS) and the external interfacing with R and Python languages data mining, classification, neural network, and fuzzy logic modules for the purpose of smart algorithm implementation. As a proof of this concept, the authors present an OLEA clustering algorithm implemented as an R external OLEA server interface and present their algorithm proof of concept experimentation on a test-bed dataset.

5.1 *Proposed OLEA Clustering Algorithm*

The authors propose a new clustering algorithm implemented as a module in the OLEA application service. This clustering algorithm selects from the AS database a number of milled virgin oil records and their attribute data uploaded by the OLEA controller (RPi3). Then it employs a clustering methodology based on *K*-means algorithm, in order to group tree cluster products of similar attributes and then initiates a selection-ensemble process of groups capable of sustaining a brand. The per-variety OLEA clustering process offers the capability to identify new products based on quality reports and cultivation conditions as well as report the accurate quantity of virgin oil availability for marketing and branding purposes. Proposed by the authors, OLEA clustering algorithm is illustrated in Fig. 3. The suggested data recording and clustering algorithm requires the following OLEA methodology steps.

Step 1: Field Selection and Cultivation Area Separation The OLEA methodology is performed in order to sustain OLEA database olive oil records per tree cluster and cluster identification records. The OLEA application service includes a per olive tree field sector identification interface with the OPEKEPE's spatial data and registration validity of the sector stored at the OPEKEPE's database of recorded olive tree fields. After successful validation, the microclimate separation process initiates where each field area is bounded to a specific microclimate area polygon accordingly (Zinas et al. 2013).

The OLEA AS includes a per olive tree field sector identification interface with the OPEKEPE's spatial data and registration validity of the sector stored at the OPEKEPE's database of recorded tree clusters.

Step 2a: Field(s) Identification Separation For each cultivation field location and microclimate, measures are kept at the OLEA database and set/updated by the farmer's mobile application or OLEA web GUI. More specifically the microclimate measurements and predictions are automatically updated with the use of agents at the Wunderground service (Wunderground API 2014). Traceability location and microclimate measures maintained by the farmer and OLEA application service agents are field altitude; field distance in Km from the sea; field solar monthly average irradiance (W/sq. m), monthly sunlight average day recordings, and flux-luminosity measurements; field soil moisture measurements; field monthly average rainfall measurements; and yearly critical meteorological

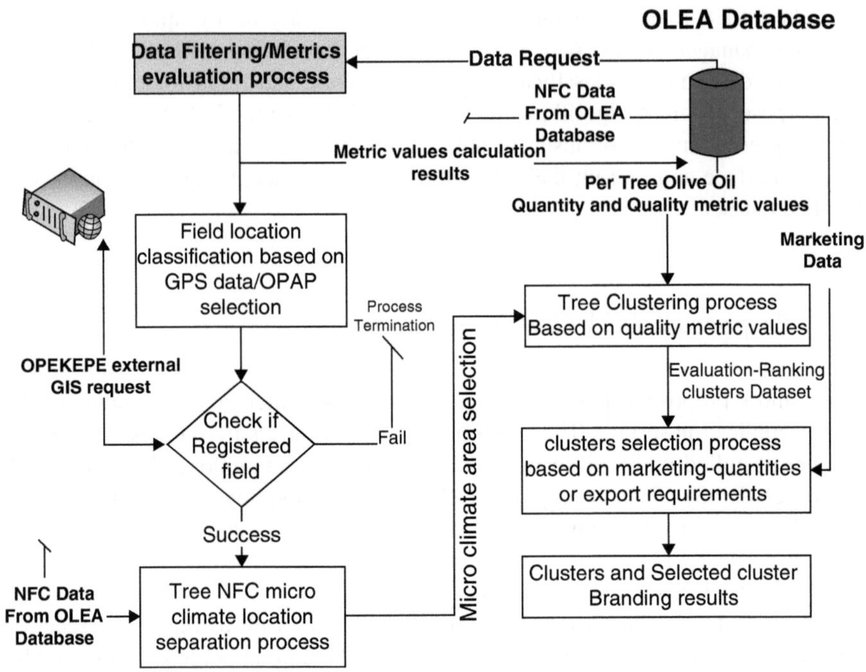

Fig. 3 OLEA clustering, data recording, and identification methodology steps

conditions recorded over each field (criticality and occurrence) such as hail or drought or frost and daily microclimate area predictions.

Step 2b: Field Tree Clustering and NFC Identification Upon first-layer cultivation area separation, a field clustering process is undertaken for each area, that is, clusters of trees of a tree area with a maximum space of 10,000 sq. m. Such areas are uniquely identified with the use of an NFC tag, which keeps track of data regarding cultivation process, pesticides used, and adjustment dates.

Step 3: Tree Cluster Product Quantity and Recorded Product Qualitative Attributes A set of metric qualitative measures of the extracted olive oil characteristics are recorded per tree cluster. This recording and data filtering process is performed separately from the clustering algorithm, and the evaluated metrics per cluster are stored at the OLEA database. That is, olives produced in each cluster are milled and measured. With the use of the OLEA sensors and controller, the following metrics can be recorded per tree cluster: olive size (number of olives per kg, 230–320 olives/kg in Europe; this is set as large to brilliant size); peroxide value (meq O_2/kg); %v/v oleic acid and free fatty acid; omega-3/6 polysaturated acids-rancidity index; UV absorbance (K232, K270, DeltaK); harvesting age (days/hours of harvesting passed); color characteristics; and organoleptic classification based on the median of any defect (Md) and median of fruity attributes (Mf) (tests based on human sensory perception).

Step 4: Clustering Algorithm Initiation After virgin oil metric evaluation of each tree, the virgin oil product clustering process is performed. This clustering process uses the *K*-means unsupervised learning algorithm. The process initializes with n clusters (the number of clusters is equal to the number of metric virgin oil values used) and is repeated by incrementing n, until market or brand metric requirements are satisfied by at least one of the generated by *K*-means clusters. Then, the cluster which satisfies the requirements and has the highest total score metric value (see Eq. 1), according to the requirements, is selected. That is, for all *K*-means-generated clusters, a total score value is calculated in order for the product clusters to be cross-examined according to market set requirements. Per-cluster total score value is equal to the sum of all qualitative virgin oil metric values multiplied by a probability index called Metric Importance Coefficient (MIC).

Step 5: Clustering Algorithm Selection Process The selection process that follows *K*-means clustering is similar to the elbow method, used for selecting the most promising number of minimum clusters generated with the minimum SSE. Selection process chooses the *K*-means number of clusters value by checking if the cluster total score (TS) value is equal to the market set TS value and continues top-down selection, until market or brand quantity needs are met, or a product cluster is reached where at least one qualitative average metric value is below the requested value for that metric by the market. The proposed clustering algorithm includes a repeating *K*-means clustering and a selection process. Proposed OLEA clustering algorithm flow diagram is illustrated in Fig. 4.

For example, let's assume OLEA clustering process (Step 4) of a set of tree clusters is based on three qualitative metrics (m_1, m_2, m_3) and the market requirement is for metric values of $U1$, $U2$, and $U3$ accordingly. Then, based on market or export brand requirements, each metric value $U1$, $U2$, and $U3$ is assigned with a statistically set probability of importance p_i, $\sum_{i=1}^{l} p_i = 1$, where l is the number of metrics used (in this case three metrics). Then the market required total score value (MIC) is calculated according to Eq. 1:

$$\mathrm{TS_M} = \sum_{i=1}^{n} p_i\ U_i,\ \ \sum_{i=1}^{n} p_i = 1 \tag{1}$$

where n is the number of metrics used. Similarly, for each *K*-means cluster, a total score (TS) metric value is calculated and compared to $\mathrm{TS_M}$ (MIC) value as shown in Eq. 2, where K is the cluster number, which the total score value corresponds to.

$$TS_k = \sum_{i=1}^{n} p_i^{'} M_i \tag{2}$$

K-means algorithm is used for the creation of product clusters where on each cluster the centroid mean metric values of M_1, M_2, and M_3 (centroid vector value) and cluster total score are calculated, based on Eqs. 3 and 4, and then compared to the market required total score value:

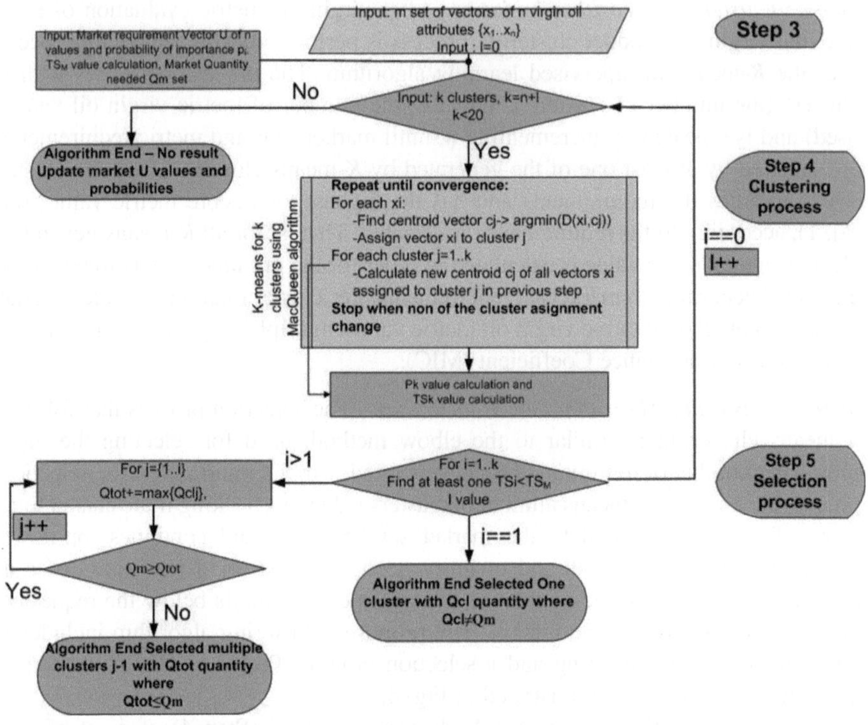

Fig. 4 OLEA clustering and branding algorithm flowchart diagram

$$\mathrm{TS_k} = p_k{}' M_k = \left(\frac{\frac{1}{n_k} \sum_{i=1}^{n_k} m_i}{\sum_{k=1}^{n} \frac{1}{n_k} \sum_{i=1}^{n_k} m_i} \right) \left(\frac{1}{n_k} \sum_{i=1}^{n_k} \mu_i \right) \tag{3}$$

where n_k is the total cluster point vectors from the *K*-means process, *i* is the number of fields in each cluster ($i = 1..n_k$), and μ_i is the qualitative metric value for each tree field ID that belongs to a cluster. Moreover, M_k is the mean metric value (centroid vector) per cluster. Probability $p_k{}'$ is expressed as the normalized M_k value, and it expresses the cluster's metric value impact on the other clusters. If equal cluster probabilities are used for all *K*-means metric values, the total score for cluster *K* is calculated based on Eq. 4.

$$\mathrm{TSk} = p_k' M_k \overset{p_k'=1/n}{\widehat{=}} \frac{1}{n} \frac{1}{n_k} \sum_{i=1}^{n_k} m_i \tag{4}$$

The *K*-means product clustering process initially starts with $K = n$ (where *n* is the number of metrics set as market requirements). This leads to the creation of three clusters with mean cluster centroid vector values *M*1, *M*2, and *M*3 accordingly per cluster (Eq. 3) and a total score value per cluster (Eq. 2).

If there is at least one cluster with $TS_k < TS_M$, then that cluster is selected. If more than one clusters with $TS_k < TS_M$, then the cluster with the maximum olive oil quantity value that fulfills metric requirements is selected first as appropriate (i.e., the cluster with the max (quantity) and $TS_k < TS_M$). If there is no cluster that fulfills the market requirements, then the clustering *K*-means process is performed repeatedly increasing each time the number of clusters by one, until the market requirements for at least one cluster mean values is met (if *K* cluster centroid vector values ($m1$,$m2$, $m3$) are less than market ($U1$,$U2$,$U3$) vector values). Based on the total score, a cluster's ranking process is acquired, and from that ranking process, the high-order quantity ranked clusters are denoted as of high exportation and branding value clusters (selection process) and are selected from top to bottom quantity until quantity requirements are met.

5.2 OLEA Clustering Algorithm Case Study Scenario I

The case study experimentation in this work includes the Manaki Greek olive variety cultivation in the area of West Greece (fields at Preveza and Lefkada). All fields are situated at the two nearby microclimate areas of Preveza and Lefkada, and the altitude variation between all is from 50 up to 250 meters. In this study, a total area of 120 square meters of Manaki olive tree cultivation is divided into 1000 square meters production fields. Each field is identified by a unique NFC ID tag that includes field's geographical location, fertilization methods used, altitude, etc. An average fields' tree-to-tree distance is about 6.5–7 m thus 25–30 olive trees are included in each field. The average field production of an NFC-tagged sector is about 1000–1200 kg/0.1 Ha.

Two metrics were used in this study in order to verify the proposed methodology: the %w/w of free acidity and the total net weight of olive production per NFC-tagged field block set to the size of 1000 sq. m. All such blocks were marked by NFC tags and square edge-bounded by GPS coordinates.

Based on the market olive oil need requests to the farmers' copartnership, the EVOO (extra-virgin olive oil) maximum acidity regulations and market requirements regarding olive oil acidity were set into a maximum of 0.5% v/v, and the market requested olive oil quantity is 7000 lt. For that purpose using our proposed OLEA methodology, a *K*-means process was performed for the selection of the fields that produced olive oil with lower acidity than the threshold set. Table 2 shows the clusters where acidity requirement is met (cluster C2) (Fig. 5).

5.3 OLEA Clustering Algorithm Case Study Scenario II

In the following scenario, the functionality of the proposed clustering algorithm is put to test using an olive oil dataset of 120 tree fields of total extracted oil quantity of 106,540 kg, from the island of Thasos, illustrated in Fig. 6. The dataset metric values

Table 2 *K*-means clusters, average quantity per cluster field in lt, and %v/v average acidity of per *K*-means cluster. Cluster 2 meets market acidity requirements and quantity requested

Cluster ID and symbol	Average cluster tree field oil quantity (*L*)	Average cluster oil acidity %v/v	Total cluster oil quantity (*L*)
C1(○)	2033	0.56	10166
C2(Δ)	1000	0.49	11000
C3(+)	1216	0.61	24322
C4(x)	769	0.59	12304
C5(◊)	661	0.55	32389
C6(∇)	861	0.56	16359

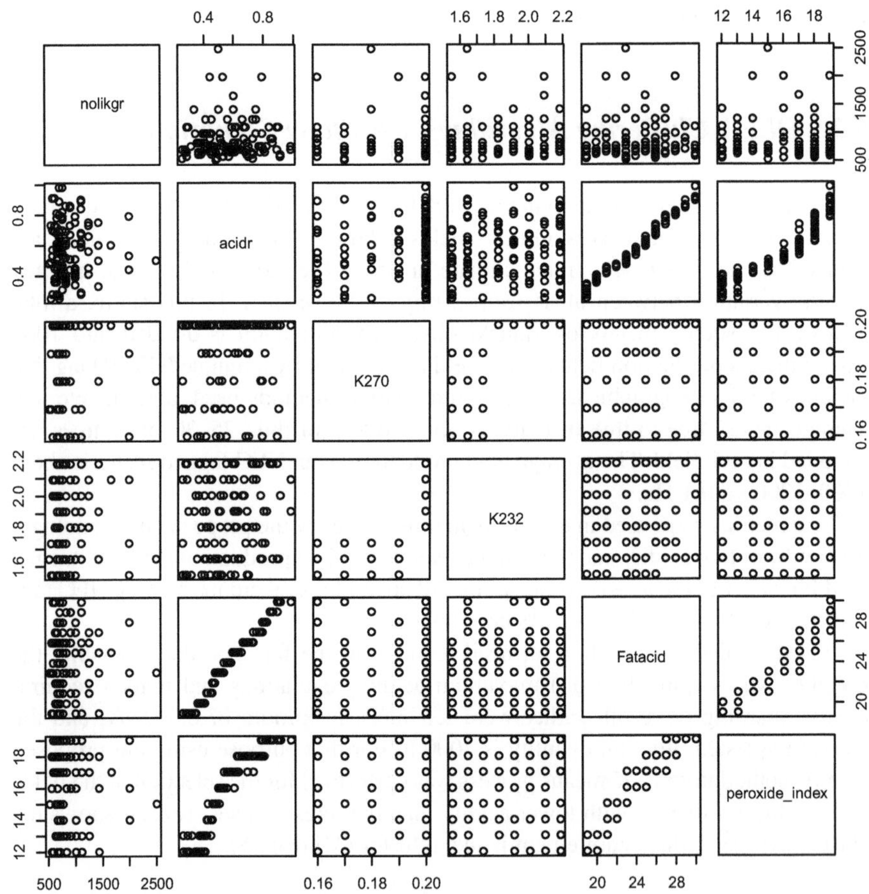

Fig. 5 Clusters returned from the OLEA algorithm. The metrics used are the quantity of extracted oil, the acidity (g/100 g), K270, K232, mean fatty acid (mg/kg) metric values and peroxide index values (mEq O2/kg). Cluster C2 is marked with the Δ symbol and is the OLEA algorithm selected cluster

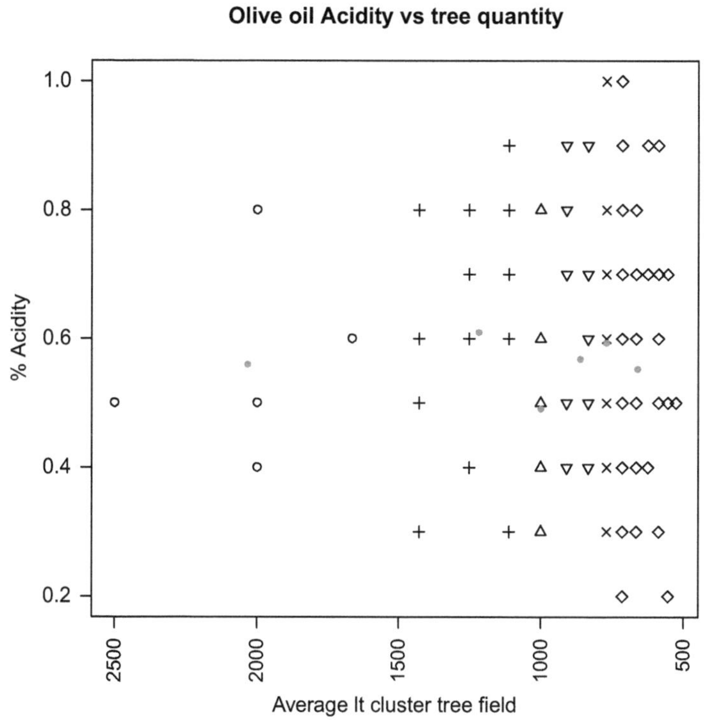

Fig. 6 Virgin olive oil experimental dataset taken from the island of Thasos, a set of 121 records of small olive tree fields, the extracted virgin oil attribute of acidity (g/100 g) over the oil quantity (in kg) per cluster the oil quantity of extracted oil (in kg), and the extracted virgin oil attributes of acidity (g/100 g), K270, K232, mean fatty acid (mg/kg), and peroxide index (mEq O_2/kg)

used are the quantity of extracted oil (kg); the acidity (g/100 g); K270, K232, and mean fatty acid (mg/kg) metric value; and peroxide index values (mEq O_2/kg).

First of all, the sum of squared error (SSE) for a cluster j is calculated as the sum of the squared distance between each member of the cluster and its centroid (Eq. 5):

$$sse_j = \sum\nolimits_{x_i \in c_j} dist(x_i, c_j)^{2 for\ all\ clusters\ j=1..k} \overset{\cdot}{\rightarrow} \langle sse_1, sse_j..sse_k \rangle \quad (5)$$

The total square error (SSE) of a clustering process is expressed as the sum of all sse{1..k} SSE vector values (Eq. 6) and the error vector ê per K-means-generated cluster equals to: $\widehat{e_i} = sse_i/SSE.$

$$SSE = \sum\nolimits_{j=1}^{k} \sum\nolimits_{x_i \in c_j} dist(x_i, c_j)^2 \quad (6)$$

The K-means is used on this dataset to cluster the set based on the five attributes (acidity, K270, K232, fatty acid, and peroxide index), and the TS_M (MIC) value is equal to $TS_M = 8.2$ according to Table 4 market requirements for virgin olive oil. Initiating *K*-means algorithm with the number of clusters equal to two gives an average error value of 0.56 > 0.05, for cluster *k* values with *k* <=5. *K*-means with *k* = 6 has an error value >0.05. *K* means results in terms of mean error; the results are shown in Table 4 (*k* = 6 clusters – *K*-means cluster results). From the *k* = 6 results, the cluster error of at least one cluster is more than 0.05%, so the clustering algorithm increases the number of clusters until the error of all clusters is less than 0.05. This happens for *K*-means *k* = 11. The results of *K*-means *k* = 11 are presented in Tables 3 and 4.

Table 3 Market requirements for the five dataset metric values for extra-virgin olive oil, probability of importance for each attribute accordingly, and total score market value TS_M

TS_M/Market req.	Ol. acid (g/100g)	K270	K232	Fatty acid (mg/kg)	Peroxide ($mEqO_2$/kg)
Req. values	0.37	0.2	2	25	15
% Prob. p_i	40%	10%	10%	20%	20%
Market importance	Total market score (MIC) $TS_{M=}8.368$ Requested quantity: 3000kg				
K=6 cluster	Cluster ID	Total score	Error value (<0.05)	Cluster size	Cluster oil quantity (Kg)
Pk=1/5	1	8.38	0.09	31	19537
	2	8.46	0.07	5	7378
	3	8.51	0.03	27	29610
	4	8.49	0.05	35	28663
	5	8.44	0.04	4	8500
	6	8.46	0.0004	18	12852

Table 4 Clustering results for *K*-means *K* = 11 clusters of the clustering algorithm

K=11 cluster	Cluster ID	Total score	Error value (<0.05)	Cluster size	Cluster oil quantity (Kg)
Pk=1/5	1	8.28	0.00066	16	10656
	2	7.18	0.000005	7	4998
	3	8.6	0.00068	16	12304
	4	9.27	0.00015	11	7854
	5	8.9	0.022	4	8500
	6	8.64	0.034	5	7378
	7	9.4	0.048	4	8500
	8	8.4	0.01	7	9996
	9	9.03	0.002	7	6363
	10	8.48	0.047	15	8881
	11	7.94	0.01	11	11000

As illustrated in Table 4, the clustering algorithm returns as an appropriate cluster with ID = 3 that has a total of 4998 kg ($TS_k < TS_M$, $TS_k = \min\{TS_i\}$). If more kgs of oil are requested by the market (6000 kg), then the quantity of cluster with ID = 11 will be allocated, setting the available quantity to 15998 Kg for this market request.

6 Conclusions

This paper presents a new monitoring system for the virgin olive oil industry called OLEA. OLEA system is capable of origin location, recording of farming methods and olive oil attributes, product traceability, and overall management of the virgin oil production and farming processes. OLEA farming monitoring methodology uses an NFC based olive tree clustering scheme that maintains cultivation and environmental attributes information stored in its database. This information is reinforced with qualitative olive product characteristics derived by the olive fruit milling process assisted by the OLEA real-time sensory system, controller, and sensory service. The attributes recorded by OLEA database are updated either by the OLEA mobile phone application for the fertilization and plant protection farming services or the OLEA mill-located sensory controller for the quantity and qualitative attributes of produced olive oil per cluster.

The OLEA system architecture is described in detail in the present work. This system includes NFC tags for tree cluster identification and sensors for the process of product measurements. OLEA system also includes an integrated application and database service where all measured NFC and sensory data are collected, allowing for application-level queries toward the database to be performed using the OLEA web interface or the OLEA mobile phone application interface.

Authors included in their OLEA implementation the capability of a modular engine for the process of developing smart agents and algorithms. In order to reinforce the necessity of this engine to contemporary management systems, a new data mining algorithm has been implemented, for processing uniform olive product characteristic clustering and group product selection in accordance to market needs. The proposed clustering algorithms uses as input a customer's product requirements in terms of threshold values and delivers back to them a selection of the most appropriate tree cluster areas product that strongly include those attributes.

To support the methodology, a case study scenario is presented, where the proposed clustering and selection algorithm is applied. From the case study results, it is clearly shown how the OLEA clustering algorithm can achieve the standardization of olive oil product quality and appropriately select production based on market needs. The use of the proposed OLEA system shall close the gap of a central information management and monitoring system facilitating the Greek virgin olive oil industry. In addition, the OLEA modular engine shall lead to the implementation of new services for the virgin oil industry.

References

Aparicio R, Harwood J (2013) Handbook of olive oil. Springer, Boston. https://doi.org/10.1007/978-1-4614-7777-8

Bettini O, Sloop C (2014) Olive Oil Annual Greece – GAIN reports, United Stated Department of Agriculture, Foreign Agricultural service, http://gain.fas.usda.gov/Recent%20GAIN%20Publications/Olive%20Oil%20Annual%202014_Rome_Greece_5-14-2014.pdf

Camarsa G, Gardner S, Jones W, Eldridge J, Hudson T, Thorpe E, O'Hara E (2010) LIFE among the olives: good practice in improving environmental performance in the olive oil sector, EU Commission Environment Directorate-General. http://ec.europa.eu/environment/life/publications/lifepublications/lifefocus/documents/oliveoil_en.pdf

Cannazza G, Cataldo A, De Benedetto E, Piuzzi E, Attivissimo F (2011) Customized system for vegetable oils quality control based on dielectric spectroscopy analysis. 2011 IEEE International Conference on Instrumentation and Measurement Technology

Codex Alimentarius Commission (2017) Standard for olive oils and olive pomace oils. CODEX STAN 33–1981, revisions and amendments

Ehaliotis C, Martinis A, Minotou C (2011) Towards a paradigm for the convergence of production, marketing and environmental protection in the field of the organic olive oil production chain- Conclusions from the Biolmed project, In: Proceedings of International Conference on organic agriculture and agroteco tourism in the Mediterranean

European Commission Agricultural Market Briefs (2012) Prospects for the olive oil sector in Spain, Italy and Greece – 2012–2020

European Commission (2015) Commission Delegated Regulation (EU) 2015/1830 amending Regulation (EEC) No 2568/91 on the characteristics of olive oil and olive-residue oil and on the relevant methods of analysis

Gao H, Zhou X, Cheng H, Zhao C, Yan X (2004) Information fusion of online oil monitoring system using multiple sensors. Tsinghua Sci Technol 9(3):258–261

Grossi M, Di Lecce G, Toschi TG, Ricco B (2014) A novel electrochemical method for olive oil acidity determination. Microelectron J Elsevier 45:1701–1707

International Olive Council (IOC) (2011) Guide for the determination of the characteristics of oil-olives. COI/OH/DOC.1

International Olive Council World (2012) Prospects for the olive oil sector – EU agricultural market briefs- International Olive council World Table Olive Figures-Greece country Profile-Policy. http://www.internationaloliveoil.org

International Olive Council (IOC) (2018) Trade standard applying to olive oils and olive-pomace oils. COI/T.15/NC No 3/Rev.12, Madrid

Kiama JW, Raman V, Patrick THH (2014) Low-cost RFID-based palm oil monitoring system (PMS): First prototype, In: 8th International Symposium of the Digital Earth (ISDE8 2014), Sarawak, vol. 18

Kontogiannis S, Tsipouras M, Kokkonis G, Valsamidis S (2016) Design and implementation of an identification productivity recording and breeding system for the sheep industry. Int J Sustain Agric Manag Inf 2(2–4):97–122. ISSN 2054-5819

Mailer R, Beckingham C (2006) Testing olive oil quality: chemical and sensory methods, Profitable & Sustainable Primary Industries, PRIMEFACT vol 231, NSW Dept. of Primary Industries, ISSN 1832-6688, pp 1–5

Minister of Agriculture (2015) National plan of action for the rational use of agricultural medicines, Greece. http://www.minagric.gr/images/stories/docs/agrotis/Georgika_Farmaka/elenxoi/NAP-GR_consolidated.pdf, K□A 6669/79087/15-7-2015

Mylonas P (2015) Olive oil: Establishing the Greek brand, National Bank of Greece sectoral report. https://www.nbg.gr/greek/.../press-office/e-spot/reports/Documents/Olive%20Oil_2015.pdf

Niaounakis M, Halvadakis CP (2006) Olive Processing Waste Management, Elsevier Literature Review and Patent Survey 2nd edn, ISBN 978-0-08-044851-0

NFC Forum NDEF (2006) NFC Data Exchange Format (NDEF), Technical specification, Rev 1. http://www.eet-china.com/ARTICLES/2006AUG/PDF/NFCForum-TS-NDEF.pdf

NFC Forum RTD (2006) NFC Forum Record Type Definition RTD v.1.0. https://www.cardsys.dk/download/NFC_Docs/NFC%2520Record%2520Type%2520Definition%2520(RTD)%2520Technical%2520Specification.pdf

OPEKEPE (2017) Greek Organization for EU CAP funding control on line web applications. http://www.opekepe.gr/applications.asp

Sakellariou M (2017) Olive trees cultivation and fertilization information, in greek, KALI GmbH technical report. http://www.kali-gmbh.com/shared/data/kali-fertiliser-broschures-pdf/broschures-gr/gr-olive-A4-1702.pdf

Scarafia L (2011) A perspective on extra virgin olive oil and premium quality: California olive oils, Agrolab Inc. http://www.agbiolab.com/files/agbiolab_California-EVOO-Premium-Quality.pdf

Shelby Z, Hartke K, Bormann C (2014) The constrained application protocol (CoAP). RFC 7252

SONY company (2014) FeliCa technical information. https://www.sony.net/Products/felica/business/tech-support/index.html

Taouil K, Chtourou Z, Kamoun L (2008) Machine vision based quality monitoring in olive oil conditioning. In: Proceedings of Image Processing Theory, Tools and Applications (IPTA 2008), Sousse, pp 1–4

Thangavel D, Ma X, Valera A, Tan HX, Tan CKY (2014) Performance evaluation of MQTT and CoAP via a common middleware. In: Intelligent Sensors, Sensor Networks and Information Processing (ISSNIP), 2014 IEEE Ninth International Conference on (pp 1–6). IEEE

Theodosiou T, Valsamidis S, Hatziliadis G, Nikolaidis M (2012) Measuring, archetyping and mining Olea europaea production data. J Syst Inf Technol 4(4):318–335. Emerald

TuryanYa I, Berezin O, Kuselman I, Shenhar A (1996) pH-metric determination of acid values in vegetable oils without titration. J Am Oil Chem Soc 73:295–301

United Nations (2015) International agreement on olive oil and table olives. United Nations conference on trade and development, Geneva. TD/OLIVE OIL11/5

Vekiari SA, Oreopoulou V, Kourkoutas Y, Kamoun N, Sallem M, Psimouli V, Arapoglou D (2010) Characterization and seasonal variation of the quality of virgin olive oil of the Throumbolia and Koroneiki varieties from Southern Greece. Grasas Aceites 61(3):221–231

Vossen P (2013) Growing olives for oil. In: Aparicio R, Harwood J (eds) Handbook of olive oil. Springer, pp 19–56. https://doi.org/10.1007/978-1-4614-7777-8_2

WBOOR (2015) World's best olive oil reports. http://www.bestoliveoils.com/page/1/

Wunderground API (2014) A Weather API designed for developers. https://www.wunderground.com/weather/api/

Zinas N, Kontogiannis S, Kokkonis G, Pikridas C (2013) A novel microclimate forecasting system architecture integrating GPS measurements and meteorological-sensor data. In: Proceedings of 6th Balkan Conference in Informatics (BCI), vol 1. ACM, Thessaloniki, pp 82–88. ISBN: 978-1-4503-1851-8

Fuzzy Control Simulation of a Smart Irrigation System

Aliki D. Muradova, Georgios K. Tairidis, and Georgios E. Stavroulakis

Abstract Centre-pivot irrigation systems, also known as flexible sprayer booms, are efficiently used in the watering process and irrigation applications. During the use of such systems, various problems can occur, due to vibrations caused either by the wind excitation or other, mainly steep, external loadings, which in turn can be caused by the uneven surface of the roads or fields, such as potholes, cavities, etc. In the present paper, fuzzy control is used for the suppression of these vibrations. The application of the control mechanism is done on a two-dimensional truss model of the structure. A hybrid, Mamdani-type, fuzzy controller is implemented and tested on the smart truss. The efficiency of the proposed simulation is shown on the numerical examples.

Keywords Smart structures · Structural control · Truss model · Finite element analysis · Fuzzy inference system · Irrigation system

Nomenclature

$\widehat{u}_{1x}, \widehat{u}_{2x}$	Nodal displacements
l	Length of an element
ε	Elastic strain
σ	Elastic stress
E	Young's modulus (modulus of elasticity)
T	Tension
A	Area of cross section
$\widehat{f}_{1x}, \widehat{f}_{2x}$	Nodal forces of a bar
f^e_{ix}, f^e_{iy}	External forces for i-th element in x,y directions

A. D. Muradova · G. K. Tairidis (✉) · G. E. Stavroulakis
Institute of Computational Mechanics and Optimization, School of Production Engineering and Management, Technical University of Crete, Chania, Greece

A. Theodoridis et al. (eds.), *Innovative Approaches and Applications for Sustainable Rural Development*, Springer Earth System Sciences,
https://doi.org/10.1007/978-3-030-02312-6_20

m_i	Mass of i-th element
$\left[\hat{k}\right]$	Stiffness matrix of an element
$[N_e]$	Number of elements
$[u]$	Displacement vector for the truss
$[\dot{u}]$	Velocity vector for the truss
$[\ddot{u}]$	Acceleration vector for the truss
$[m]$	Mass matrix for the truss
$[c]$	Damping matrix for the truss
T_r	Transformation matrix for the truss
$[k]$	General stiffness matrix for the truss
β,γ	Newmark's parameters
T_f	Final time of the simulation
Δt	Time step
K_t	Number of time steps

1 Introduction

Centre-pivot irrigation systems, also called waterwheels or sprinkler irrigation systems, are very popular for the watering of crops. Such systems are highly efficient, as they combine perfect irrigation and maximum water conservation. However, the use of these systems can confront several problems due to the vibrations which are caused by the wind and/or other external loadings. Vibrations and vibration damping using passive and active techniques of flexible spray booms have been studied in several works. A detailed study has been presented in the doctoral thesis Anthonis (2000), for horizontal vibrations, and in the subsequent paper Anthonis et al. (2005), for vertical vibrations. Active force control is proposed in Tahmasebi et al. (2013) and multi-rate control in Arvanitis et al. (2003).

In the present investigation, the reduction of the vibrations by means of simulation of a smart truss model is sought. This structure can be considered as a simplified model of the most commonly used centre-pivot irrigation systems. The smart trusses embody piezoelectric elements along with control mechanisms that provide the intelligent behaviour. Linear systems for control of vibrations in trusses and other structures can be studied by classical control methods. However, classic mathematical theory of control can meet many restrictions, as nonlinearity in the system and/or the controllers increases dramatically the complexity of the problem. In this case, fuzzy and hybrid neuro-fuzzy controllers can be used instead.

Different control strategies or mixtures of them are presented in the work of Abe (1996). More specifically, a rule-based control algorithm for active tuned mass dampers, exploiting the capabilities of fuzzy logic, along with classic control tools, such as the linear quadratic regulator (LQR) feedback gains, is proposed. In general, fuzzy inference rules systematize existing experience and can be used for the rational formulation of nonlinear controllers (Driankov et al. 1996). In the present investigation, a collocated controller in the sense of Preumont (2002) is developed. Several controllers have been developed in previous investigations. For example, a Mamdani-type fuzzy controller for the control of a smart composite beam is developed in Tairidis et al. (2009). The verbal fuzzy rules are written in the sense of the motion of a pendulum, while the numerical results indicate the efficiency of the proposed control scheme. Fuzzy vibration control of a smart thin elastic rectangular plate, with representative numerical examples, is given in Muradova and Stavroulakis (2013, 2015). The development and tuning of a neuro-fuzzy control scheme are presented in detail in Stavroulakis et al. (2011).

A literature review on the static and dynamic shape control of structures by means of piezoelectric actuation is conducted in Irschik (2002).The significant work which was done on active and semi-active vibration control of structures during the previous years is reviewed in Fisco and Adeli (2011). Namely, modified LQR and LQG, neural network-based, fuzzy logic, sliding mode and wavelet-based controllers are reviewed. Finally, a brief survey on industrial applications of fuzzy control is given in the work of Precup and Hellendoorn (2011).

Strong computational tools for the simulation of smart structural systems can be considered among others MatLab, Simulink or other similar software (Tairidis et al. 2017; Muradova et al. 2018 etc.). Once the controllers are built, some fine tuning may be necessary and can be based on several concepts, which can be among others the trial-and-error method or global optimization methods, such as particle swarm optimization (Marinaki et al. 2011; Tairidis et al. 2015) or genetic algorithms (Lu et al. 2003; Tairidis et al. 2016; Chang 2011; Karavas et al. 2017).

In this study, a Mamdani-type hybrid fuzzy controller is implemented on a smart truss model, and it is tested within the MatLab environment. The investigation is based on the study of a smart truss with piezoelectric sensors and actuators. The examined system consists of a truss which is supported at two nodes, and it is subjected to an external loading at one node. The main contribution of the work is the introduction and incorporation of control actions through thermal stresses on a truss finite element model. This technique can be extended for more complex, two- and three-dimensional systems. Moreover, the use of fuzzy control within this setting can be considered a novel aspect.

The present work is organized as follows. In Sect. 2 a mechanical model is presented. The irrigation system is modelled as a smart truss, and the equations of motion are derived. Section 3 is devoted to a computational algorithm. The Newmark-β method is employed for the time integration of the linear system of ordinary differential equations. Section 4 focuses on the construction of the control scheme, which is based on a Mamdani-type fuzzy inference system. Numerical examples are illustrated in Sect. 5, while in Sect. 6 the main results are discussed.

2 Mechanical Model

In this section, the simplified mechanical model with embedded piezoelectric components, along with the control system, is presented. The truss model is considered to have similar behaviour in vibrations with each part (right and left) of a small centre-pivot irrigation system (Fig. 1), like the ones used in Greek agriculture and in other places of the world.

The simplified model of the smart truss structure consists of a finite number of parts (bars), connected with each other. Each bar of the structure corresponds to one element. The elements are connected with each other by nodes. The truss is fixed at two nodes, namely, at nodes 1 and 2, as shown in Fig. 2. The structure is subjected to a vertical time-dependent loading, and it is equipped with a fuzzy controller.

A dynamic truss equation with stiffness, mass and damping matrices is derived by applying a finite element approach (Logan 2007).

It is assumed that the displacement at each bar element is a linear function

$$u = \frac{\widehat{u}_{2x} - \widehat{u}_{1x}}{l} x + \widehat{u}_{1x},$$

where $\widehat{u}_{1x}, \widehat{u}_{2x}$ are nodal displacements of the bar element and l is the length of the element. According to the classical theory, the strain displacement and the stress-strain relationship are given as

Fig. 1 A typical small centre-pivot irrigation system in Greece

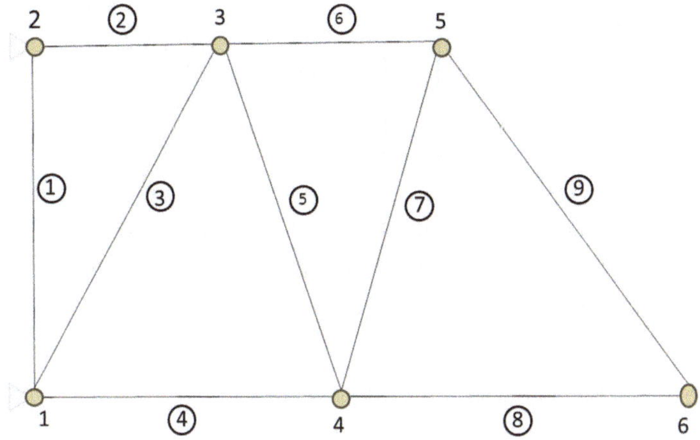

Fig. 2 Truss with 6 nodes, 9 elements and 2 fixed supports at nodes 1 and 2

$$\varepsilon = \frac{du}{dx}, \quad \sigma = E\varepsilon,$$

where ε is the strain (deformation) tensor, σ is the stress tensor and E is Young's modulus (modulus of elasticity). Hence

$$T = A\,\sigma = \frac{AE}{l}\left(\widehat{u}_{1x} - \widehat{u}_{2x}\right),$$

where T is the tension and A is the area of cross section of a bar.

Furthermore, for the nodal forces, we have $\widehat{f}_{1x} = -T, \ \widehat{f}_{2x} = T.$

Since we consider a dynamic case, the Newton's second law of motion is applied to each node. Therefore, for each node we have the external (applied) force f_x^e, which is equal to the sum of the internal force and nodal mass times acceleration. Thus, for each element in x and y directions, we have

$$f_{1x}^e(t) = \widehat{f}_{1x}(t) + m_1 \frac{\partial^2 \widehat{u}_{1x}}{\partial t^2}, f_{2x}^e(t) = \widehat{f}_{2x}(t) + m_2 \frac{\partial^2 \widehat{u}_{2x}}{\partial t^2}, \tag{1}$$

$$f_{1y}^e(t) = \widehat{f}_{1y}(t) + m_1 \frac{\partial^2 \widehat{u}_{1y}}{\partial t^2}, \ f_{2y}^e(t) = \widehat{f}_{2y}(t) + m_2 \frac{\partial^2 \widehat{u}_{2y}}{\partial t^2}. \tag{2}$$

or in the matrix form

$$\begin{Bmatrix} f_{1x}^e \\ f_{1y}^e \\ f_{2x}^e \\ f_{2y}^e \end{Bmatrix} = \begin{Bmatrix} \widehat{f}_{1x} \\ \widehat{f}_{1y} \\ \widehat{f}_{2x} \\ \widehat{f}_{2y} \end{Bmatrix} + \begin{bmatrix} m_1 & 0 & 0 & 0 \\ 0 & m_1 & 0 & 0 \\ 0 & 0 & m_2 & 0 \\ 0 & 0 & 0 & m_2 \end{bmatrix} \begin{bmatrix} \ddot{\widehat{u}}_{1x} \\ \ddot{\widehat{u}}_{1y} \\ \ddot{\widehat{u}}_{2x} \\ \ddot{\widehat{u}}_{2y} \end{bmatrix}. \quad (3)$$

Distributed mass of the bar elements can be neglected or considered as lumped mass on the nodes of the elements. Replacing $\left\{\widehat{f}\right\}$ with $\left[\widehat{k}\right]\{\widehat{u}\}$ from (3), we obtain a system of equations for the bar element

$$\{f^e(t)\} = \left[\widehat{k}\right]\{\widehat{u}\} + [m]\left\{\ddot{\widehat{u}}\right\}, \quad (4)$$

where $\left[\widehat{k}\right] = \frac{AE}{l}\begin{bmatrix} 1 & 0 & -1 & 0 \\ 0 & 0 & 0 & 0 \\ -1 & 0 & 1 & 0 \\ 0 & 0 & 0 & 0 \end{bmatrix}$ is the stiffness matrix. Furthermore, $[m]$ is the mass matrix in (3), and $\{\ddot{u}\}$ is the acceleration vector in (1) and (2).

Introducing also the damping matrix, we can write down Eq. (4) in the following form:

$$\{f^e(t)\} = \left[\widehat{k}\right]\{\widehat{u}\} + [c]\left\{\dot{\widehat{u}}\right\} + [m]\left\{\ddot{\widehat{u}}\right\}. \quad (5)$$

At this point, a transformation matrix for the bar element needs to be defined. According to the rules for the nodal displacements, we have

$$\{\widehat{u}\} = [T_{\mathrm{r}}]\{u\}, \quad (6)$$

i.e.

$$\begin{bmatrix} \widehat{u}_{1x} \\ \widehat{u}_{1y} \\ \widehat{u}_{2x} \\ \widehat{u}_{2y} \end{bmatrix} = \begin{bmatrix} c & s & 0 & 0 \\ -s & c & 0 & 0 \\ 0 & 0 & c & s \\ 0 & 0 & -s & c \end{bmatrix} \begin{bmatrix} u_{1x} \\ u_{1y} \\ u_{2x} \\ u_{2y} \end{bmatrix}$$

where $T_{\mathrm{r}} = \begin{bmatrix} c & s & 0 & 0 \\ -s & c & 0 & 0 \\ 0 & 0 & c & s \\ 0 & 0 & -s & c \end{bmatrix}$ is the transformation matrix.

Analogously for the nodal forces

$$\left\{\widehat{f}\right\} = [T_{\mathrm{r}}]\{f\}. \quad (7)$$

Using the previous results from (6), (7) one can obtain

$$[T_{\rm r}]\{f\} = \left\{\widehat{f}\right\} = \left[\widehat{k}\right]\{\widehat{u}\} = \left[\widehat{k}\right][T_{\rm r}]\{u\}.$$

Hence,

$$\{f\} = \left[T_{\rm r}^T\right]\left[\widehat{k}\right][T_{\rm r}]\{u\},$$

or

$$\{f\} = [k]\{u\},$$

where the global stiffness matrix $[k]$ is defined as

$$[k] = \left[T_{\rm r}^T\right]\left[\widehat{k}\right][T_{\rm r}],$$

i.c.

$$[k] = \begin{bmatrix} c & -s & 0 & 0 \\ s & c & 0 & 0 \\ 0 & 0 & c & -s \\ 0 & 0 & s & c \end{bmatrix} \frac{AE}{l} \begin{bmatrix} 1 & 0 & -1 & 0 \\ 0 & 0 & 0 & 0 \\ -1 & 0 & 1 & 0 \\ 0 & 0 & 0 & 0 \end{bmatrix} \begin{bmatrix} c & s & 0 & 0 \\ -s & c & 0 & 0 \\ 0 & 0 & c & s \\ 0 & 0 & -s & c \end{bmatrix}$$
$$= \frac{AE}{l} \begin{bmatrix} c^2 & cs & -c^2 & -cs \\ cs & s^2 & -cs & -s^2 \\ -c^2 & -cs & c^2 & cs \\ -cs & c^2 & cs & s^2 \end{bmatrix}.$$

In order to obtain the global stiffness, mass and damping matrices for the whole structure (as the truss is composed of more than one elements), we must assemble these matrices for each element of the structure. Thus, the global stiffness, mass and force matrices are defined as

$$\mathbf{K} = \sum_{e=1}^{N_{\rm e}} k^{(e)}, \quad \mathbf{M} = \sum_{e=1}^{N_{\rm e}} m^{(e)}, \quad \mathbf{C} = \sum_{e=1}^{N_{\rm e}} c^{(e)}, \quad \mathbf{F} = \sum_{e=1}^{N_{\rm e}} f^{(e)},$$

where $k^{(e)} \equiv k$ for each bar element and $N_{\rm e}$ is the number of elements.

After assembling the stiffness, mass and damping matrices for all the elements, the equation of motion (5) is written as

$$\mathbf{M}\ddot{\mathbf{u}}(t) + \mathbf{C}\dot{\mathbf{u}}\,(t) + \mathbf{K}\mathbf{u}(t) = \mathbf{F}(t). \tag{10}$$

3 Newmark-β Method for the Equations of Motion

In order to solve the system of equations of motion (10), the second order direct numerical integration Newmark-β method (Newmark 1959) is applied. This method is flexible, since it does not require the smoothness of the second derivative. According to Newmark-β method, the formulas for nodal displacement, velocity and acceleration hold

$$\mathbf{u}_k = \mathbf{g}_{k,0} + \beta\Delta t^2\ddot{\mathbf{u}}_{k+1},$$

$$\dot{\mathbf{u}}_k = \mathbf{g}_{k,1} + \gamma\Delta t\ddot{\mathbf{u}}_{k+1},$$

where

$$\mathbf{g}_{k,0} = \mathbf{u}_k + \Delta t\dot{\mathbf{u}}_k + \Delta t^2(0.5 - \beta)\ddot{\mathbf{u}}_{k+1},$$

$$\mathbf{g}_{k,1} = \dot{\mathbf{u}}_k + \Delta t(1 - \gamma)\ddot{\mathbf{u}}_{k+1},$$

$$\mathbf{u}_k = \mathbf{u}(t_k), \quad t_k = k\Delta t, \quad where\ k = 0, 1, \ldots, K_T, \quad \Delta t = \frac{T}{K_T},$$

$$\mathbf{u} = \left(u_{1x}, u_{1y}, u_{2x}, u_{2y}, \ldots., u_{Nx}, u_{Ny}\right)^T$$

and N is the number of nodes. Substituting the expressions for $\mathbf{u}_k$ and $\dot{\mathbf{u}}_k$ into (10) and adding the control function to the right-hand side, we obtain

$$\ddot{\mathbf{u}}_{k+1} = \bar{\mathbf{M}}^{-1}(\mathbf{F}_{k+1} + \mathbf{Z}_{k+1} - \mathbf{C}\mathbf{g}_{k1} - \mathbf{K}\mathbf{g}_{k0}),$$

where $\bar{\mathbf{M}}=\mathbf{M}+\Delta t\gamma\mathbf{C}+\Delta^2 t\beta\mathbf{K}$.

For the initial values of the nodal displacements, velocity and acceleration, we have $\mathbf{u}_0 = \mathbf{u}(t_0) = \mathbf{u}(0)\dot{\mathbf{u}}_0 = \dot{\mathbf{u}}(t_0) = \dot{\mathbf{u}}(0)$, $\ddot{\mathbf{u}}_{k+1}=\bar{\mathbf{M}}^{-1}(\mathbf{F}_0(t) - \mathbf{C}\dot{u}_0 - \mathbf{K}u_0)$.

The integration constants are selected to be $\beta = 0.25$ *and* $\gamma = 0.5$, which corresponds to the case of unconditionally stable constant average acceleration method.

4 Control Simulation Procedure

A nonlinear fuzzy controller is developed using the Fuzzy Toolbox of MatLab. Namely, a Mamdani-type fuzzy inference system with two inputs and one output is employed. Through fuzzification, a set of given "mappings" of both input and output variables are turned into membership functions. Then, the controller can take decisions, i.e. supply an output for given inputs, through a decision-making system which is based on a set of "if-then" verbal rules (Muradova and Stavroulakis 2013).

Table 1 The fuzzy inference rules (e.g. IF displacement is "far up" AND velocity is "up" THEN control force is "max")

Vel.\Disp.	Far up	Close-up	Equil	Close Dn	Far Dn
Up	Max	Med+	Low+	Null	Low−
Null	Med+	Low+	Null	Low−	Med−
Down	Low+	Null	Low−	Med−	Min

In the present investigation, the inputs of the system are the displacement and the velocity of a bar element, which coincides with the one where the control is applied, while the output is the control force. Since the system of equations (10) is given at each node, a transformation of the obtained element control on the nodes is necessary, i.e. if e_{ct} is the control element with starting node $(x_1(e_{ct}), y_1(e_{ct}))$ and final node $(x_2(e_{ct}), y_2(e_{ct}))$, the displacement of the control element will be computed as a product, $u_{e_{ct}} = \delta_{ct} \cdot \mathbf{u}$, where δ_{ct} is a vector with zero components except the ones which correspond to the control element (with starting and final node contribution) $\delta_{ct} = (0, 0, \ldots, 0, -cos\theta, -sin\theta, cos\theta, sin\theta, 0, 0, \ldots, 0)$. The angle between the bar element of control and the x–axis is computed as $\theta = \text{atan}((y_2 - y_1)/(x_2 - x_1))$. The application of control on the truss is similar with the one of introducing thermal stresses, cf. Logan (2007), Stavroulaki et al. (1997).

The controller has 15 rules (Table 1), which use the AND operator. Triangular- and trapezoidal-shaped membership functions have been chosen for the inputs and for the output (Figs. 3, 4 and 5). The selection of polygonal functions has some serious advantages. First, they can be defined only with a small amount of data. Moreover, the condition of unity of the partition is easier to be met with polygonal shapes instead of curves (e.g. Gaussian or sigmoid functions), as the sum of degrees of membership of the involved parameters can easier amount to 1. The implication and aggregation methods have been set to minimum and maximum, respectively. The mean of maximum method (MOM) has been selected for defuzzification.

5 Numerical Results

In this section, the truss model, which is presented in Fig. 2, is simulated using the procedures described in Sects. 3 and 4. The truss, which is fixed at nodes 1 and 2, is subjected to an external force in the vertical direction at node 6. The control is applied on the nearby element 9. The purpose of control is to reduce the oscillations of the structure. The collocated controller takes as inputs the displacement and the velocity at element 9 and returns the control force, which is applied on the truss at each time step of the simulation. The membership functions of fuzzy variables are given in Figs. 3, 4 and 5. Constant and sinusoidal loadings, caused by wind and/or other external sources, are considered. In the examples below, the elasticity modulus is $E = 100{,}000$, and the cross area is $A = 0.1$. The simulation time is $T_f = 50$ and the number of time steps is $K_t = 500$.

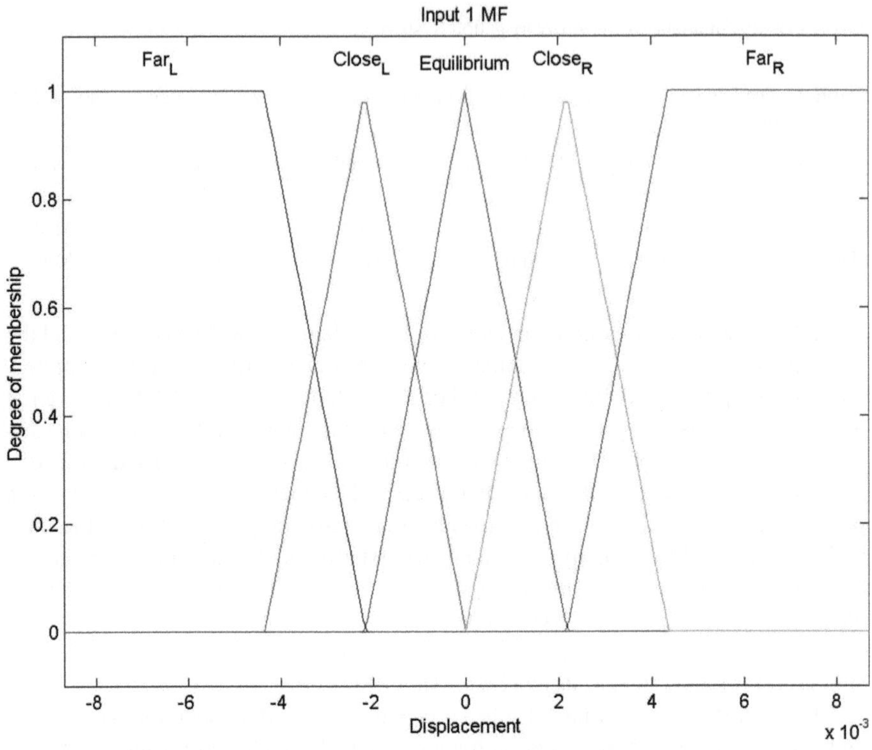

Fig. 3 The membership functions for the input 1 (displacement) of the fuzzy inference system

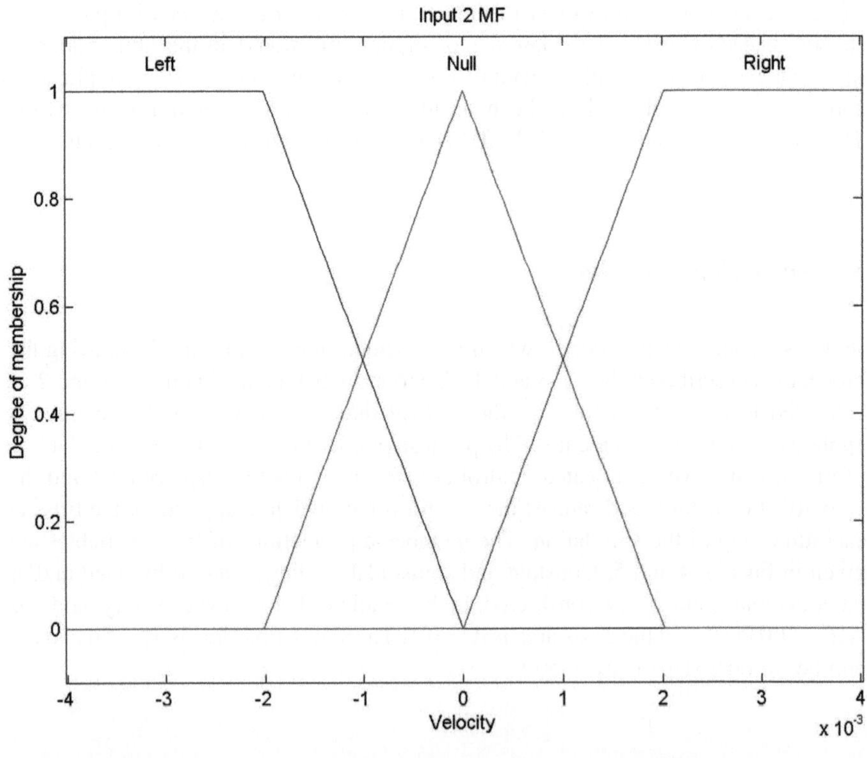

Fig. 4 The membership functions for the input 2 (velocity) of the fuzzy inference system

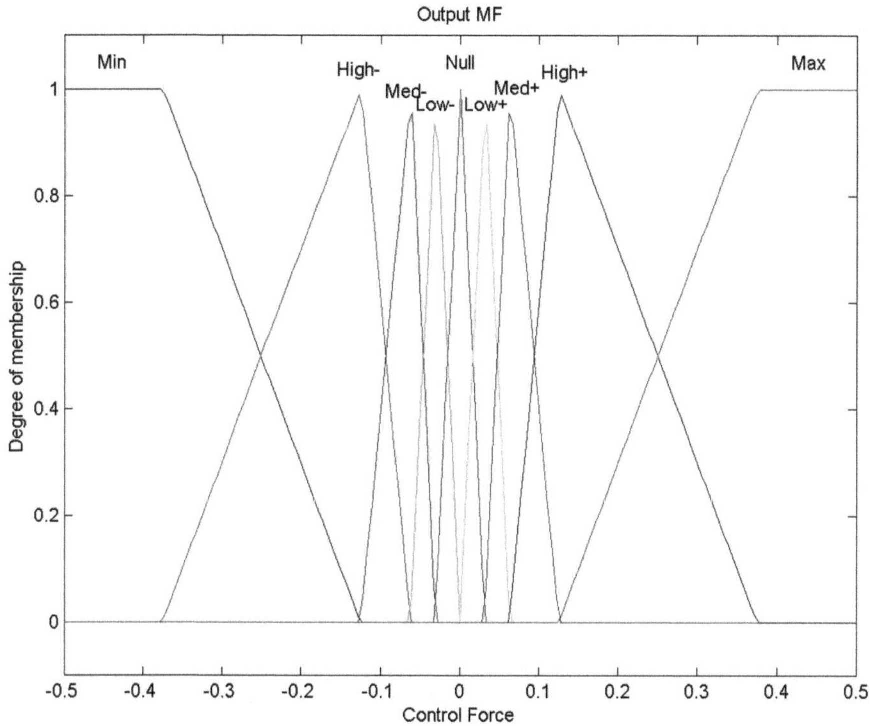

Fig. 5 The membership functions for the output (control force) of the fuzzy inference system

Example 1 For the external loading $\mathbf{F}(t)$, the component $f_{2N}(t) = -0.5$ and the other components $f_i(t)$, $i = 1,\ldots 2N-1$, of $\mathbf{F}(t)$ are zero. The results are presented in Figs. 6, 7 and 8. Namely, initial coordinates of nodes are given in Fig. 6. In Figs. 7 and 8, displacement and velocity at node 9 are shown, respectively.

Example 2 For the external loading $\mathbf{F}(t), f_{2N}(t) = -2sin\omega t$, $\omega = 10\pi$, and $f_i(t) = 0$, $i = 1,\ldots 2N-1$. The results are presented in Figs. 9, 10 and 11.

From the examples, one can conclude that the harmonic loading influences the control of vibrations. It should be emphasized that fuzzy control is rule-based, which does not correspond to an optimal control problem. However, at the same time, the proposed methodology is applicable to a wider spectrum of problems. The effectiveness of the control depends on several parameters, with most important the selection of the fuzzy rules, which in turn have been chosen empirically. The results in some cases can be improved by using an adaptive neuro-fuzzy inference system (ANFIS), which have been introduced for a simplified beam model in the recent work of Tairidis et al. (2017). The details of ANFIS for different problems are also described in the works of Papachristou et al. (2011), Stavroulakis et al. (2011) and Muradova et al. (2017).

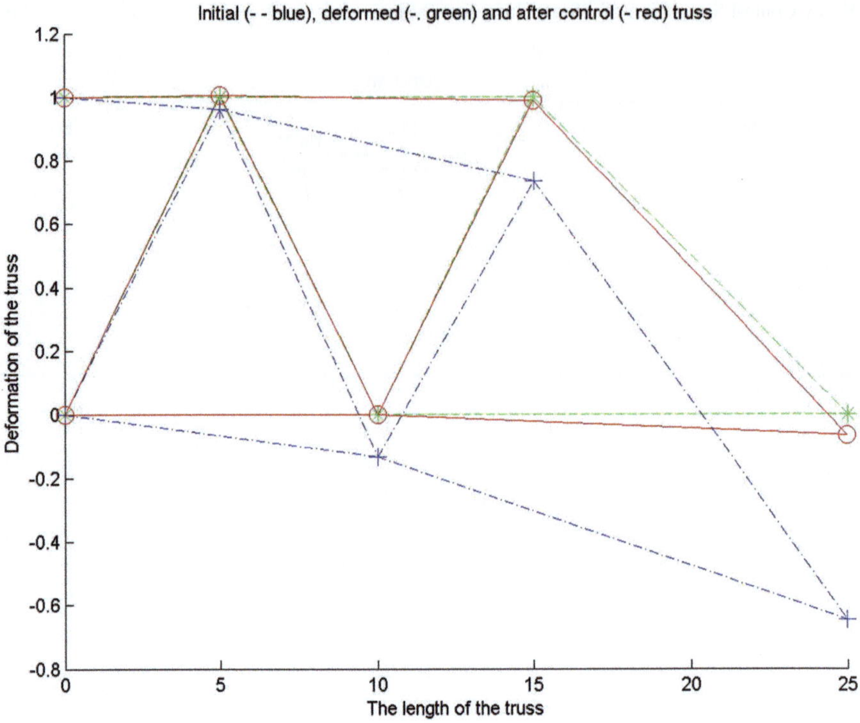

Fig. 6 The initial, deformed and after control truss at the end of simulation for example 1

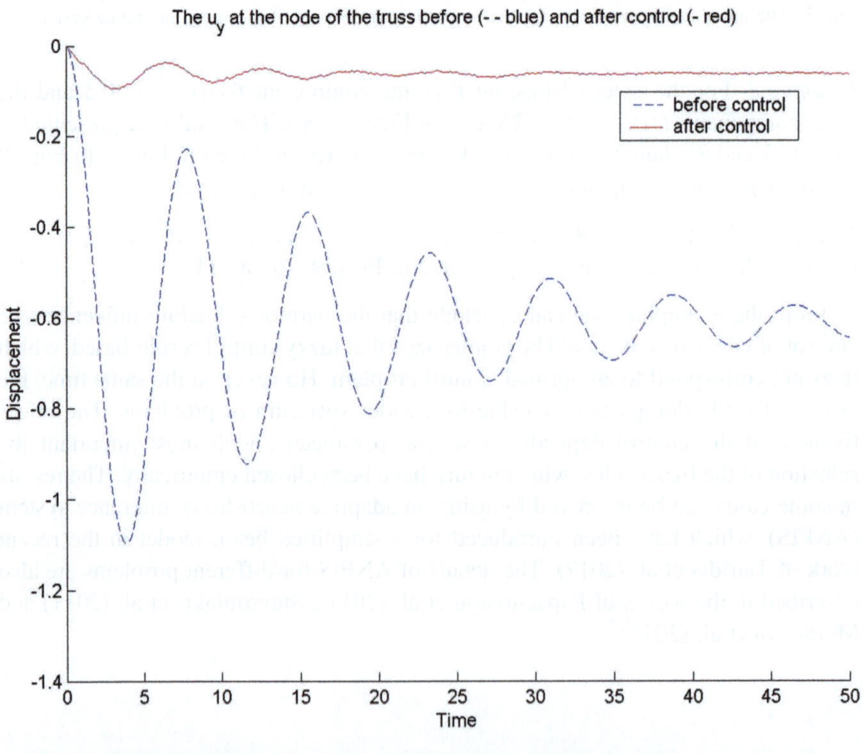

Fig. 7 The displacement before and after control at the node 9 for example 1

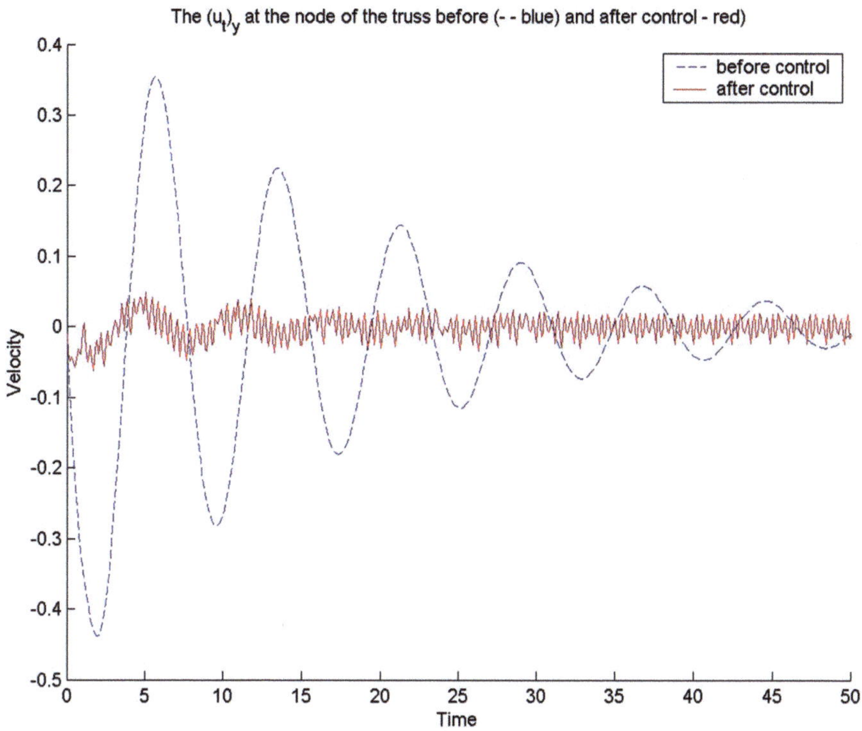

Fig. 8 The velocity before and after control at the node 9 for example 1

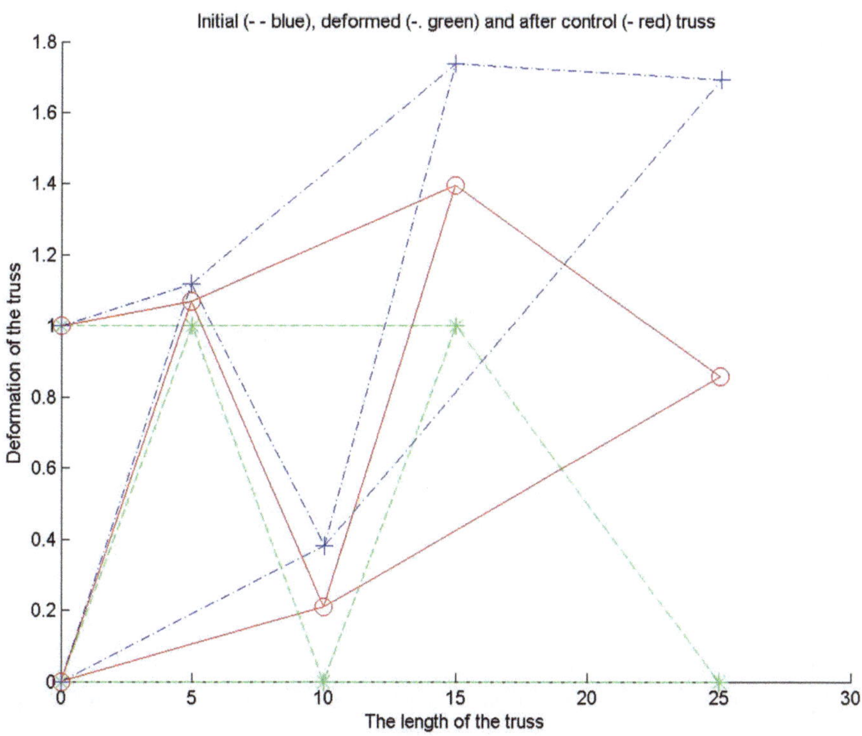

Fig. 9 The initial, deformed and after control truss at the end of simulation for example 2

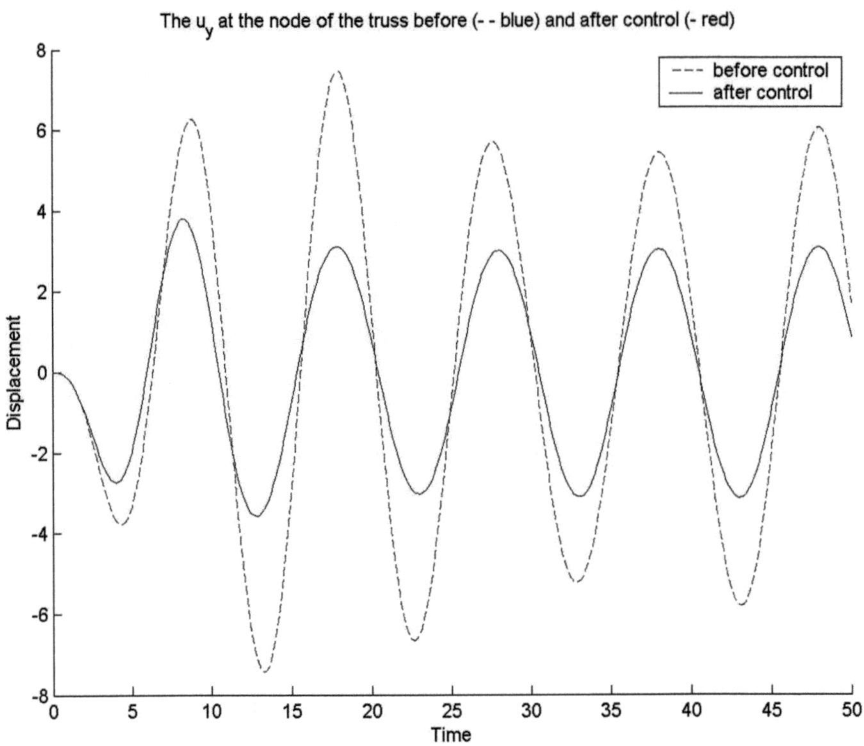

Fig. 10 The displacement before and after control at the node 9 for example 2

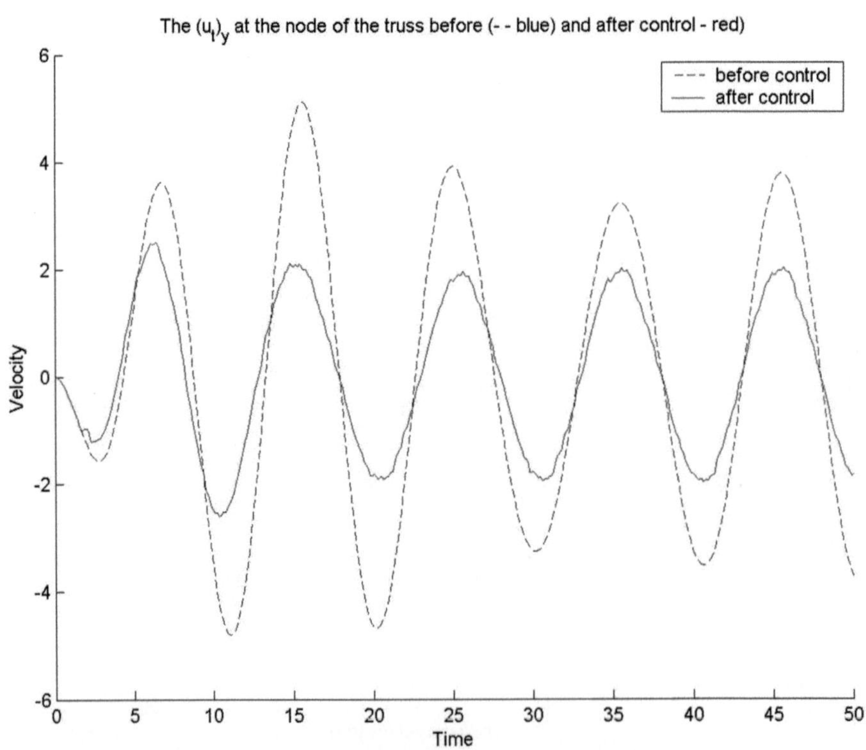

Fig. 11 The velocity before and after control at the node 9 for example 2

6 Conclusion

A control of centre-pivot irrigation system has been simulated with the use of a smart truss model. A finite element analysis has been applied for the investigation of the behaviour of the truss. The control has been performed on the base of fuzzy logic rules. Namely, a Mamdani-type fuzzy inference system has been used for the creation of the controller. The numerical experiments have shown that a location of the controller depends on the location of external forces. Moreover, it is shown that the proposed methodology is general, thus it can be used for various loading scenaria, as well as for different control implementations, as with the present formulation, the control can be placed at any element of the structure. Finally, the obtained numerical results have shown the efficiency of the introduced techniques; however, optimization can help in the design of more complex systems based on the proposed concept.

Robust controllers like the ones which are developed by using fuzzy techniques are suitable for the fusion of different kinds of information, which is a demand on modern precision agriculture. In this direction, sensors of various types, GIS systems and database data are used together for automation purposes. A concrete application on irrigation spray booms is presented herein, although the fuzzy control concepts are quite general and can be applied on different types of dynamical systems.

References

Abe M (1996) Rule-based control algorithm for active tuned mass damper. J Eng Mech 122 (8):705–713

Anthonis J (2000) Design and development of an active horizontal suspension for agricultural spray booms. PhD Thesis, KU Leuven. Available at https://lirias.kuleuven.be/bitstream/123456789/246205/2/doct+Jan+Anthonis.pdf

Anthonis J, Audenaert J, Ramon H (2005) Design optimization for the vertical suspension of a crop sprayer boom. Biosyst Eng 90(2):153–160

Arvanitis KG et al (2003) Chapter 8. New trends in optimal structural control. In: Belyaev A, Guran A (eds) Selected topics in structronic and mechatronic systems. World Academic Publishers, Singapore, pp 321–416

Chang H-H (2011) Genetic algorithms and non-intrusive energy management system based economic dispatch for cogeneration units. Energy 36:181–190

Driankov D, Hellendoorn H, Reinfrak M (1996) An introduction to fuzzy control, 2nd edn. Springer, Munich

Fisco NR, Adeli H (2011) Smart structures: part II, hybrid control systems and control strategies. Sci Iran 18(3):285–295

Irschik H (2002) A review on static and dynamic shape control of structures by piezoelectric actuation. Eng Struct 24:5–11

Karavas CS, Arvanitis K, Papadakis G (2017) A game theory approach to multi-agent decentralized energy management of autonomous polygeneration microgrids. Energies 10(11):1756

Logan DL (2007) A first course in the finite element method, 4th edn. Thomson, Toronto

Lu Q, Zh P, Chu F, Huang J (2003) Design of fuzzy controller for smart structures using genetic algorithms. Smart Mater Struct 12(6):979–986

Marinaki M, Marinakis Y, Stavroulakis GE (2011) Fuzzy control optimized by a multi-objective particle swarm optimization algorithm for vibration suppression of smart structures. Struct Multidisc Optim 43:29–42

Muradova AD, Stavroulakis GE (2013) Fuzzy vibration control of a smart plate. Int J Comput Methods Eng Sci Mech 14(3):212–220

Muradova AD, Stavroulakis GE (2015) Hybrid control of vibrations of smart von Kármán. Acta Mech 226(10):3463–3475

Muradova AD, Tairidis GK, Stavroulakis GE (2017) Adaptive neuro-fuzzy vibration control of a smart plate. Numer Algebra Contr Optim 7(3):251–271

Muradova AD, Tairidis GK, Stavroulakis GE (2018) Fuzzy vibration suppression of a smart elastic plate using graphical computing environment. Soft Comput Civil Eng 2(1):1–17

Newmark NM (1959) A method of computation for structural dynamics. ASCE J Eng Mech Division 85:N. EM3

Papachristou I, Salonikidis S, Tairidis G, Stavroulakis G (2011) Adaptive fuzzy control of smart structures – ANFIS. Athens, 7th GRACM international congress on computational mechanics

Precup R-E, Hellendoorn H (2011) A survey on industrial applications of fuzzy control. Comput Ind 62:213–226

Preumont A (2002) Vibration control of active structures. Springer, Newyork

Stavroulaki ME, Leftheris B, Stavroulakis GE (1997) Optimal prestress in modal analysis via induced temperature modelling. Struct Optimization 13(2–3):95–103

Stavroulakis GE, Papachristou I, Salonikidis S, Papalaios I, Tairidis GK (2011) Neurofuzzy control for smart structures. In: Tsompanakis Y, Topping B (eds) Soft computing methods for civil and structural engineering. Saxe-Coburg Publications, Stirlingshire, pp 149–172

Tahmasebi M et al (2013) Active force control applied to spray boom structure. Appl Mech Mater 315:616–620

Tairidis GK, Stavroulakis GE, Marinova DG, Zacharenakis EC (2009) Classical and soft robust active control of smart beams. In: Papadrakakis M, Charmpis DC, Tsompanakis Y, Lagaros ND (eds) Computational structural dynamics and earthquake engineering. CRC Press, London, pp 165–177

Tairidis GK, Foutsitzi G, Koutsianitis P, Stavroulakis GE (2015) Fine tuning of fuzzy controllers for vibration suppression of smart plates using particle swarm optimization. Volos, 8th GRACM international congress on computational mechanics proceedings

Tairidis GK, Foutsitzi G, Koutsianitis P, Stavroulakis GE (2016) Fine tuning of a fuzzy controller for vibration suppression of smart plates using genetic algorithms. Adv Eng Softw 101:123–135

Tairidis GK, Koutsianitis P, Muradova A, Stavroulakis GE (2017) Neuro-fuzzy control for the reduction of the vibrations on smart irrigation systems, proceedings of 8th international conference on information and communication Technologies in Agriculture. Food Environ 2030:533–540

Processing of Digested Pulp from Agricultural Biogas Plant

Wojciech Czekała

Abstract In the European Union, renewable energy sources (RES) constitute an important alternative to fossil fuels. Solid biomass is mainly used for the production of renewable energy. One of the RES types is the production of agricultural biogas as a result of the anaerobic digestion process (AD). This process generates biogas and digestate. The second product is a residue from the fermentation process, consisting mainly of nondigested organic and mineral components. The market offers many devices for the management of the digestate (digested pulp), e.g., separators that separate the raw material into solid and liquid fractions. The aim of the present book chapter was to discuss the possibility of improving the management efficiency of biogas plant by using the digested pulp as a fertilizer and for other purposes. For a designed 1 MWel biogas plant installation, modern technological solutions have been proposed. It can be concluded that a proper digested pulp management can bring additional profits to the biogas plant.

Keywords Waste management · Fertilizer production · Digestate · Digested pulp · Renewable energy · Environmental protection

1 Introduction

With the increase in energy demand, it is necessary to seek new technologies for its production (Bhattacharyya 2011). It should be cheap, efficient, and, what is most important, safe for the environment. Biofuels are renewable energy sources (RES) that are produced from biomass (Wang et al. 2009). For many years in Europe, the growth of the biogas plant market and the growth of installations producing electricity and heat have been observed. At the same time, the development of technologies related to biofuels is seen, and one of the types of renewable energy

W. Czekała (✉)
Institute of Biosystems Engineering, Poznań University of Life Sciences, Poznań, Poland
e-mail: wojciech@up.poznan.pl

A. Theodoridis et al. (eds.), *Innovative Approaches and Applications for Sustainable Rural Development*, Springer Earth System Sciences,
https://doi.org/10.1007/978-3-030-02312-6_21

installations that has been becoming more and more significant for a few years are agricultural biogas plants. The aim of the present book chapter was to discuss the possibility of improving the management and economic revenue of biogas plant by further use of the digested pulp.

2 Agricultural Biogas as a Source of Renewable Energy

With the decline of fossil fuels, it has become necessary to look for new sources of energy, and RES are one of the solutions (Demirtas 2013). The greatest advantages for RES include inexhaustibility and distributed production location. This is contrary to energy coming from fossil fuels or nuclear energy. The energy production based on renewable resources is one of the priorities of the European Union in the field of environmental protection. The main source of renewable energy in Poland is biomass (Igliński et al. 2011). Depending on the applied technologies, biomass can be transformed into liquid, solid, and gas biofuels among which agricultural biogas can be distinguished. Agricultural biogas plants are type of installations producing stable and environmentally friendly energy at a similar level throughout the year. Agricultural biogas plants use various substances rich in biodegradable organic matter for energy production. They are part of the energy policy of each country, producing energy from renewable sources. Unstable legal regulations in Poland were the reason for their small development in recent years. Despite this, the conditions for construction and operation of plants in Poland and most countries worldwide are satisfactory.

The process of biogas production based on AD and the main product of which is biogas, a mixture of gases in which methane dominates (Weiland 2010), generally ranges from 50 to 65%. The second important constituent is carbon dioxide as well as remaining gases present in the mixture in trace amounts. The products of biogas combustion in a combined heat and power generator unit (CHP) are electricity and heat. The biogas plant is an installation that produces energy in a recovery process while managing waste (Dach et al. 2016; Kozłowski et al. 2016) (Fig. 1). The most common criterion for division of an installation is the substrate used. Therefore, one may distinguish biogas plants in agricultural landfills or municipal waste and sewage treatment plants.

Currently, apart from choosing the right production technology, more and more attention is paid to the substrate selection for energy production. There are many substrates that can be used for agricultural biogas production, having different properties such as state of aggregation, chemical composition, and presence of impurities (Waliszewska et al. 2018). From the biogas plant point of view, the most important are their availability in a given area, biogas and methane yield, as well as price. In practice, the price of the raw material is extremely important, because it can account for up to 40% of the biogas plant cost. Having this in mind, it is necessary to look for substrates with both high energy efficiency and a relatively low price.

Fig. 1 Agricultural biogas plant. (Author: Andrzej Lewicki)

For the proper functioning of installation, by reaching nominal power, it is necessary to ensure a certain daily amount of substrates. It depends primarily on the power of the biogas plant and the methane yield of individual substrates. The substrate being used in biogas plants for many years has been animal slurry, which is a highly available material (Yadav and Garg 2016). Slurry is an inefficient substrate; however, it acts as a carrier for microorganisms necessary in the AD process. Complement to the natural slurry there is often maize silage, the yield of which is high and reaches more than 200 m^3 of biogas from 1 Mg of fresh state substrate. For economic reasons, there was a search for other alternative substrates for biogas production. Suitable substrates for biological waste transformation processes, including biogas production, are waste and residues from the agri-food industry (Fig. 2), which is a well-developed and diversified sector both in Poland and Europe. In addition to maize silage, slurry, and residues from agricultural sources, the most frequently used are also distillers' grains and beet pulp.

Agricultural biogas plants are safer from the environmental perspective because the substrates are from agricultural origin which often undergo through high sanitation standard procedures as it is designed for human consumption (Czekała et al. 2015). Agricultural biogas is produced in biogas plants that use agricultural substrates to produce energy (Budzianowski 2016). It is a mixture of gases with high methane content, so it can be used for energy production (Cieślik et al. 2016). In addition to the selection of substrates, the management of fermentation process products should also be remembered. Another product than biogas, often overlooked substrate from the AD process of agricultural products, is digested pulp (digestate) (Börjesson and Berglund 2007).

Fig. 2 Food waste for biogas production. (Author: Wojciech Czekała)

3 Digested Pulp: Production and Properties

In agricultural biogas plants, the main product of the AD process is biogas. Biogas production is the most important from a financial point of view, due to the methane presence allowing for energy production. The second main product is digestate, also known as digestate pulp (Koszel and Lorencowicz 2015; Arthurson 2009) (Fig. 3). One of the biggest challenges for agricultural biogas plants is the utilization of the by-product, which in fact is the digestate itself. Moreover, many studies and articles confirm the suitability of digestate for soil fertilization due to the high content of available nutrients needed for plant growth. There is no precise definition of digestate in Polish legislation. Most often, it is defined as the residue resulting from the fermentation process. The digested pulp consists primarily of organic compounds, minerals, and biomass of living organisms undecomposed in the AD process.

The composition of digestate is mainly dependent on the substrates used for the AD process. Other important factors affecting the composition and properties of the digested pulp depend on the process conditions and its correctness, mixing in fermentation chambers and fermentation time. As mentioned before, the substrates in each area are widely available, so the digestate after AD may have different properties in each case. Other factors affecting the residue properties include

Fig. 3 Digested pulp for laboratory research. (Author: Wojciech Czekała)

substrate fragmentation, the course and technology of the fermentation process, or technologies used in agricultural biogas plants (Sogn et al. 2018). The discussed residue from agricultural biogas plants can be treated as waste, by-product, as well as organic fertilizer. Regardless of the legal point of view, the methods of proceeding of the mentioned material are the same.

Digestate is characterized by high moisture content. Thus, the dry matter content of the digestate is rather low than 6%. In wet technologies dehydration is carried out by means of separation or drying (Kaparaju and Rintala 2008). In case of separation, solid and liquid fraction are the products which can be used for different purposes. For this reason, separation is an important process and is carried out on many biogas plants. Dry matter content in solid fraction is usually 20–35%, whereas liquid fraction is about 1.5–4%. One advantage of drying is the efficiency of the process because it allows the substrate to be processed to a specific moisture content. However, the drawback is the need to supply a significant amount of energy in the operation. Moreover, in case of the use of digestate as a fertilizer, it is recommended to maintain certain level of moisture that can favor the absorption of nutrients by plants (Table 1).

In Europe, the market for separators and digestate driers is developed, so it is possible to improve agricultural biogas installations to include such technologies. For example, separators primarily used for sewage sludge were installed in agricultural biogas plants. The digestate should be stored until it can be used as a fertilizer. The method of storage of both raw digestate and liquid fraction is the same, due to the liquid consistency of both types of material. The popular solution for temporary

Table 1 Average composition of solid and liquid fractions of digestate from selected Polish biogas plants (Kowalczyk-Juśko and Szymańska 2015)

Parameters	Solid fraction	Liquid fraction
Dry matter [%]	22–27	2.7–4.3
Dry organic matter [%]	89–94.5	58–62
Total nitrogen [%]	0.4–0.8	0.2–0.75
$N\text{-}NH_4$ [%]	0.08–0.52	0.28–0.38
P [%]	0.1–0.28	0.03–0.05
K [%]	0.12–0.69	0.5–0.62
Ca [%]	0.22–0.43	0.05–0.07
Mg [%]	0.06–0.17	0.01–0.02

Fig. 4 Open lagoons for digested pulp. (Author: Wojciech Czekała)

storage of the raw digestate is concrete or steel tanks. Their construction resembles fermentation chambers of biogas plants or slurry tanks. Another solution is storage in open lagoons (Fig. 4) or closed lagoons. They are tanks in the ground, sealed with a special membrane protecting against leaks. Open lagoons are cheaper to build but have more disadvantages. The biggest problem is the possibility of rain penetration and higher gas emissions, especially ammonia. The solid fraction, characterized by a loose structure, can be stored on flooring, preferably with a roofing.

The digestate treatment is important both from the environmental and economic side. This is caused by the lack of solutions favoring the development of the discussed residue, causing the need of longer or improper storage, adversely affecting the environment. On the other hand, the introduction of solutions allowing digestate application may result in financial benefits.

4 Digestate Processing and Directions of Use

The substrate used in the agricultural biogas plant is subject to numerous transformations. Regarding elemental composition, the greatest changes occur in relation to carbon. If organic compounds are degraded, consequently, the amount of organic matter decreases, and methane is produced. In relation to macroelements, another relevant aspect is the conversion of organic matter into mineral compounds. Digestate is characterized by better parameters than the unprocessed substrate, e.g., slurry, because nutrients are more available for plants. The intensity of all those transformations depends primarily on the type of substrate used (Czekała et al. 2016a). In addition, the process conditions are an important element, for example, in the parameters: type of technology used, process temperature, substrate size and type, and intensity of the process.

After obtaining the digestate, it must be managed in accordance with the law. By far, the most popular method in Europe is its direct use for fertilizing purposes. However, there are more ways to digestate management (Fig. 5).

It can be concluded that there are many ways to utilize digestate. Generally, the selection of elements required for waste management should be preceded by:

1. Analysis of the scientific literature and legislation in force in the country.
2. Industry market analysis of available equipment and technologies.
3. Calculating the amount of digestate available.

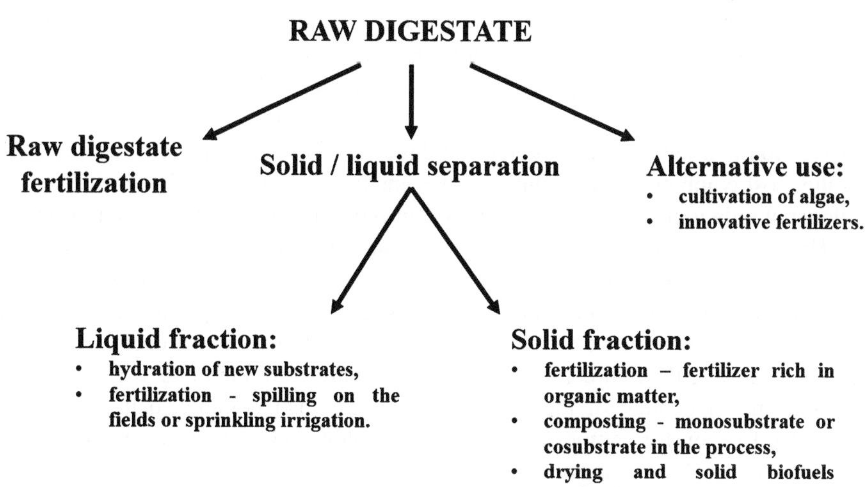

Fig. 5 The possibility of using raw and processed digestate. (Author: Wojciech Czekała)

4. Determining the preferred directions of using the digestate.
5. Development of a biogas effluent processing plant that is compatible with the biogas plant.

The most common way to use the digested pulp is use it directly as a fertilizer (Tampio et al. 2016). This is due to the ease and high efficiency of its application in the soil. This solution is practiced mainly due to the lack of implementation of other activities such as separation, drying, or confectioning. The only requirement is to mix it beforehand to equalize the nutrient content. In many countries in Europe and, for example, in China, digestate is perceived as a valuable fertilizer. This is a desirable substrate and apparently presents no environment concerning related to its management. In Poland, there is some resistance about the use of digestates as fertilizers mostly due a misperception and understanding about the biogas and the use of digested pulp. Currently, some biogas plants in Poland have problems with the sale of digested pulp. However, with the understanding of the benefits of digestate, the perception of its application as fertilizer by farmers is favorably shifting, and digestate started to be treated as a valuable fertilizer.

Fertilizer application of the digestate brings many benefits. The most important is to provide macro- and microelements necessary for the development of plants. The liquid form facilitates applications that can be carried out using machines for fertilizing with liquid form natural fertilizers, e.g., slurry (Cavalli et al. 2016). Among the most common hazards are those concerning the content of heavy metals and pathogens. In cases when the feedstock for biogas plant is agricultural or food substrates of known origin, the permissible standards should not be exceeded. However, it is recommended to check the digestate before its application to the soil at least twice a year. Research conducted by Sogn et al. (2018) were intended to determine the fertilizer properties of pulp depending on the type of soil (loam, sand, silt). The authors highlighted the positive properties allowing for the use of digestate in agriculture. Kataki et al. (2017) tested the possibilities of using four types of digestate for fertilizing purposes: whole (before separation), solid, liquid, and ash from solid digestate. The authors conclude that integrated production of bioenergy and use of by-product as fertilizer could be very significant in terms of energy production and fertilizers. Although the direct application of digestate as fertilizer is the most commonly used solution, other methods are also sought for that would allow not only to manage it but also to obtain economic and environmental benefits.

However, in case of separation, the digestate liquid and solid fraction can be either used as fertilizer in the fields. Liquid fraction also can be used to decrease the dry matter content of the new substrate for the biogas plant (Sigurnjak et al. 2017). The solid fraction biomass can be used as well as a fertilizer or for energy purposes (Czekała et al. 2017). Direct distribution on the surface of the fields will enrich the organic matter content of the top soil. And, it is also possible to use the solid digestate fraction for compost, both as a stand-alone substrate or as an additive in combination with other ingredients.

One possible solution for this problem is the separation (dehydration) of the digestate (Wu et al. 2017). The effect of such action is to obtain a minimum of two significantly differentiated fractions. This will allow easier management of the product after fermentation process. Another advantage will be the possibility of increased revenue of the biogas plant.

The technology which can even increase the fertilizing value of digestate solid fraction is called composting. Because of intensive temperature growth during composting, the obtained material is characterized by high fertilizing value, sanitary safety, lack of odor, and chemical stability (Waszkielis et al. 2013; Wolna-Maruwka and Dach 2009).

The liquid fraction is most often used directly as a fertilizer. Lower dry matter content than in unprocessed digestate contributes fertilization, above all, due to better ingredient mixing. Another option is to use it as a substrate for hydration of a new batch of solid feed (e.g., silage or pomace). The liquid fraction allows to increase the degree of hydration, which will enable the flow of feedstock through pumps in the installation. In addition, there are more organisms in the liquid fraction that are responsible for inoculating the new batch of feedstock. In scientific research carried out around the world, it was used, among others, in the barley cultivar (Haraldsen et al. 2011) which was proved by Sigurnjak et al. (2017). The research was carried out in three subsequent years, and the authors came to a conclusion, among others, that liquid fraction of digestate has a potential to be used as N fertilizer.

Definitely more directions of use are characterized by solid fractions. The solid fraction of the digested pulp formed as a result of separation is characterized by porosity and high content of organic matter, despite partial degradation under AD. Due to this fact, it can be utilized to achieve the fertilizer and energy goals. The mentioned fraction is rich in organic matter (most often 60–90%); therefore, it is a suitable fertilizer for Polish soils, usually poor in organic matter. At the same time, in order to increase the fertilizer value of this fraction, composting is advised. Composting is an oxygen decomposition of organic matter that results in a good quality organic fertilizer called compost. Own research of Czekała et al. (2017) has indicated that even the solid fraction itself can be used in the composting process. It should be noted that using additives for the composting process is more popular. A good example is the use of sewage sludge. The solid fraction is characterized by a high content of organic matter, and sewage sludge is rich in nitrogen. This combination allows to improve the C:N ratio, which will favorably affect the process and thus the quality of the obtained fertilizer.

The water-free solid fraction becomes a valuable substrate with a wide range of applications in the solid biofuel sector. The residue from the biogas plant can be used in the mechanical molding process to obtain pellets or briquettes or can be directly thermally transformed. The energy value in terms of dry matter is usually in the range of 14.5–18 $MJ \cdot kg^{-1}$. Some objections have been raised by scientists within the very course of the combustion process; hence it is suggested to conduct research in this topic.

Another possibility of raw digestate management is to use it as feeding medium for algae production as the substrate for biogas plant (Cerbin et al. 2012). The experimental installation (greenhouse of 3000 m^2) for algae production with usage of digestate solution for microalgae nutrition was built in Central Poland in 2015. However, the real scale investment has shown much lower microalgae production (mainly because of colder climate and technical problems) than promising results obtained in laboratory scale (Lewicki et al. 2013).

Ideally, a part of the digestate could be used directly to fertilize the fields that provide substrate to the biogas plant. This will keep the nutrients within the production chain of the biogas plant. Another advantage is related to economic and environmental aspects. The direct use of digestate as fertilizer can decrease the cost of crop production, while the production of mineral fertilizers, in some cases, can pose threat to the environment. Therefore, this alternative is economically and environmentally ideal.

5 Case Study for Selected Biogas Plant

In the present case study, the energetic use of digestate was the objective. The production of solid biofuels, i.e., briquettes and pellets, was considered. Both briquettes and pellets are widely used in Poland for heat production (Szmigielski et al. 2014; Kowalczyk-Juśko et al. 2015a). The design of the agricultural biogas plant with power 1 MW will use the substrates shown in Table 2. The presented data are from studies conducted in Laboratory of Ecotechnologies in the Poznań University of Life Sciences (PULS). The research was carried out according to the modified German standard DIN 38 414 – S8.

It is estimated that about 90% of the initial substrate matter still remain in the digestate. These values are the assumed values often used in calculation in Poland. Consequently, a typical biogas plant with 1 MW electric power produces, daily, a few tens of tons of digestate. Such a considerable amount of digestate often becomes a liability for biogas plants operating in Poland.

In order to calculate the amount of digestate material produced on a biogas plant, it is necessary to estimate the dry matter and water content of the substrates. The following formulas were used (1):

Table 2 Substrates use for biogas production

Substrate	Quantity [$Mg \cdot year^{-1}$]	dm [%]	odm [% dm]	Biogas efficiency [$m^3 \cdot Mg^{-1} FM$]	Methane content [%]
Maize silage	11,000	32	95	210	55
Brewers grains	11,000	24	90	140	60
Slurry	22,000	4	70	10	60

Non-published data from Laboratory of Ecotechnologies at PULS

$$M_{\mathrm{dm}} = M_{\mathrm{fm}} \cdot \mathrm{dm} \ \left[\mathrm{Mg} \cdot \mathrm{year}^{-1}\right] \quad (1)$$

where:

M_{dm} – mass of dry substance mass [Mg·year^{-1}]
M_{fm} – mass of fresh mass [Mg·year^{-1}]
dm – dry matter content [%]

The most important parameter when planning a separation installation is the water content of the raw digestate and its fractions. The quantity and characteristics of the substrates are shown in Table 2. The mass of substrates used for biogas production in the analyzed mix was 44,000 Mg (Table 2). Assuming that the amount of the digestate constitutes 90% of the feed substrate, the annual amount of digestate is calculated (2):

$$M_{\mathrm{dig}} = M_{\mathrm{s}} \cdot n \ \left[\mathrm{Mg} \cdot \mathrm{year}^{-1}\right] \quad (2)$$

where:

M_{dig} – digestate mass [Mg·r^{-1}]
M_{s} – mass of input to biogas plant [Mg·r^{-1}]
n – mass retention coefficient in the fermentation process −0.9 [%]

At the present case study, an annual production of 39,600 Mg of digestate was expected, which gives about 108 Mg digestate per day. The quantity is relatively large, since half of the load is a slurry characterized by high hydration. The addressed project is planned to use solid fractions for the production of solid biofuels. And, the liquid fraction will be directed to the hydration of new substrates and partly for the production of microorganism-enriched fertilizers. This is an innovative solution that complies with both scientific and entrepreneurial interest in relation to research on microbial inoculation of the liquid fraction to enhance fertilizer properties. The following assumptions were made for calculations (Table 3).

The generated solid biofuel from the digestate has a slightly lower calorific value than the wood pellet treated as the standard solid biofuel. The determined calorific value was 16.58 MJ·kg^{-1} (Czekała et al. 2016b). In the analyzed example, it was assumed that the daily production of the solid fraction is 20 Mg with a dry mass of

Table 3 Design assumptions

Assumption	Value of assumption
Quantity of digestate production [m^3·year^{-1}]	39,600
Mass of 1 m^3 of fermentation pulp [Mg]	1
Dry matter of digestate [%]	8
Amount of liquid fraction after separation [Mg·year^{-1}]	28,800
Amount of solid fraction after separation [Mg·year^{-1}]	10,800

Non-published data based on real scale biogas plants

23%. Generally, water content can affect the strength of materials (Chocyk et al. 2015; Kowalczyk-Juśko et al. 2015b). Thus, to achieve appropriate level of water in the substrate, a belt dryer will operate in the biogas plant. Then, with the resulting solid fraction, it will be possible to produce about 4 Mg of briquettes from the digestate. The produced briquette can be sold to external buyers for 100–130 Euro·Mg^{-1}.

Necessary installations for digestate processing in the biogas plant are as follows:

- Separator EUR 44,000
- Belt dryer EUR 60,000
- Line for the production of briquettes (0.5 Mg·h^{-1}) EUR 14,000

The cost of installation of the above listed separation and treatment devices is EUR 118,000. However the biogas plant building cost is about five million Euros, consequently, the above device cost seems economically acceptable.

6 Environmental and Economic Benefits as a Result from the Use of Digestate

Digested pulp can be used as a fertilizer on fields; however, certain conditions must be observed. It is forbidden i.a. to use fertilizer on frozen soils, water flooded areas, covered with snow and in liquid form during the vegetation of plants intended for direct human consumption. Digestate is considered to be environmentally friendly as a fertilizer under the condition that typical agricultural substrates were used for fermentation, including slurry, manure, or by-products of the agri-food industry. The use of digestate will not only close the circulation of nutrients within the field but also reduce the amount of mineral fertilizers.

As mentioned before, digestate due to its properties and chemical composition can and should be used in agriculture as a fertilizer. In many countries, it is treated as a valuable organic fertilizer. In Poland, the possibilities of managing the pulp are not very beneficial, although in recent years slow changes have been observed heading toward improvement of situation. When considering the issue of agricultural use of digestate, it is worth keeping in mind that it can be an additional source of income for installations. Counting only the NPK fertilizer content according to market prices, the average value of 1 m^3 varies between EUR5 and 10, depending on the content of nutrients.

In addition, the presence of microelements necessary for the plant development (copper, zinc, manganese, iron, and others) that increase the fertilizer and economic value of the digestate should be taken into account. In the AD, a biomass of known origin is used. This means that the digestate produced in the process does not contain any harmful compounds in relation to the input substrates (natural fertilizers, silage, food waste). The only conditions necessary for this type of process is proper fermentation and, if necessary, pasteurization of the load before sending to the digester.

The use of digestate carries a beneficial effect on the environment, mainly by management of organic waste in a biogas plant. Not processed biodegradable waste is a source of hazardous leachate and gas emissions. The digestate application itself has allowed to reduce, among others, emissions of noxious gaseous compounds, as described in Riva et al. (2016). Often, it is also highlighted that fertilization with digestate increases soil fertility not only in nutrients but also organic matter (Möller and Müller 2012). The pH is usually around 7–9. The use of digestate allows to reduce the negative impact on the environment resulting from acidification of soils. The information about digestate suggest that the material is beneficial for ecosystems.

7 Conclusion

Numerous world studies clearly indicate the suitability of digestate in the process of plant growth and development. In practice, owners of agricultural biogas plants usually dispose off the digestate for small amounts of money, inadequate to the fertilizer value characterizing the residue in question. The lack of specific regulations and social resistance is responsible for limiting the use of digestate at field. Moreover, biogas plant owners often do not have enough farmland to distribute the entire produced digestate. Consequently, this situation creates barriers to the development of the biogas sector in Poland. Additionally, it is recommended to carry out physical, chemical, and microbiological analyses of the digestate before its application in the soil. A better characterization of the fertilizer can increase the trustiness of farmers and favor the utilization of digestate at fields. Investments in an installation for industrial separation and processing of digestate create opportunities for the use of digestate with additional revenue.

References

Arthurson V (2009) Closing the global energy and nutrient cycles through application of biogas residue to agricultural land – potential benefits and drawbacks. Energies 2:226–242

Bhattacharyya SC (2011) Understanding and analysing energy demand. In: Energy economics. Springer, London, pp 41–76

Börjesson P, Berglund M (2007) Environmental system analysis of biogas system – Part II: the environmental impact of replacing various systems. Biomass Bioenergy 31:326–344

Budzianowski M (2016) A review of potential innovations for production, conditioning and utilization of biogas with multiple-criteria assessment. Renew Sust Energ Rev 54:1148–1171

Cavalli D, Cabassi G, Borrelli L, Geromel G, Bechini L, Degano L, Gallina PM (2016) Nitrogen fertilizer replacement value of undigested liquid cattle manure and digestates. Eur J Agronomy 73:34–41

Chocyk D, Gładyszewska B, Ciupak A, Oniszczuk T, Mościcki L, Rejak A (2015) Influence of water addition on mechanical properties of thermoplastic starch foils. Int Agrophysics 29 (3):267–275

Cerbin S, Nowakowski K, Dach J, Pilarski K, Boniecki P, Przybył J, Lewicki A (2012) Possibilities of neural image analysis implementation in monitoring of microalgae production as a substrate for biogas plant. 4th international conference on digital image processing (ICDIP 2012). Proceedings of SPIE, vol 8334. Article Number: 83342A. https://doi.org/10.1117/12.954164

Cieślik M, Dach J, Lewicki A, Smurzyńska A, Janczak D, Pawlicka-Kaczorowska J, Boniecki P, Cyplik P, Czekała W, Jóźwiakowski K (2016) Methane fermentation of the maize straw silage under meso- and thermophilic conditions. Energy 115(2):1495–1502

Czekała W, Dach J, Czekała J (2015) Operational possibilities of a biogas plant at the brewery under polish conditions. Proceedings of the 2nd international conference on energy & environment: bringing together engineering and economics. Guimaraes, Portugal 520–525

Czekała W, Smurzyńska A, Cieślik M, Boniecki P, Kozłowski K (2016a) Biogas efficiency of selected fresh fruit covered by the Russian embargo. In: Energy and clean technologies conference proceedings SGEM. International Multidisciplinary Scientific GeoConferences SGEM - Sofia, Bulgaria 3:227–233

Czekała W, Bartnikowska S, Lewicka A, Bugała A, Zbytek Z, Lewicki A (2016b) Economic and energy efficiency of the solid biofuels produced from digested pulp. MATEC web of conferences 60, 04005 ICCBS 2016. https://doi.org/10.1051/matecconf/20166004005

Czekała W, Dach J, Dong R, Janczak D, Malińska K, Jóźwiakowski K, Smurzyńska A, Cieślik M (2017) Composting potential of the solid fraction of digested pulp produced by a biogas plant. Biosyst Eng 160:25–29

Dach J, Jóźwiakowski K, Kowalczyk-Juśko A, Kozlowski K, Neugebauer M (2016) Biogas plant exploitation under the low-subsidies market conditions: maize silage versus biowaste scenarios. In: Energy and clean technologies conference proceedings, SGEM. International Multidisciplinary Scientific GeoConferences SGEM - Sofia, Bulgaria 3:221–226

Demirtas O (2013) Evaluating the best renewable energy technology for sustainable energy planning. Int J Energy Econ Policy 3(Special Issue):23–33

Haraldsen TK, Andersen U, Krogstad T, Sørheim R (2011) Liquid digestate from anaerobic treatment of source-separated household waste as fertilizer to barley. Waste Manag Res 29:1271–1276

Igliński B, Iglińska A, Kujawski W, Buczkowski R, Cichosz M (2011) Bioenergy in Poland. Renew Sust Energ Rev 15(6):2999–3007

Kaparaju PLN, Rintala JA (2008) Effects of solid-liquid separation on recovering residual methane and nitrogen from digested dairy cow manure. Bioresour Technol 99:120–127

Kataki S, Hazarika S, Baruah DC (2017) Assessment of by-products of bioenergy systems (anaerobic digestion and gasification) as potential crop nutrient. Waste Manag 59:102–117

Kowalczyk-Juśko A, Szymańska M (2015) Poferment nawozem dla rolnictwa. http://ksow.pl/uploads/tx_library/files/Poferment_nawozem_dla_rolnictwa_01.pdf. Accessed 3 Aug 2008

Kowalczyk-Juśko A, Kowalczuk J, Szmigielski M, Marczuk A, Jóźwiakowski K, Zarajczyk K, Maslowski A, Slaska-Grzywna B, Sagan A, Zarajczyk J (2015a) Quality of biomass pellets used as fuel or raw material for syngas production. Przemysl Chemiczny 94(10):1835–1837

Kowalczyk-Juśko A, Koscik B, Jóźwiakowski K, Marczuk A, Zarajczyk J, Kowalczuk J, Szmigielski M, Sagan A (2015b) Effects of biochemical and thermochemical conversion of sorghum biomass to usable energy. Przemysl Chemiczny 94(10):1838–1840

Kozłowski K, Lewicki A, Sołowiej P, Neugebauer P, Smurzyńska A (2016) Usage of waste whey as mono-substrate in continuous fermentation process. In: Energy and clean technologies conference proceedings, SGEM. International Multidisciplinary Scientific GeoConferences SGEM - Sofia, Bulgaria 3:345–350

Koszel M, Lorencowicz E (2015) Agricultural use of biogas digestate as a replacement fertilizers. Agric Agric Sci Procedia 7:119–124

Lewicki A, Dach J, Janczak D, Czekała W (2013) The experimental macro Photoreactor for microalgae production. 6th international conference on information and communication technologies in agriculture, food and environment (HAICTA 2013). Book Series: Procedia Technology 8:622–627

Möller K, Müller T (2012) Effects of anaerobic digestion on digestate nutrient availability and crop growth: a review. Eng Life Sci Special Issue Biogas 12(3):242–257

Riva C, Orzi V, Carozzi M, Acutis M, Boccasile G, Lonati S, Tambone F, D'Imporzano G, Adani F (2016) Short-term experiments in using digestate products as substitutes for mineral (N) fertilizer: agronomic performance, odours, and ammonia emission impacts. Sci Total Environ 547:206–214

Sigurnjak I, Vaneeckhaute C, Michels E, Ryckaert B, Ghekiere G, Tack FMG, Meers E (2017) Fertilizer performance of liquid fraction of digestate as synthetic nitrogen substitute in silage maize cultivation for three consecutive years. Sci Total Environ 599–600:1885–1894

Sogn TA, Dragicevic I, Linjordet R, Krogstad T, Eijsink VGH, Eich-Greatorex S (2018) Recycling of biogas digestates in plant production: NPK fertilizer value and risk of leaching. Int J Recycling Organic Waste Agric 7(1):49–58

Szmigielski M, Zarajczyk J, Kowalczyk-Juśko A, Kowalczuk J, Rydzak L, Slaska-Grzywna B, Krysiak Z, Cycan D, Szczepanik M (2014) Quality of biomass briquettes as stock for thermo-chemical conversion and syngas production. Przemysl Chemiczny 93(11):1986–1990

Tampio E, Marttinen S, Rintala J (2016) Liquid fertilizer products from anaerobic digestion of food waste: mass, nutrient and energy balance of four digestate liquid treatment systems. J Clean Prod 125:22–32

Yadav A, Garg VK (2016) Vermiconversion of biogas plant slurry and parthenium weed mixture to manure. Int J Recycl Org Waste Agricult 5:301–309

Waliszewska H, Zborowska M, Waliszewska B, Borysiak S, Antczak A, Czekała W (2018) Transformation of Miscanthus and Sorghum cellulose during methane fermentation. Cellulose 25(2):1207–1216

Wang F, Xiong XR, Liu CZ (2009) Biofuels in China: opportunities and challenges. In Vitro Cell Dev Biol Plant 45:342–349

Waszkielis KM, Wronowski R, Chlebus W, Białobrzewski I, Dach J, Pilarski K, Janczak D (2013) The effect of temperature, composition and phase of the composting process on the thermal conductivity of the substrate. Ecol Eng 61:354–357

Weiland P (2010) Biogas production: current state and perspectives. Appl Microbiol Biotechnol 85:849–860

Wolna-Maruwka A, Dach J (2009) Effect of type and proportion of different structure-creating additions on the inactivation rate of pathogenic bacteria in sewage sludge composting in a cybernetic bioreactor. Arch Environ Prot 35(3):87–100

Wu D, Lü F, Shao L, He P (2017) Effect of cycle digestion time and solid-liquid separation on digestate recirculated one-stage dry anaerobic digestion: use of intact polar lipid analysis for microbes monitoring to enhance process evaluation. Renew Energy 103:38–48

List of Reviewers

Eleni Abraham, Aristotle University of Thessaloniki, Greece
Vasileios Ambas, Regional Government of Western Macedonia, Greece
Konstantinos Arvanitis, Agricultural University of Athens, Greece
Chintan Bhatt, Chandubhai S Patel Institute of Technology, India
Sandro Bimonte, National Research Institute of Science and Technology for Environment and Agriculture, France
Boško Blagojević, University of Novi Sad, Serbia
Evangelia Daskalakou, Hellenic Agriculture Organization "Demeter", Greece
Pavlos Delias, University of Kavala Institute of Technology, Greece
Yucheng Dong, Xi'an Jiaotong, University, China
Domenica Farinella, Universita' degli Studi di Messina, Italy
Athanasios Gertsis, Perrotis College, Greece
Safwat H. Shakir Hanna, Prairie View A&M University, USA
Mina Kafkaletou, Hellenic Agriculture Organization "Demeter", Greece
Paraskevi Karanikola, Democritus University of Thrace, Greece
Paata Kervalishvili, Georgian Technical University, Georgia
Kateryna Kononova, V. N. Karazin Kharkiv National University, Ukraine
Georgia Koutouzidou, University of Macedonia, Greece
Stavriani Koutsou, Alexander Technological Educational Institute of Thessaloniki, Greece
Alina Kowalczyk-Juśko, University of Life Sciences in Lublin, Poland
Tomas Kusio, Cracow University of Economics, Poland
Grigorios Kyriakopoulos, National Technical University of Athens, Greece
Georgios Markou, Hellenic Agriculture Organization "Demeter", Greece
Ioannis Mitsopoulos, Ministry of Environment & Energy, Greece
Radmehr Monfared, Loughborough University, UK
Giovanni Ottomano, Mediterranean Agronomic Institute of Bari, Italy
Anastasia Pantera, University of Applied Sciences of Central Greece, Greece
Vasileios Papatsiros, University of Thessaly, Greece

A. Theodoridis et al. (eds.), *Innovative Approaches and Applications for Sustainable Rural Development*, Springer Earth System Sciences,
https://doi.org/10.1007/978-3-030-02312-6

Georgia Papoutsi, Hellenic Agriculture Organization "Demeter", Greece
Zoi Parissi, Aristotle University of Thessaloniki, Greece
Andrea Pisanelli, National Research Council, Italy
Emanuele Schimmenti, University of Palermo, Italy
Alexandra Solomou, Hellenic Agriculture Organization "Demeter", Greece
Eleni Tani, Agricultural University of Athens, Greece
Dimitrios Tseles, University of West Attica, Greece
Eleni Tzika, Aristotle University of Thessaloniki, Greece
Laszlo Varallyai, University of Debrecen, Hungary
Konstantinos Zaralis, Technological Educational Institute of Western Macedonia, Greece

Index

A. Theodoridis et al. (eds.), *Innovative Approaches and Applications for Sustainable Rural Development*, Springer Earth System Sciences,
https://doi.org/10.1007/978-3-030-02312-6

K

MIX
Papier aus verantwortungsvollen Quellen
Paper from responsible sources
FSC® C105338
FSC
www.fsc.org

If you have any concerns about our products,
you can contact us on
ProductSafety@springernature.com

In case Publisher is established outside the EU,
the EU authorized representative is:
Springer Nature Customer Service Center GmbH
Europaplatz 3, 69115 Heidelberg, Germany

Printed by Libri Plureos GmbH
in Hamburg, Germany